RUSSIAN–ENGLISH, ENGLISH–RUSSIAN FORESTRY AND WOOD DICTIONARY

2nd Edition

Russian–English, English–Russian Forestry and Wood Dictionary

2nd Edition

W. Linnard
Freelance Translator
Radyr
Cardiff
UK

and

D. Darrah-Morgan
Freelance Translator
Pembridge
Herefordshire
UK

CABI *Publishing*

CABI is a trading name of CAB International

CABI Head Office
Nosworthy Way
Wallingford
Oxfordshire OX10 8DE
UK

Tel: +44 (0)1491 832111
Fax: +44 (0)1491 833508
Email: info@cabi.org
Web site: www.cabi.org

CABI North American Office
38 Chauncy St
Suite 1002
Boston, MA 02111
USA

Tel: +1 617 395 4051
Fax: +1 617 354 6875
Email: cabi-nao@cabi.org

A catalogue record for this book is available from the British Library, London, UK.

Library of Congress Cataloging-in-Publication Data
Linnard, William
Russian-English, English-Russian forestry and wood dictionary / W. Linnard and D. Darrah-Morgan.
p. cm.
ISBN 0-85199-321-4 (alk. paper)
1. Forests and Forestry—Dictionaries—Russian. 2. Forests and Forestry—Dictionaries. 3. Russian Language—Dictionaries—English. 4. English Language—Dictionaries—Russian. I. Darrah-Morgan, D. II. Title.
SD126.L56 1999 99-19672
634.9'039171—dc21 CIP

ISBN-13: 978-0-85199-321-8

First printed 1999
Transferred to print on demand 2014

Printed and bound by CPI Group (UK) Ltd, Croydon, CR0 4YY

Contents

Preface

The first edition of the *Russian-English Forestry and Wood Dictionary*, published by CAB in 1966, was well received but has long been out of print.

Since then, great changes have taken place in Russian forestry, and Russia and the other republics of the former USSR are now more accessible than at any time in history. The forest resource of Russia, easily the greatest of any country in the world, will become even more globally important, both environmentally and commercially, in the future. Increasing interest is being shown in Russian forests by international organizations such as the World Bank, FAO and WWF, as well as by a host of bilateral joint-venture schemes. With economic recovery in the forest sector in Russia, and the increasing internationalization of research, the scope and importance of Russian forestry literature will increase.

This new edition of the dictionary is an updated and considerably enlarged version of the first Russian-English edition, containing many new terms, species-names, acronyms and abbreviations to take account of the great developments that have taken place in Russian forestry, not only in forest mechanization and woodworking technology, but also in forest management, economics, environmental pollution, conservation and concern for the environment.

All the terms in the dictionary are from classical Russian forestry textbooks that form the bedrock of Russian forest terminology, and the stream of Russian forestry periodicals and books that have appeared over the last thirty years. The new English-Russian section makes this a handy specialist two-way dictionary, designed as a practical aid in reading, translating and abstracting Russian literature on forestry and forest products. A list of the botanical names of trees and shrubs, with their Russian and English equivalents, has also been included.

We gratefully acknowledge the support and encouragement of Mrs Gillian Petrokofsky (Forestry, CABI), and the practical assistance provided during compilation by Mrs Caroline Darrah-Morgan.

William Linnard
David Darrah-Morgan

email: darrahmorgan@compuserve.com

Common Russian abbreviations for the main tree species

Ак	acacia (usually *Robinia*)	Б	birch
Бк	beech	Бр	elm
В	elm	Г	hornbeam
Гр	hornbeam	Д	oak
Е	spruce	Ив	willow
Ил	elm	К	*Pinus sibirica*
Кл	maple	Кл	*Acer platanoides*
Кп	*Acer campestre*	Кт	*Acer tataricum*
Л	larch	Лп	lime
Лщ	hazel	Ол	alder
Ос	aspen	Оск	*Populus nigra*
П	fir	С	pine
Т	poplar	Я	ash
Яв	*Acer pseudoplatanus*	Яс	ash

Russian classification of site conditions (условия произрастания): four main groups of sites classified by nutrient status, subdivided by six soil moisture classes

		poor	fairly poor	fairly rich	rich
		A	B	C	D
very dry, xerophile	0	A_0	B_0	C_0	D_0
dry, meso-xerophile	1	A_1	B_1	C_1	D_1
fresh, mesophile	2	A_2	B_2	C_2	D_2
moist, meso-hygrophile	3	A_3	B_3	C_3	D_3
wet, hygrophile	4	A_4	B_4	C_4	D_4
very wet, ultra hygrophile	5	A_5	B_5	C_5	D_5

А

аапа aapa (eutrophic wet peatland in northern taiga)

абапол slab

аблактировка approach grafting

абрикос apricot

абрикос, маньчжурский Manchurian apricot, *Prunus mandshurica* (Maxim.) Koehne

абрикос, обыкновенный apricot, *Prunus armeniaca* L.

абрикос, сибирский Siberian apricot, *Prunus sibirica* L.

абрис outline, sketch, rough plan

авария average (in insurance), breakage, breakdown

авиалес aircraft timber

авиапиломатериалы aircraft lumber

авиация, лесная forestry aviation

автограф *Dryocoetus autographus* Ratz.

автоклав autoclave

автокубатурник automatic volume calculator

автолесовоз lorry, timber carrier, straddle carrier

автолесовоз, портальный straddle carrier

автомобиль automobile, lorry

автопогрузчик automatic loader, lift truck

автопогрузчик с вилочным захватом fork-lift truck

автопоезд 'train' of lorry-plus-trailer(s), road train

автотранспорт lorry transport

агрегат, фрезерно-пильный chipping headrig, Chip-N-Saw type of unit

агролесомелиоративная опытная станция experiment station for improvement of agriculture by forestry measures

агролесомелиорация improvement of agriculture by forestry measures

агротехника agricultural techniques

агрофон cultivation history (of a site)

адамово дерево paulownia, *Paulownia tomentosa* (Thunb.) Steud.

ажурность openness (e.g. of canopy or shelterbelt)

ажурный open

азалеа azalea (see also рододендрон)

азалия azalea (see also рододендрон)

азот nitrogen

азотобактерин azotobacterin (a preparation of *Azotobacter* spp. used as a bacterial fertilizer)

айва quince, *Cydonia oblonga* Mill.

айлант tree of heaven, *Ailanthus altissima* Swingle

акатник Amur acacia, *Maackia amurensis* Rupr.

акация acacia (but see below), *Acacia*

акация, амурская Amur acacia, *Maackia amurensis* Rupr.

акация, белая black locust, *Robinia pseudoacacia* L.

акация, желтая pea tree, *Caragana arborescens* Lam.

акация, песчаная *Ammododendron* Fisch.

акация, серебристая silver wattle, mimosa, *Acacia dealbata* Link.

акация, сибирская pea tree, *Caragana arborescens* Lam.

акация, шелковая silk tree, *Albizzia julibrissin* Dur.

акватория water (as opposed to land surface), area of water

акклиматизация acclimatization, conditioning

алас wet tundra-type area

албедо, интегральный integrated albedo

аллея avenue

АЛОС experiment station for improvement of agriculture by forestry measures

алтей лекарственный marsh mallow, *Althaea officinalis* L.

алыча plum, damson, *Prunus divaricata* Led.

амбар storehouse, warehouse

аммиак ammonia

аммоний, сернокислый ammonium sulphate

амортизация depreciation (allowance), amortization, repayment
аморфа, кустовидная *Amorpha fruticosa* L.
анализ analysis
анализ ствола stem analysis
ангелика angelica, *Angelica* L.
андромеда, дубровниколистная marsh andromeda, *Andromeda polifolia* L.
анизотропия anisotropy
анкер anchor, stay
антипирен fire retardant
антипка mahaleb cherry, *Prunus mahaleb* L.
антисептик preservative
антропоген anthropogenic, influenced by human activity
аншпуг lever, crowbar
аралия, маньчжурская Manchurian aralia, *Aralia mandshurica* Rupr. et Max.
арболит wood concrete
арборикультура arboriculture
ареал area, range
арена sandy terrace
аренда lease, rent
арендатор lessee
арендодатель lessor
арка arch
арочный arch, arched (adj.)
артель artel
артель, лесная forestry artel
артель, рабочих gang of workers
архив, клоновый clone bank
арча Central Asian juniper (several species) (see also можжевельник)
арчевник Central Asian juniper woodland
арык irrigation ditch
асбокартон asbestos board
ассектатор 'assectator' (natural plant associate of an 'edificator' species - see эдификатор)
ассессор, лесной junior (subordinate) forester
ассистент, лесной assistant forester
ассоциация association
астрогал, сладколистный milk vetch, *Astragalus glycyphyllos* L.
атом, меченый labelled atom, tracer
ауксибласт auxiblast (elongated one-year shoot)
аэробный aerobic
аэрозоль aerosol
аэросев aerial sowing
аэростат balloon
аэрофотография aerial photography
аэрофотоснимок aerial photograph
аэрофотосъемка aerial photography

Б

бабочка butterfly
баган horizontal wood pile (in charcoal burning)
багасса bagasse
багет strip, fillet, moulding
багон Scots pine forest on waterlogged soil
багор floating hook, pike pole, cant hook
багрянник katsura tree, *Cercidiphyllum japonicum* Sieb. et Zucc.
багульник, болотный *Ledum palustre* L.
бадан, толстолистный *Bergenia crassifolia* Fritsch.
база base
база, лесоперевалочная place where wood is transferred from water to land transport
база, лесосырьевая forest land leased for logging
базирование providing a flat base, positioning (for working in a machine), planing to give the required angle between two adjacent faces
байбак bobak marmot, *Marmota bobac* L.
байрак wooded ravine
бак tank, container
бакаут lignum vitae, *Guaiacum officinale* L.
бактерия bacterium
баланс древесины recovery

балансы pulpwood
балка beam, cant; ravine
балх-арча Turkestan juniper, *Juniperus turkestanica* Kom.
бальзам balsam
бальзам, пихтовый fir balsam
бальзамин-недотрога touch-me-not, *Impatiens nolitangere* L.
БАМ Baikal/Amur railway
бамбук bamboo
бандаж bandage (wood preservation)
банкет bench (of land)
барабан drum
барбарис barberry, *Berberis* L.
барбарис, обыкновенный common barberry, *Berberis vulgaris* L.
барвинок, малый lesser periwinkle, *Vinca minor* L.
барда spent liquor
баржа barge
баррас scrape (in resin tapping)
барсук badger, *Meles meles* L.
бархан barkhan (crescent-shaped shifting sand dune)
бархат, амурский Amur cork tree, *Phellodendron amurense* Rupr.
бархатовник *Phellodendron amurense* forest
бассейн basin, log pond
бастард hybrid
батрак hired labourer
баттенсы battens
бахрома fringe
башмак shoe
башня tower
башня, водонапорная water tower
башня, силосная silo
башня, сторожевая watch tower, look-out tower
баялыш *Salsola arbuscula* Pall.
бедра дерева root swelling
без коры under bark, u.b.
безвершинник pollard (tree)
бездорожье absence of roads
безлесица absence of forests
безлесный without forests, treeless
безлесье absence of forests
безопасность safety
безсучковый branchless, knot-free
безымянка short thin board
белена, черная henbane, *Hyoscyamus niger* L.
беленый bleached
белесый bleached (of soil)
белизна whiteness
белка squirrel, *Sciurus* L.
белка, закавказская *Sciurus persicus* Erx.
белка, обыкновенная red squirrel, *Sciurus vulgaris* L.
белкование squirrel hunting
белковый protein, albumin(ous)
белладонна deadly nightshade, *Atropa belladonna* L.
белодушка stone marten, *Martes foina* Erx.
белок protein, albumin
белокудренник, черный black horehound, *Ballota nigra* L.
белолоз willow, *Salix alba* L. or *Salix triandra* L.
беломошник area of *Cladonia* (see also бор беломошник)
белотал willow, *Salix viminalis* L. or *Salix triandra* L.
белоус, торчастый mat-grass, *Nardus stricta* L.
беляк mountain hare, *Lepus timidus* L.
бензин petrol, benzine
бензол benzene
бензопила petrol-driven power saw
берег bank, shore
береза birch, *Betula* L.
береза, бородавчатая silver birch, *Betula verrucosa* Ehrh.
береза, даурская black birch, *Betula dahurica* Pall.
береза, железная iron birch, *Betula schmidtii* Reg.
береза, желтая (далневосточная) Siberian yellow birch, *Betula costata* Traut.
береза, каменная stone birch, *Betula ermanii* Cham.
береза, карельская Karelian birch, *Betula verrucosa* f. *carelica*

береза, карликовая dwarf birch, *Betula nana* L.
береза, кустарниковая dwarf birch, *Betula fruticosa* Pall.
береза, низкая *Betula humilis* Schrank.
береза, повислая silver birch, *Betula pendula*
береза, пушистая downy birch, *Betula pubescens* Ehrh.
береза ребристая Siberian yellow birch, *Betula costata* Traut.
береза, черная black birch, *Betula dahurica* Pall.
береза Шмидта iron birch, *Betula schmidtii* Reg.
березка dwarf birch
березняк birch forest, *Betuletum*
березовик, обыкновенный birch mushroom, *Boletus scaber* Bull.
березовица birch sap
березовый birch
берека, лечебная wild service, *Sorbus torminalis* Crantz
бересклет spindle tree, *Euonymus* L.
бересклет, бородавчатый rough spindle tree, *Euonymus verrucosus* Scop.
бересклет, европейский European spindle tree, *Euonymus europaeus* L.
берест elm, *Ulmus* L., often smooth elm, *Ulmus carpinifolia* Gled. (*Ulmus foliacea* Gilib.) (see also вяз)
береста birch bark
беркут golden eagle, *Aquila chrysaetos* L.
берма berm
бессмертник, лесной wood cudweed, *Gnaphalium sylvaticum* L.
бессортный unsorted
бессучковый branchless, knot-free
бетон, древесный wood concrete
бешеница hemlock, *Conium maculatum* L.
билет ticket, pass, permit
билет, лесорубочный felling permit, felling licence
биогеоценоз biogeocoenosis
биогеоценология biogeocoenology
биом biome
биоопыт bioassay
биоразнообразие biodiversity
биосфера biosphere
биостойкость resistance to biological agencies
биота arbor-vitae, *Thuja* L.(see also туя)
биотест bioassay
биотип biotype
биотоп biotope
биоценоз biocoenosis
биржа **1.** timber yard, wood storage place **2.** exchange, commodity exchange
биржа, лесная wood exchange
бирка tally
бирючина, обыкновенная common privet, *Ligustrum vulgare* L.
биюргун *Anabásis salsa* C. A. May
б/к under bark, u.b.
блеск lustre
блик sunfleck, fleck (of light)
блок block, pulley, log
блок, оконный window frame
блок, подцепочный fall block
блошак, дубовый oak flea beetle, *Haltica saliceti* Wse.
блошник polypody, *Polypodium vulgare* L.
блуд play (of saw)
блуждание пилы saw wobble
бобовые Leguminosae
бобовый leguminous
бобр beaver, *Castor fiber* L.
богара dry to arid, non-irrigated s(i)erozem or montane chestnut soils in Central Asia, usually on slopes and eroded or threatened by erosion; hence, the system of sowing (or planting) without irrigation
богарный without irrigation (see богара)
бодяк thistle, *Cirsium* Mill.
бодяк, болотный *Cirsium palustre* Scop.
бодяк, ланцетолистный *Cirsium lanceolatum* Scop.
бодяк, овощный *Cirsium oleraceum* Scop.
боек head (of a hammer)
бой face (of a hammer)
боккаут Persian iron wood, *Parrotia persica* C.A.M.

боковник staves (for casks)
болванка blank, hewn blank
болезнь disease, illness
болезнь, голландская (береста, ильма, вяза) Dutch elm disease
болезнь, патологическая pathological disease
болиголов hemlock, *Conium maculatum* L.
болонь sapwood (see also заболонь)
болотистый boggy, marshy, swampy
болото bog, fen, marsh, swamp, wet moor
болото, верховое bog, upland bog
болото, висячее hanging bog
болото, выпуклое raised bog
болото, зыбунное quaking bog
болото, моховое moss bog, moss fen
болото, низинное eutrophic peatland, fen
болото, низовое eutrophic peatland, fen
болото, переходное transitional bog
болото, плоское blanket bog
болото, склоновое hanging bog
болото, торфяное peat bog
болотоведение science of bogland (see болото)
большеболонность excessively wide sapwood development
бон boom (see also запань)
бон, задерживающий holding boom, storage boom
бон, направляюший guiding boom, glance boom
бон, однобревенный single-log boom
бон, поперечный transverse boom, holding boom
бон, продольный longitudinal boom, glance boom
бон, реевный glance boom, wing boom, fin boom
бон, шпоночный keyed boom
бондарная cooperage, cooper's workshop
бондарня cooperage, cooper's workshop
бондарь cooper
бонитет site class
бонитирование determination of site class
бонитировка determination of site class
бор **1.** pine forest, infertile pine site type, *Pinetum* **2.** boron
бор, багульниковый *Pinetum ledosum*
бор -беломошник *Pinetum cladinosum*
бор, болотный *Pinetum uliginosum*
бор -брусничник *Pinetum vacciniosum*
бор, вересковый *Pinetum callunosum*
бор -долгомошник *Pinetum polytrichosum*
бор -зеленомошник *Pinetum hylocomiosum*
бор, кисличный *Pinetum oxalidosum*
бор, ленточный ribbon-like pine forest in Siberia
бор, лишайниковый *Pinetum cladinosum*
бор, папоротниковый *Pinetum dryopteridosum*
бор, развесистый wood millet, *Milium effusum* L.
бор, разнотравный *Pinetum mixtoherbosum*
бор, сложный *Pinetum compositum*
бор, сфагновый *Pinetum sphagnosum*
бор, травяной *Pinetum herbosum*
бор -черничник *Pinetum myrtillosum*
бордсы boards
боровик *Boletus edulis* Bull.
бородавник, обыкновенный nipplewort, *Lapsana communis* L.
борозда furrow
бороздка drill, small furrow
бороздка, посевная seed drill
бороздник drill marker
борона harrow
боронение harrowing
бортник forest bee-keeper
бортничество forest bee-keeping
борщевик cow parsnip, *Heracleum* L.
борьба struggle, control, prevention
борьба, химическая chemical control
бочар cooper
бочка barrel, cask
бочка, заливная wet barrel
бочка, сухотарная dry barrel
боярышник hawthorn, *Crataegus* L.
боярышник, восточный eastern hawthorn, *Crataegus orientalis* Pall.

боярышник, колючий Midland hawthorn, *Crataegus oxyacantha* L.
боярышник, обыкновенный Midland hawthorn, *Crataegus oxyacantha* L.
боярышник, однопестичный hawthorn, quickthorn, *Crataegus monogyna* Jacq.
боярышник, пятипестичный hawthorn, *Crataegus pentagyna* W.K.
боярышница *Aporia crataegi* L.
БПК BOD, biological oxygen demand
бражник sphingid, hawk moth
бражник, сиреневый *Sphinx ligustri* L.
бражник, сосновый *Sphinx pinastri* L.
брак rejects, rejected material
бракер grader, sorter, inspector
бракованный rejected
браковать reject, grade, inspect
браковка rejection, grading, inspection
браконьер poacher
браконьерство poaching
брахибласт brachyblast (short shoot)
брашпиль windlass
бревенник tall tree, timber tree
бревно log
бревно долготье long log
бревно, комлевое butt log
бревно, пиловочное sawlog
бревнокатка log rollway
бревносбрасыватель log ejector
бревносвал log unloader
бревноспуск log chute
бревнотаска log conveyor
бредина goat willow, sallow, *Salix caprea* L.
бригада team, gang, crew
бригада, комплексная mechanized gang
бригадир ganger, team leader
брикет briquette
бровка edge, border, 'Chinese beard' (inverted V-shaped swelling on stems, associated with branch stub occlusions)
брожение fermentation
бронзовка, золотистая rose chafer, *Cetonia aurata* L.
брус balk, square, cant, flitch
брус, двухкантный cant (log with two opposite slabs removed)
брус, переводный crossing sleeper
брус с обзолом waney balk
брус, трехкантный cant (with three straight sides)
брус, четырехкантный square
брусника cowberry, red whortleberry, *Vaccinium vitis-idaea* L.
брусничник area of *Vaccinium vitis-idaea* (see also бор брусичник)
брусовка removal of two opposite slabs, slabbing, breaking down a log so as to leave a baulk or baulks in the centre
брусок scantling, batten
брусок, мерный measuring lath
брусок, сажальный lining-out board
брусок, сеяльный sowing lath
брусоперекладчик cant re-setter
бугор hill(ock), protuberance, mound
будра, плющевидная ground ivy, *Glechoma hederacea* L.
будра, шерстистая *Glechoma hirsuta* W. et K.
бузина elder, *Sambucus* L.
бузина, красная red elder, *Sambucus racemosa* L.
бузина, черная common elder, *Sambucus nigra* L.
бук beech, *Fagus* L.
бук, восточный oriental beech, *Fagus orientalis* Lipsky
бук, европейский beech, *Fagus sylvatica* L.
бук, кавказский oriental beech, *Fagus orientalis* Lipsky
бук, крымский *Fagus taurica* Popl. (a Crimean hybrid between *F. sylvatica* and *F. orientalis*)
бук, лесной beech, *Fagus sylvatica* L.
букашник, горный sheep's bit, *Jasione montana* L.
букняк beech forest, fertile beech forest type, *Fagetum*
буковый beech (adj.)
буксир tug (boat)
буксировка towing
буль boule
бульдозер bulldozer
бумага paper

бумага, влагопрочная wet-strength paper
бумага, газетная newsprint
бумага, жиростойкая grease-proof paper
бумага, лакмусовая litmus paper
бумага, мешочная bag or sack paper
бумага, оберточная wrapping paper
бумага, обойная wallpaper
бумага, текстурная decorative paper overlay
бумага, фильтровальная filter paper
Бумимпэкс Paper Import/Export Agency
бур borer, drill, auger
бурав borer, drill, auger
бурав, возрастный increment borer
бурав, почвенный soil auger
бурав Пресслера Pressler increment borer
бурав, приростный increment borer
буравить drill, bore
буравчик gimlet
буран heavy snow storm
буревал windthrow
бурелом wind breakage, wind-broken trees
бурозем burozem (brown earth)
бурундук Siberian chipmunk, *Eutamias sibiricus* Laxm.
бурьян weeds
буря storm
буссоль compass
бутон flower bud
бутонизация flower-bud formation
бухарник, мягкий creeping soft-grass, *Holcus mollis* L.
бухарник, шерстистый Yorkshire fog, *Holcus lanatus* L.
бучина beech forest, fertile beech forest type, *Fagetum*
бык pier, bull, ox
бьеф reach (of a river)
бьеф, нижний lower reach, downstream

В

вагон wagon
вагонетка trolley, cart
важенка female deer
вайма glue press, clamp
вайя frond
вал 1. roll, roller, shaft 2. bank, embankment
вал, ножевой cutter-block
вал отбора мощности power take-off
валежник windthrown tree, windfall
валериана, лекарственная valerian, *Valeriana officinalis* L.
валик 1. roll, roller, shaft 2. bank, embankment
валить to fell, to cut out, to cut down
валка felling, cutting
валка, направленная directional felling
валок roll, roller, cylinder
валокант timber square on 4 sides but still having waney edges
валочно-пакетирующая машина feller-buncher
валочно-трелевочная машина feller-skidder
валун boulder
вальдшнеп woodcock, *Scolopax rusticola* L.
вальцовка rolling, tensioning (e.g. saws)
вальщик feller
ванна bath, trough, vat
ванчес wainscot billet, flitch
вар pitch
варакушка bluethroat, *Luscinia svecica* L.
варка cooking, boiling
варка, гидротропная hydrotropic cooking (cooking with Na xylenesulphonate solution)
василек *Centaurea* L.
вахта 1. watch, duty 2. logging camp, woods camp
вахта, трехлистниковая buckbean, *Menyanthes trifoliata* L.
ваяние carving
ввоз import, importation
вдоль волокон along the grain, parallel to grain
ведение 1. knowledge 2. management
ведение рубки felling system

ведомость report, list
ведомость рубкам felling plan
ведомство department, office
ведро pail, bucket
ведьмина метла witches' broom
вездеход all-terrain vehicle, amphibious vehicle
вейник *Calamagrostis* Roth
вейник Лангсдорфа *Calamagrostis langsdorffii* Trin.
вейник, ланцетный purple smallreed, *Calamagrostis lanceolata* Roth
вейник, лесной *Calamagrostis arundinacea* Roth
вейник, наземный bush grass, *Calamagrostis epigeios* Roth
вейник, тростниковидный *Calamagrostis arundinacea* Roth
вексель bill of exchange
векша red squirrel, *Sciurus vulgaris* L.
венец, зубчатый sawing chain
веник thin twig with leaves
венчик corolla
вепренок young wild boar
вепрь wild boar, *Sus scrofa* L.
верба willow, *Salix daphnoides* Vill.
верба, красная willow, *Salix acutifolia* Willd.
вербейник, обыкновенный yellow loosestrife, *Lysimachia vulgaris* L.
вербена, лекарственная vervain, *Verbena officinalis* L.
вербняк stand of willow
верболоз willow, *Salix viminalis* L. or *Salix acutifolia* Willd.
вереск heather, *Calluna vulgaris* L. (Hull.)
вереск, болотный cross-leaved heath, *Erica tetralix* L.
вереск, древовидный tree heath, *Erica arborea* L.
вересковые Ericaceae
веретено spindle
веретенообразный fusiform
верещатник area overgrown with heather
вероника, лекарственная common speedwell, *Veronica officinalis* L.
верста verst (= 1.07 km)
верстак workbench
вертишейка wryneck, *Jynx torquilla* L.
вертлюг shackle
вертун cant hook, peavy
вертун, сосновый pine twisting rust, *Melampsora pinitorqua* Rost.
верховодка perched water-table
верховой crown thinning (see прореживание, верховое)
верхолаз high-rigger, tree climber
верхушечный terminal, tip, apical
верхушка top, tip, summit, apex
вершалина harrow made of a spruce top
вершина top, tip, summit, apex
вершинник top log
вершины и сучья lop and top
вершняк upper storey, overwood
вершок **1.** top, tip, summit, apex **2.** vershok (= 4.4 cm)
вес weight
вес, воздушно-сухой air-dry weight
вес, объемный density
вес, удельный **1.** specific gravity **2.** share, proportion
вес, чистый net weight
весовые части parts by weight
весы scales, balance
весы, гидростатические scales for weighing wood in air and then in water
ветверезка, древолазающий tree-climbing branch-pruner
ветвистость branchiness
ветвление branching
ветвь branch, twig; a branched shoot formed of ветки (see ветка below)
ветка branch, twig; unbranched shoot over 1 year old
ветла white willow, *Salix alba* L.
ветлянник willow stand
веточка small twig
ветреница **1.** shake (visible at butt end only) **2.** anemone
ветреница, белая wood anemone, *Anemone nemorosa* L.
ветреница, лесная *Anemone sylvestris* L.
ветреница, лютиковая yellow wood anemone, *Anemone ranunculoides* L.

ветренница 1. shake (visible at butt end only) 2. anemone
ветровал windthrow, windthrown trees
ветролом wind breakage, wind-broken trees
ветроопыляемость wind-pollination, anemophily
ветроопыляемый wind-pollinated, anemophilous
ветроупорный wind-resistant
ветроустойчивый wind-resistant
ветхость dilapidation, decrepitude
вех, ядовитый cowbane, *Cicuta virosa* L.
веха 1. boundary mark 2. surveying rod, surveying pole
вечнозеленый evergreen
вешенка, обыкновенная *Pleurotus ostreatus*
вещество substance, material
вещество, горючее fuel
вещество, дубильное tanstuff, tanning material
вещество, кормовое fodder, feed, forage
вещество, красящее dye, dyestuff, pigment
вещество, отпугивающее repellent
вещество, питательное nutrient
вещество, ростовое growth substance
вещество, сухое dry matter
вещество, усвояемое минеральное available nutrient
взаимодействие interaction, reaction
вздымка freshening, streaking, streak (in resin tapping)
вздымщик resin-tapper
взрослый adult
взрыв explosion
взрывание blasting
взрывать to blow up, to blast
взрыхление loosening (of soil)
взрыхлитель soil loosener, scarifier
взрыхлять to loosen
вид 1. appearance, form 2. species 3. kind, sort
визир sight line, sighting line, survey line
визировать to sight
вика vetch, *Vicia* L.
вилка fork
вилка, валочная push pole (in felling)
вилка, мерная caliper
вилообразный forked
вилохвостки Collembola
вильчатый forked
винт screw
вискоза viscose
витютень wood pigeon, *Columba palumbus* L.
вишарник ground cherry, *Prunus fruticosa* Pall.
вишня cherry
вишня, кустарниковая ground cherry, *Prunus fruticosa* Pall.
вишня, обыкновенная sour cherry, *Prunus cerasus* L.
вишня птичья wild cherry, *Prunus avium* L.
вишня, степная ground cherry, *Prunus fruticosa* Pall.
вишняк ground cherry, *Prunus fruticosa* Pall.
в/к o.b., over bark
вкладыш bearing, bush
в коре o.b., over bark
влага moisture
влага, свободная free moisture
влага, связанная bound moisture
влагалище sheath
влагоемкость moisture capacity
влагоемкость, максимальная адсорбционная maximum adsorption capacity
влагоемкость, наименьшая field capacity
влагоемкость, полевая field capacity
влагоемкость, полная maximum water-holding capacity
влагомер moisture meter, hygrometer
влагопоглощение moisture absorption, hygroscopicity
влагопроводность moisture conduction
влажность moisture, moisture content, humidity
влажность, относительная relative humidity

влажность, равновесная equilibrium moisture content
влажность, транспортная shipping-dry condition
влажность устойчивого завядания permanent wilting point
влажность, эксплуатационная service moisture content
влажность почвы soil moisture
ВЛТИ Voronezh Forestry Institute
вместилище receptacle, container
вместилище, смоляное resiniferous cavity
вместимость capacity, volume
внедрение introduction
внешний external, outer
ВНИАЛМИ All-Union Research Institute for Forestry in Agricultural Improvement
ВНИИЛМ All-Union Research Institute for Forestry and Forestry Mechanization
внутренний internal, inner
внутривидовой intraspecific
вода water
вода, грунтовая groundwater
вода, подсмольная tar water
вода, сточная effluent
вода, талая melt water, thaw water
водонепроницаемый watertight, waterproof, impermeable to water
водоносный слой aquifer
водопоглощение water absorption
водопой watering place
водопроницаемость permeability to water, infiltration capacity
водораздел watershed, divide
водоросль alga
водосбор water catchment
водосбор, обыкновенный columbine, *Aquilegia vulgaris* L.
водослив 1. spillway, overflow 2. water bombing (aerial fire fighting)
водослой wetwood, watercore
водосток drain, outlet
водохранилище reservoir
водяника, черная crowberry, *Empetrum nigrum* L.
водянка, бактериальная bacterial disease of oak (wilt)
воз cart-load, cart
возбудитель 1. pathogen 2. exciter, inducer
воздух air
воздушная трелевочная установка aerial cableway
воздушно-сухой air-dry
возка transport, transporting, hauling
возмужалость maturity
возобновление regeneration, renewal
возобновление, вегетативное vegetative regeneration
возобновление, дополнительное supplementary regeneration
возобновление, естественное natural regeneration
возобновление, искусственное artificial regeneration
возобновление, каемчатое strip regeneration
возобновление, куртиное group regeneration
возобновление, порослевое coppice regeneration, sucker regeneration
возобновление, последовательное regeneration established after felling of the main crop
возобновление, последующее regeneration established before felling of the main crop
возобновление, предварительное regeneration established before felling of the main crop
возобновление, семенное regeneration by seed
возобновляемый renewable
возобновлять to regenerate
возраст age
возраст, господствующий prevailing age, age of the majority of the trees forming the stand
возраст, оборотный rotation age, exploitable age
возраст, предельный age limit
возраст рубка rotation age, exploitable age

возраст спелости maturity, rotation age (see also спелость)
возраст, средний mean age
войлок **1.** felt, tomentum **2.** mat (of dry grass etc.)
войлочник, буковый felted beech coccus, *Cryptococcus fagi* Bär.
волдырник, ягодный berry catchfly, *Cucubalus baccifer* L.
волк **1.** wolf, *Canis lupus* L. **2.** wolf tree
волнистость waviness
волнистый wavy
волнолом wave-breaker
волнянка lymantriid
волок skidding trail, skidding track
волокнистый fibrous
волокно **1.** fibre **2.** grain
волокно, вырванное torn grain
волокно, древесное wood fibre
волокно, задранное chipped grain
волокно либриформа libriform fibre
волокно, лубяное phloem fibre
волоконце fibril
волокуша **1.** braking drag **2.** drag harrow
волоснец, песчаный lyme grass, *Elymus arenarius* L.
волосность capillarity
волосок, корневой root hair
волочение dragging
волчеягодник mezereon, *Daphne mezereum* L.
волчок epicormic shoot
волчье лыко mezereon, *Daphne mezereum* L.
ВОМ power take-off
воробей sparrow
воробей, домовый house sparrow, *Passer domesticus* L.
воробей, полевой tree sparrow, *Passer montanus* L.
ворон raven, *Corvus corax* L.
ворона crow
ворона, серая hooded crow, *Corvus cornix* L.
ворона, черная carrion crow, *Corvus corone* L.
ворот winch, windlass
ворота gate, gateway
ворсистость woolliness (of grain)
воск wax
воск, древесный grafting wax
восковник, обыкновенный sweet gale, *Myrica gale* L.
воспитание education, training, raising, tending
воспламеняемость inflammability
воспроизведение reproduction
воспроизводство reproduction
восстановление restoration, re-establishment, regeneration
В/О ЭКСПОРТЛЕС All-Union Wood Exporting Agency
впадина **1.** hollow **2.** gullet (of saw tooth)
ВПМ feller-buncher
впотай flush, even
впритык butt, flush
вразбежку alternate, staggered
вразброс broadcast
вразвал through and through (in sawing)
вразнокомелицу 'butt to top' (i.e. describing loads having butts and tops at both ends)
врастание growing in, ingrowing
вращение rotation, revolution
вред harm, damage, injury
вредитель injurious organism, pest
вредительный injurious, harmful
вредный injurious, harmful
врезать to cut in, to notch, to score
время time, season
всасывание suction, absorption
вспахивание ploughing, tillage
вспахивать to plough, to till
вспашка ploughing, tillage
вспашка, плантажная deep ploughing
вспомогательный auxiliary
вспышка outbreak, insect outbreak
ВСРПМ feller-trimmer-buncher
встречаемость frequency, incidence
всход young seedling (usually first-year)
всходить to germinate
всхожесть germinative capacity, viability
всхожесть, грунтовая field germination
всхожий viable
всхожный viable

втаскивание dragging in
ВТМ feller-skidder
ВТУ aerial cableway
втулка bush, sleeve
в.ч. parts by weight
выбор **1.** choice, selection **2.** removal
выбор пород species choice
выборка **1.** choice, selection **2.** removal
выборка, случайная random sample
выборочный selective, selection
выбраковать to sort out, to cull
выбросы, промышленные industrial emissions
вываливание windthrow
выверка adjustment (e.g. of circular saw)
выветривание weathering
вывоз export
вывозить to extract, to export
вывозка extraction, transport, hauling (of timber from the forest)
вывозка, безтрелевочная direct extraction (without skidding)
выгон pasture
выгонка distillation
выгрузка unloading, discharging
выдел subcompartment
выдел, таксационный subcompartment
выдел участка division of an area into subcompartments
выделение separation, emission, secretion, excretion
выдра otter, *Lutra lutra* L.
выдувание blowing away, wind erosion
выемка cutting, excavation, notch
выжег main burn (in shifting cultivation)
выживаемость capacity for survival
выживание survival
выжигание burning, prescribed burning
выжимание frost heaving
вызревание ripening
выкапывание digging up, lifting
выкапывать to dig up, to lift
выкопка digging up, lifting
выкорчевывание rooting out, grubbing up, uprooting
выкорчевывать to grub up, to uproot
выкрошивание breaking into crumbs, crumbling
вылущивать to extract (seed)
вымачивание steeping, soaking, wetting
вымерять to measure
вымирание dying off, extinction
вымокание 'drowning' (of seedlings)
вымывание washing out, leaching
выносливость endurance, hardiness, resistance
выносливый resistant, hardy
вынос, селевой mud flow, rock flow
выпадение precipitation
выпалывание weeding
выпиливание sawing, sawing out
выпревание rotting (e.g. of seedlings)
выпуск **1.** number (of journal) **2.** manufacture, production, output **3.** outlet
выпь bittern, *Botaurus stellaris* L.
выработка manufacture, production, output
выработка, комплексная annual output (in m^3 per worker, usually in logging)
выравнивать to level off
выращивать to raise, to rear, to cultivate
вырезать to cut (out)
вырождение degeneration
вырост excrescence, outgrowth
вырубание felling, cutting
вырубать to fell, to cut out, to cut down
вырубка (see also рубка) **1.** felling, cutting **2.** felling site, felled area
вырубка, сметная planned cut
вырубленный cut, felled
высадка planting out
высаживать to plant out
высверливание drilling, boring
выслеживание tracking, stalking
высокосортный high-grade
высокоствольник high forest (of seedling origin)
высокоствольный high-forest (of seedling origin) (adj.)
высокотравие tall herbaceous cover
высота height
высота вершины height to top
высота, видовая form height
высота груди breast height (1.3 m)
высота линейки pressure-bar lead
высота насаждения stand height

высота пропила depth of cut (height of kerf), board thickness
высота, срединная mid height
высота, средняя mean height
высотомер hypsometer
выставить на свободу to release
выставка 1. exhibition 2. projection (e.g. of a blade)
выстрагивать to plane
выстроганный planed
выступ-ограничитель depth-gauge (in chain-saw teeth)
высушенный dried, seasoned
высушивание drying, drying up, desiccation, dying
высушивание, зимнее winter drying
высыхание drying, drying up, desiccation, dying
вытаптывание trampling
вытаптывать to trample
вытаска extraction, dragging out
вытеснение displacement, crowding out
вытрелевка skidding out
вытяжка extract
выход yield
вычисление calculation
вышка tower, watch tower
вышка, пожарная fire tower, look-out
выщелачивание leaching, lixiviation
выщепление segregation (genetics)
вьюга blizzard
вьюнок, полевой bindweed, *Convolvulus arvense* L.
вьюрок brambling, *Fringilla montifringilla* L.
вьющийся twining
вяз elm, *Ulmus* L.
вяз, гладкий European white elm, *Ulmus laevis* Pall.
вяз, густой *Ulmus densa* Litv.
вяз, лиственный smooth elm, *Ulmus carpinifolia* Gled. (*U. foliacea* Gilib.)
вяз, листоватый smooth elm, *Ulmus carpinifolia* Gled. (*U. foliacea* Gilib.)
вяз, мелколистный usually Siberian elm, *Ulmus pumila* L. var. *arborea* Litv.; sometimes Chinese elm (*U. parvifolia* Jacq.)
вяз, обыкновенный 'common elm', usually European white elm, *Ulmus laevis* Pall.
вяз, перистоветвистый Siberian elm, *Ulmus pumila* L., var. *arborea* Litv. (*U. pinnato-ramosa* Dieck.)
вяз, приземистый Siberian elm, *Ulmus pumila* L.
вяз, пробковый *Ulmus suberosa* Moench
вяз, шершавый wych elm, *Ulmus glabra* Huds. (*U. scabra* Mill.)
вязание binding
вязанка bundle
вязель, разноцветный crown vetch, *Coronilla varia* L.
вязка binding
вязкость 1. viscosity 2. toughness
вязник elm stand, *Ulmetum*
вязовник elm stand, *Ulmetum*
вяхирь wood pigeon, *Columba palumbus* L.

Г

га ha (hectare)
габарит overall dimensions, external dimensions
габитус habit
гавань port
газ gas
газ, генераторный producer gas
газификация древесины production of gas from wood
газогенератор gas generator
газопроницаемость permeability to gas
газоустойчивость resistance to gas
газочурка wood block for producer gas
галечник pebble-bed
галипот galipot
галка jackdaw, *Corvus monedula* L.
галл gall
галлица cecidomyid, gall-gnat
галлица, еловая семенная *Plemeliella abietina* Seitn.
галлица, ивовая *Rhabdophaga salicis* Schr.

галлица, лиственничная *Dasyneura laricis* Loew.
галлица, сосновая красная *Thecodiplosis brachyntera* Schwaeg.
галька pebbles
гарпиус colophony, rosin
гаршнеп jack snipe, *Lymnocryptes minimus* Brünn.
гарь burnt area, burn
гаситель скорости течения current-reducer
гать **1.** dam **2.** brushwood or corduroy road
гваякол guaiacol
гвоздодер nail claw, nail pullers
гвоздь nail
гейстер sapling, large transplant (2-2.5 m tall)
гексахлоран BHC
гектар hectare
гектограф *Dryocoetes hectographus* Reitt.
гемицеллюлоза hemicellulose
ген gene
генеративный reproductive
геном genome
генотип genotype
генофонд gene resource(s), gene bank
геоботаника plant-sociology
геоценоз geocoenosis
герань *Geranium* L.
герань, болотная *Geranium palustre* L.
герань, кровянокрасная *Geranium sanguineum* L.
герань, лесная *Geranium sylvaticum* L.
герань, луговая *Geranium pratense* L.
герань, Роберта herb Robert, *Geranium robertianum* L.
герань, темная *Geranium phaeum* L.
гербарий herbarium
гербицид herbicide
гибель mortality, dying
гибкость flexibility
гибридизация hybridization
гигроскопичность moisture absorption, hygroscopicity
гидроклин hydraulic felling wedge
гидрокран hydraulic crane
гидролес wood used in freshwater and marine constructions
гидролиз hydrolysis
гидрология hydrology
гидроманипулятор grapple
гидростанция hydroelectric station
гидротехника hydraulic engineering
гидротропный hydrotropic (see варка)
гидрофильность hygroscopicity
гидрофобизация water-repellent treatment
ГИЛФ State Forest Inventory
гильза **1.** bush, sleeve **2.** case, hull
гинкго maidenhair tree, *Ginkgo biloba* L.
ГИС geographic information system, GIS
ГК gibberellic acid
г.к. hydrolytic acidity, pH units
главный chief, principal, main
глаз eye
глаз вороний herb Paris, *Paris quadrifolia* L.
глаз птичий bird's-eye figure
глазок **1.** eye, eyelet **2.** undeveloped intergrown dormant bud **3.** bud shield
глазомер ocular estimation
глазомерный by eye, ocular
гледичия honey-locust, *Gleditsia* L.
гледичия, каспийская Caspian honey-locust, *Gleditsia caspica* L.
гледичия, обыкновенная honey-locust, *Gleditsia triacanthos* L.
глина clay
глинистый clayey
глинозем clay soil
глог dogwood, *Cornus sanguinea* L.
глоговина wild service tree, *Sorbus torminalis* Crantz
глубина depth
глухарь capercaillie, *Tetrao urogallus* L.
глушь thicket, wilderness
глыба clod, ball (of earth), lump
гнездо **1.** nest, clump **2.** mortise, pocket (e.g. for hinge)
гнездовка, настоящая bird's-nest orchid, *Neottia nidus-avis* Rich.
гнездовье nesting place
гниение rotting, decay, rot
гнилой rotten, decayed

гниль rot, decay
гниль, белая white rot
гниль, бурая brown rot
гниль, бурая (дуба) brown rot of oak, *Daedalea quercina* Fr.
гниль, бурая (лиственницы) brown rot of larch, *Fomes officinalis* Fr.
гниль, вершинная top rot
гниль, деструктивная brown rot
гниль, заболонная sap rot
гниль, комлевая butt rot
гниль, корневая root rot
гниль, коррозионная white rot
гниль, красная (сосны) red rot of pine and larch, *Trametes pini* Fr.
гниль, мягкая soft rot
гниль, периферическая sap rot
гниль, пестрая mottled rot
гниль, пестрая (ели) mottled rot of spruce, *Trametes abietis* Karst.
гниль, пестрая напенная mottled butt rot, (*Fomes annosus* (Fr.) Cke.)
гниль, плесневая soft rot
гниль, призматическая cubical rot
гниль, сердцевинная heart rot
гниль, стволовая trunk rot, stem rot
гниль, умеренная soft rot
гниль, центральная heart rot
гниль, ямчатая pocket rot
гнус biting flies, blood-sucking flies
гнуть to bend
гнутье bending
гнутье, торцовое end-grain bending
год year
год, семенной seed year, mast year
год, урожайный seed year, mast year
годичный annual, yearly
годный suitable
годовалый yearling, one year old (adj.)
годовик yearling
голец bare mountain
головка **1.** small head **2.** end, tip
головка, ножевая cutter head
головка, фрезерная cutter head
голодание deficiency (nutrient)
гололедица glazed frost, ice crust
голосеменные gymnosperms
голосеменный gymnospermous
голосемянные gymnosperms
голосемянный gymnospermous
голубика bog whortleberry, *Vaccinium uliginosum* L.
голубь pigeon, dove, *Columba* L.
голый naked, bare
голье bare patch, gap
гольцы screes
гон rut (zool.)
гонка **1.** distillation **2.** river driving **3.** chase
гонт shingle
гонять **1.** to drive, to chase **2.** to distil
гора mountain
горбина slab
горбыль slab
гордовина wayfaring tree, *Viburnum lantana* L.
горельник fire-damaged forest
горение combustion, burning
гореть to burn
горец, вьюнковый black bindweed, *Polygonum convolvulus* L.
горец, песчаный *Polygonum arenarium* W.K.
горец, почечуйный *Polygonum persicaria* L.
горец, узловатый *Polygonum nodosum* Pers.
горечавка gentian, *Gentiana* L.
горизонт horizon
горизонт, пахотный ploughed horizon
горизонт вмывания illuvial horizon
горизонт вымывания eluvial horizon
горизонт почвы soil horizon
горизонталь horizontal, level, contour line
горимость **1.** fire danger **2.** forest fire statistics
горихвостка-лысушка redstart, *Phoenicurus phoenicurus* L.
горихвостка-чернушка black redstart, *Phoenicurus ochruros* Gmel.
горлец snake-root, *Polygonum bistorta* L.
горлица turtle dove, *Streptopelia turtur* L.
горностай stoat, *Mustela erminea* L.
горный mountain, montane

городить to hedge, to fence, to enclose
городчатый crenate
городьба enclosure, fence
горошек vetch, *Vicia* L.
горошек, заборный *Vicia sepium* L.
горошек, лесной *Vicia sylvatica* L.
горошек, мышиный tufted vetch, *Vicia cracca* L.
горючее fuel
горючесть combustibility
Госзаповедник State reserve
Гослесхоз State forest enterprise
Гослесфонд state forest resource, the total area of all land and water in the charge or control of the state forestry service (comprising non-forest area and forest area, the latter consisting of forested and non-forested areas)
Госплан State plan
господство dominance
господствование dominance
господствовать to dominate
господствующий dominant, dominating
ГОСТ State Standard
готовальня set of drawing instruments
гофр corrugation, crimp
ГПС global positioning system, GPS
граб hornbeam, *Carpinus* L.
граб, восточный eastern hornbeam, *Carpinus orientalis* Mill.
граб, кавказский Caucasian hornbeam, *Carpinus caucasica* Grossh.
граб, обыкновенный hornbeam, *Carpinus betulus* L.
граб, сердцелистный *Carpinus cordata* Blume
грабинник eastern hornbeam, *Carpinus orientalis* Mill.
грабли rake
гравер 1. engraver, carver 2. *Pityogenes chalcographus* L.
гравер, двузубый *Pityogenes bidentatus* Hrbst.
гравер, обыкновенный *Pityogenes chalcographus* L.
гравилат, городской wood avens, *Geum urbanum*
гравилат, речной water avens, *Geum rivale* L.
град hail
градация 'gradation', population build-up (of pests)
градина hailstone
градобитие hail damage
градобой hail damage
гранат pomegranate, *Punica granatum* L.
гранатник pomegranate, *Punica granatum* L.
граница limit, boundary
граница древесной растительности, верхняя tree line
грань 1. face, side 2. arris, edge
грань, задняя back (of saw tooth)
грань, передняя face, front (of saw tooth)
график graph, diagram
грач rook, *Corvus frugilegus* L.
гребенщик tamarisk, *Tamarix* L.
гребенщик, ветвистый branchy tamarisk, *Tamarix ramosissima* Ldb.
гребенщик, вытянутый *Tamarix elongata* Ldb.
гребенщик, жестковолистный Kashgar tamarisk, *Tamarix hispida* Willd.
гребенщик, изящный graceful tamarisk, *Tamarix gracilis* Willd.
гребенщик, опушенный Kashgar tamarisk, *Tamarix hispida* Willd.
гребенщик, рыхлый limp tamarisk, *Tamarix laxa* Willd.
гребенщик, четырехтычинковый tetrandrous tamarisk, *Tamarix tetrandra* Pall.
гребни, направляющие rails (of chainsaw guide bar)
грейфер grab
гречиха, сахалинская *Polygonum sachalinense* L.
гриб (see also грибы) fungus, mushroom
гриб, белый *Boletus edulis* Bull.
гриб, домовый белый *Poria vaporaria* Pers.
гриб, домовый настоящий *Merulius lacrymans* Schum.

гриб, домовый пластинчатый *Paxillus pannoides* Fr.

гриб, домовый пленчатый *Coniophora cerebella* Schr.

гриб, зимний *Collybia velutipes* Quél.

гриб, столбовой *Lenzites saepiaria* Fr.

гриб, шпальный чешуйчатый *Lentinus lepideus* Fr.

грибница mycelium

грибокорень mycorrhiza

грибы, базидиальные Basidiomycetes

грибы-водоросли Phycomycetes

грибы гименомицеты Hymenomycetes

грибы, деревоокрашивающие wood-staining fungi

грибы, дереворазрушающие wood-destroying fungi

грибы, дрожжевые Saccharomycetes, yeasts

грибы, мучнисторосяные powdery mildews, Erisyphaceae

грибы, несовершенные Fungi imperfecti

грибы, низшие lower fungi

грибы, пластинчатые Agaricaceae

грибы, пленчатые Thelephoraceae

грибы, плесневые Mucorales, mould fungi

грибы, ржавчинные Uredinales, rust fungi

грибы, сумчатые Ascomycetes

грибы, трубчатые Polyporaceae

грибы, трутовые Polyporaceae

грибы, шляпочные pileate fungi

гроза thunder storm

громобойна lightning shake

груббер grubber, cultivator

груд 1. fertile (usually hardwood) site type 2. oak/spruce forest type

грудь (высота груди) breast (breast height)

груз load, cargo

грузить to load

грузовик truck, lorry

грузооборот traffic load, freight turnover (e.g. of a road)

грузоподъемность load capacity, carrying capacity

грузчик loader

грунт 1. ground, bottom, earth, subsoil 2. priming, first coat

грунт, защищенный 'protected ground' i.e. greenhouse

грунтование priming

грунтомет soil-thrower (fire-fighting)

группа group, clump

группа I (первая) group I forests (shelterbelts, amenity and greenbelt forests, etc., where tending and regeneration are the prime objects of management)

группа II (вторая) group II forests (forests of commercial and water-conservation importance in less well-forested regions, where cut should not exceed m.a.i.)

группа III (третья) group III forests (all other forests, i.e. production forests, where logging is the prime concern)

группа просветов pore cluster

груша pear, *Pyrus* L.

груша, земляная Jerusalem artichoke, *Helianthus tuberosus* L.

груша, иволистная willow-leafed pear, *Pyrus salicifolia* Pall.

груша, обыкновенная common pear, *Pyrus communis* L.

груша, лохолистная *Pyrus elaeagnifolia* Pall.

груша, уссурийская Ussuri pear, *Pyrus ussuriensis* Max.

грушанка wintergreen, *Pyrola* L.

грушанка, круглолистная larger wintergreen, *Pyrola rotundifolia* L.

грушанка, малая common wintergreen, *Pyrola minor* L.

грушанка, однобокая *Pyrola (Ramischia) secunda* L.

грушанка, одноцветковая *Pyrola (Moneses) uniflora* L.

грызун rodent

грызун, мышевидный mouse-like rodent (any small species of the Muridae)

гряда 1. layer, bed, stratum 2. ridge

гряда, посадная plant bed

гряда, семенная seed bed

грядиль plough beam

грядка 1. small bed 2. small hardwood log
губка 1. fungus (see also гриб; трутовик) 2. sponge 3. jaw (of vise)
губка, березовая *Polyporus betulinus* Fr.
губка, дубовая brown rot of oak, *Daedalea quercina* Fr.
губка, еловая *Trametes abietis* Sacc.
губка, кленовая *Fomes connatus* Fr.
губка, корневая mottled butt rot, (*Fomes annosus* (Fr.) Cke)
губка, лиственничная *Fomes officinalis* Bress.
губка, окаймленная *Fomes pinicola* (Gill.) Fr.
губка, пихтовая *Fomes hartigii* All.
губка, сосновая red rot of pine and larch, *Trametes pini* Fr.
гуж tug
гужом by land, by vehicle
гумбрин bleaching clay (waste product of the petroleum industry used as a growth promoter)
гуммоз gummosis
гумус humus
гумус, грубый mor, raw humus
гумус, кислый mor, raw humus
гумус, мягкий mull, mild humus
гумус, нейтральный mull, mild humus
гумус, сладкий mull, mild humus
гумус, сырой mor, raw humus
гурт drove, flock, herd
гусеница larva, caterpillar
густой dense
густота density
густота насаждения stand density, crop density, stocking (expressed in trees/ha)
гутта(перча) gutta-percha
гуща 1. thicket 2. sediment

Д

давление pressure
ДальНИИЛХ Far East Forest Research Institute
дальномер range finder, distance meter, perambulator
дальность distance, remoteness
дамба dam
датчик gauge
датчик сопротивления resistance strain gauge
дауервальд Dauerwald, continuous forest
дача, лесная forest estate, forest district (subdivision of a лесничество)
дача, учебная demonstration forest, training forest
дверь door
двигатель motor, engine
движение movement
движитель forwarder, skidding tractor
двойник *Ips duplicatus* Sahlb.
двойчатка forked tree
двор yard, court
ДВП (древесноволокнистые плиты) fibreboards
двудольные dicotyledons
двудомный dioecious
двукрылые Diptera
двуполый bisexual
двусеменодольный dicotyledonous
двухлетний two-year, biennial
двухэтажный two-storeyed
двухярусный two-storeyed
дебри jungle, thicket
девственный virgin
девясил *Inula* L.
деготь tar
деградация degradation
дегтеотделитель tar separator
дегтярня tar works
дезинфекция disinfection
действие effect, action
декар decare
деление division
дело business, matter, thing, act
дело, землемерное surveying
дело, лесное forestry, forest management
дело, опытное experimentation
дело, охотничье hunting
дело, пожарное fire service
дело, сплавное floating
дело, строительное building

делянка plot
демутация succession, recovery, development, regeneration
дендрарий arboretum
дендрология dendrology
дендроклиматология dendroclimatology
дендрометр instrument for tree measurement, dendrometer
дендрометрия tree measurement, dendrometry
дендрохронология dendrochronology
дендрошкала tree-ring chronology
денитрификация denitrification
день, рабочий working day
Деплесбумпром Department of Forest, Pulp and Paper, and Woodworking Industry
дербенник, иволистный purple loosestrife, *Lythrum salicaria* L.
дербенник, прутовидный *Lythrum virgatum* L.
дербянка *Blechnum* L.
деревенение lignification
деревенеть to lignify
дерево 1. tree 2. wood
дерево, адамово paulownia, *Paulownia tomentosa* (Thunb.) Steud.
дерево, бакаутовое lignum vitae, *Guaiacum officinale* L.
дерево, бархатное Amur cork tree, *Phellodendron amurense* Rupr.
дерево будущего final crop tree
дерево, вечнозеленое evergreen tree
дерево, господствующее dominant tree
дерево, гуттаперчевое gutta-percha tree, *Eucommia ulmoides* Oliv.
дерево, деловое timber tree
дерево, дровяное firewood tree
дерево, железное Persian iron wood, *Parrotia persica* C.A.M.
дерево, зависшее lodged tree
дерево, заглушенное suppressed tree (dead or dying)
дерево, заклейменное marked tree
дерево, закомлеватое tree with large butt swelling
дерево, защитное shelter tree, protective tree
дерево, земляничное strawberry tree, *Arbutus* L.
дерево, иудино Judas tree, *Cercis siliquastrum* L.
дерево, карликовое dwarf tree
дерево, кевевое *Pistacia mutica* F. et M.
дерево, корявое stunted tree, scrub tree
дерево, краевое marginal tree, border tree
дерево, красное 'red wood' (usually mahogany)
дерево, лаковое varnish tree, *Rhus verniciflua* Stokes
дерево, лесное forest tree
дерево, лиственное broadleaved tree
дерево, ловчее trap tree
дерево, мамонтовое giant sequoia, *Sequoia gigantea* Decne.
дерево, мастиковое mastic, *Pistacia lentiscus* L.
дерево, материнское mother tree, parent tree
дерево, маточное mother tree, parent tree
дерево, модельное sample tree
дерево, окраинное marginal tree, border tree
дерево, оставленное residual tree, standard tree, reserved tree
дерево, отдельное individual tree
дерево, отмирающее dying tree
дерево, охлестивающее whip
дерево, перезрелое overmature tree
дерево, переспелое overmature tree
дерево, плодовое fruit tree
дерево, пограничное boundary tree
дерево, подсечное lopped tree
дерево, подчиненное subordinate tree
дерево, разросшееся wolf tree
дерево, родовое family tree, pedigree
дерево, рожковое carob, *Ceratonia siliqua* L.
дерево, свежесрубленное freshly felled tree
дерево, сверхгосподствующее predominant tree

дерево, семенное seed tree
дерево, согосподствующее co-dominant tree
дерево, среднее mean tree
дерево, суховершинное stagheaded tree
дерево, сухостойное dead standing tree
дерево, тиковое teak, *Tectona grandis* L.
дерево, тутовое mulberry, *Morus* L.
дерево, тюльпанное tulip tree, *Liriodendron tulipifera* L.
дерево, уксусное staghorn sumac, *Rhus typhina* L.
дерево, фаутное defective tree
дерево, хвойное conifer, coniferous tree
дерево, черное ebony, *Diospyros ebenum* König
дерево, эбеновое ebony, *Diospyros ebenum* König
деревобетон wood-concrete
деревообработка woodworking (e.g. sawmilling, planing, plywood manufacture)
деревопереработка wood conversion (e.g. furniture manufacture, hydrolysis, paper making)
деревцо sapling, small tree
деревянение lignification
деревянистый woody, ligneous
деревянный 1. woody, ligneous 2. wooden, wood
дерен, красный dogwood, *Cornus sanguinea* L.
дерен, татарский Tatarian dogwood, *Cornus tatarica* Mill.
держи-дерево Christ-thorn, *Paliurus spina-christi* Mill.
дерн turf, sward, sod
дернина turf, sward, sod
дернорез turf cutter
дерносним turf-remover, scalper, skim coulter
дерноснимaтель turf-remover, scalper, skim coulter
деррик derrick
деряба mistle-thrush, *Turdus viscivorus* L.
десантник, пожарный smoke-jumper
десятина desyatina (=1.09 ha)
деталь detail, piece, component, member
дефибратор defibrator
дефибрер grinder
дефибрирование grinding, defibration
дефицитный deficient, in short supply
дефляция blowing away, wind erosion
деформация deformation, strain
Децис Decis (insecticide)
дешифрирование interpretation (of photos)
джида oleaster, *Elaeagnus angustifolia* L.
джузгун *Calligonum* L.
джузгун, безлистный *Calligonum aphyllum* Gurke
джузгун голова медузы *Calligonum caput-medusae* Schr.
джузгун, щетинистый *Calligonum setosum* Litw.
джунгли jungle
дзельква, граболистная Caucasian elm, *Zelkova carpinifolia* Dipp.
диаметр diameter, thickness
диаметр без коры diameter inside bark
диаметр в коре diameter outside bark
диаметр верхнего отруба top diameter
диаметр на высоте груди d.b.h., breast-height diameter
диаметр нижнего отруба butt diameter, diameter at thick end of log
диаметр с корой diameter outside bark
диаметр, срединный mid diameter
диаметр, средний mean diameter
диаметр, средний квадратический diameter corresponding to mean basal area, quadratic mean diameter
диапазон range
дивала, многолетняя *Scleranthus perennis* L.
дивала, однолетняя *Scleranthus annuus* L.
дикарь wilding
дикий wild
дикобраз porcupine, *Hystrix leucura* Sykes
дикорастущий wild, growing wild
дилены ends (pl.)
дильсы deals

диморфант prickly castor-oil tree, *Kalopanax pictus* (Thunb.) Nakai
динамика численности population dynamics
диоптр diopter
диск disk, blade (of a circular saw)
дисперсия variance
дистанционный distant, remote
дистанция distance
дичание becoming wild (see одичание)
дичок wilding
дичь 1. wildlife, game 2. wilderness
дичь, боровая Tetraonidae (grouse, ptarmigan, capercaillie, hazel hen)
длина length
длинноволокнистый long-fibred
длинноствольный long-stemmed
дно bottom
добротность quality
добротность места произрастания site quality
добротность насаждения stand quality
добывание getting, procuring, harvesting, extraction
добыча output, production, gain, yield
довсходовый pre-emergence
дождевание sprinkling, sprinkler irrigation
дождемер rain gauge
дождь rain
дождь, золотой golden-chain, *Laburnum anagyroides* Med.
дождь, кислый acid rain
дозревание ripening, after-ripening
долбеж mortise, groove, slot
долбежка mortise, groove, slot
долбление chiselling, slotting, mortising
долговечность longevity
долгомошник area of *Polytrichum commune* (see also бор долгомошник)
долгоножки Tipulidae
долгоносик curculionid, weevil (see also смолевка)
долгоносик, большой сосновый pine weevil, *Hylobius abietis* L.
долгоносик, дубовый *Curculio glandium* Marsh.
долгоносик, еловый *Pissodes harcyniae* Hbst.
долгоносик, малый сосновый banded pine weevil, *Pissodes notatus* F.
долгоносик, синий сосновый *Magdalis frontalis* Gyll.
долгоносик, средний сосновый *Hylobius pinastri* Gyll.
долгоносик, шишковый *Pissodes validirostris* Gyll.
долготье long log, long timber
долина valley
долгота longitude
долото chisel
долото, полое hollow mortising chisel
дом house
дом, сборный prefabricated house
доминант dominant, dominating
доминирующий dominant, dominating
домкрат jack
донник headings (for casks)
донник, лекарственный *Melitotus officinalis* L.
донце basal scar (of an acorn)
дополнение beating up
допуск tolerance, allowance (e.g. for shrinkage)
доработка after-treatment, finishing
дорога road, path, way
дорога, автомобильная lorry road
дорога, бревенчатая corduroy road
дорога, воздушная канатная aerial cableway
дорога, грунтовая earth road
дорога, железная railway
дорога, зимняя winter road
дорога, канатная cableway
дорога, канатно-подвесная aerial cableway
дорога, канатно-рельсовая cable railway
дорога, колейная strip road, road consisting of two parallel tracks
дорога, ледяная ice road
дорога, лежневая fore-and-aft road, plank road
дорога, лесная forest road

дорога, лесовозная extraction road, timber transport road
дорога, подвесная cableway, overhead line
дорога, рельсовая railway
дорога, сланевая road made of logs covered with layer of earth
дорога, снежная snow road
дорога, узкоколейная narrow-gauge railway
дорога, фашинная fascine road
дорожка 1. path, track 2. groove
дорожка и шпонка tongue and groove
дорожник grooving plane
доска board
доска, байдачная pine or spruce board (6.4 m long, 1.5-3.7 cm thick, variable width)
доска, боковая the board next to the slab
доска, выстроганная planed board
доска, краевая the board next to the slab
доска, кровельная roofing board
доска, ларевая sawn timber 7.6-15.2 cm thick, 30 cm wide, variable length
доска, лафетная board 8.9 cm thick, 38-42 cm wide
доска, мерная measuring rod
доска, настилочная floor-board
доска, необрезная unedged board
доска, обзольная waney board
доска, обрезная edged board
доска, палубная decking plank
доска, паркетная parquet board
доска пола floor-board
доска, половая floor-board
доска, посадная lining-out board
доска, посадочная lining-out board
доска, сеяльная sowing lath
доска, толстая plank
доска, чертежная drawing board
доставка delivery, supply
доступность accessibility
доход 1. income, revenue 2. return, profit
доход, валовой gross income
доход, чистый net income
доходность profitability
дощечка slat
дощечка, карандашная pencil slat
ДП (древесина прессованная) compressed wood
ДПК (древпресскрошки) wood particles (compressed into mouldings with resin)
дранка slat, lath, batten
дранка, коробочная box slat, box veneer
дранка, кровельная shingle
дранка, штукатурная lath
драть slat, lath, batten (see дранка)
древесина wood, timber, lumber (see also лес)
древесина, аварийная flotsam, wood floating loose
древесина, весенняя early wood, spring wood
древесина, воздушно-сухая air-dry wood
древесина, вторичная secondary xylem
древесина, гнилая rotten wood
древесина, деловая timber (all assortments except fuelwood)
древесина, дровяная firewood, fuelwood
древесина, здоровая sound wood
древесина, измельченная comminuted wood
древесина, кольцепоровая ring-porous wood
древесина, кондиционная standard timber
древесина, кондовая high-quality softwood
древесина, консервированная preservative-treated wood
древесина, косослойная spiral-grained wood
древесина, креневая compression wood
древесина, крупная large timber (usually over 7 cm diam.)
древесина, крупнослойная wide-ringed wood
древесина, летняя late wood, summer wood
древесина, ликвидная 'realizable' wood, merchantable wood
древесина, лиственная hardwood

древесина, мелкослойная narrow-ringed wood
древесина, мелкотоварная small wood
древесина, молевая loose-floated wood
древесина, мяндовая wide-ringed pine wood
древесина на корню standing timber
древесина, некондиционная non-standard timber
древесина, окоренная barked wood
древесина, осенняя late wood, autumn wood
древесина, пневая stump wood
древесина, поделочная timber
древесина, поздняя late wood
древесина, порочная defective wood
древесина, посиневшая bluestained wood
древесина, прелая rotten wood
древесина, прессованная compressed wood
древесина, пропаренная steamed wood
древесина, пропитанная impregnated wood
древесина, ранняя early wood
древесина, рассеяннопоровая diffuse-porous wood
древесина, резонансовая resonance wood
древесина, рудовая high-quality softwood
древесина, спелая ripewood, mature wood
древесина, спрессованная compressed wood
древесина, стволовая stemwood
древесина, строганная planed wood
древесина, сырая green wood
древесина, твердая hard wood
древесина, тяговая tension wood
древесина, узкослойная narrow-ringed wood
древесина, узорчатая figured wood
древесина, фаутная defective wood
древесина, хвойная softwood
древесина, цельная solid wood
древесина, ядровая heartwood
древесинник, многоядный *Trypodendron signatum* Ol.
древесинник, полосатый *Trypodendron lineatum* Ol.
древесиноведение wood science
древесница, въедливая leopard moth, *Zeuzera pyrina* L.
древесностружечные плиты chipboard, particleboards
древесный wood, woody, tree
древо tree (poetical)
древовал tree-feller (machine)
древоведение wood science
древовидный woody, tree-like
древогнезд carpenter bee (Xylocopidae)
древогрыз lyctid, powder-post beetle
древогрыз, бороздчатый *Lyctus linearis* Goeze
древогрыз, одноцветный *Lyctus brunneus* Steph.
древогрыз, опушенный *Lyctus pubescens* Panz.
древоизмерение wood measurement
древонасаждение stand
древопласт wood/plastic composite
древопластик wood/plastic composite
древоруб feller
древоспуск slide, chute
древостой stand (see also лес)
древосушка wood drying plant, kiln, timber drier
древоточец cossid
древоточец, ивовый goat moth, *Cossus cossus* L.
древоточец, пахучий goat moth, *Cossus cossus* L
древоточина frass
дренаж draining, drainage
дренирование draining, drainage
дробилка chipper, hog
дробление chipping, hogging
дробленка chip, particle
дрова firewood, fuelwood
дрова, колотые cleft fuelwood
дрова, швырковые fuelwood cut into billets *c.* 12 in (= *c.* 30 cm) long
дровосек cerambycid, longhorn beetle (see also усач)

дрожжи, кормовые fodder yeasts
дрозд thrush, *Turdus* L.
дрозд, белозобый ring ouzel, *Turdus torquatus* L.
дрозд, певчий song thrush, *Turdus philomelos* L.
дрозд, черный blackbird, *Turdus merula* L.
дрок, германский *Genista germanica* L.
дрок, красильный dyer's greenweed, *Genista tinctoria* L.
дрофа great bustard, *Otis tarda* L.
дрючок thin pole
дряблость incipient decay
ДСП particleboards, chipboard
ДСП (древеснослойстые пластики) laminated densified sheets
дуб oak, *Quercus* L.
дуб, араксинский Araks oak, *Quercus araxina* (Trautv.) Grossh.
дуб, аракский Araks oak, *Quercus araxina* (Trautv.) Grossh.
дуб, восточный Caucasian oak, *Quercus macranthera* F. et M.
дуб, Гартвиса Hartwiss oak, *Quercus hartwissiana* Stev.
дуб, грузинский Georgian oak, *Quercus iberica* Stev.
дуб, длинноножковый *Quercus longipes* Stev.
дуб, зимний durmast oak, sessile oak, *Quercus sessiliflora* Salisb. (*Quercus petraea* (Matt.) Liebl.)
дуб, иволистный willow oak, *Quercus phellos*
дуб, имеретинский *Quercus imeretina* Stev.
дуб, каменный holm oak, *Quercus ilex* L.
дуб, каштанолистный chestnut-leafed oak, *Quercus castaneifolia* C.A.M.
дуб, красный red oak, *Quercus rubra* L. (*Quercus borealis* Michx.)
дуб, крупнопыльниковый Caucasian oak, *Quercus macranthera* F. et M.
дуб, летний pedunculate oak, *Quercus robur* L.
дуб, монгольский Mongolian oak, *Quercus mongolica* Fisch.
дуб, моренный bog oak
дуб, ножкоцветный *Quercus pedunculiflora* C.Koch
дуб, понтийский Pontine oak, *Quercus pontica* Koch
дуб, пробковый cork oak, *Quercus suber* L.
дуб, пушистый downy oak, *Quercus pubescens* Willd.
дуб, сидячецветный durmast oak, sessile oak, *Quercus sessiliflora* Salisb. (*Quercus petraea* (Matt.) Liebl.)
дуб, скальный durmast oak, sessile oak, *Quercus sessiliflora* Salisb. (*Quercus petraea* (Matt.) Liebl.)
дуб, стебельчатый pedunculate oak, *Quercus robur* L.
дуб, черешчатый pedunculate oak, *Quercus robur* L.
дубильня tannery
дубина club, cudgel
дубитель tanstuff, tanning material
дубление tanning
дубло oak tanbark
дубняк oak forest, *Quercetum*
дубовый oak, oaken
дубонос hawfinch, *Coccothraustes coccothraustes* L.
дубрава oak forest, fertile oak forest type, *Quercetum*
дубровник germander, *Teucrium* L.
дуга arch, arc
дугласия Douglas fir, *Pseudotsuga menziesii* Mirb.
дудник wild angelica, *Angelica sylvestris* L.
дуплистый hollow, empty
дупло hollow, cavity
дурман thorn-apple, *Datura stramonium* L.
душевик, обыкновенный wild basil, *Clinopodium vulgare* L.
душица, обыкновенная marjoram, *Origanum vulgare* L.
дым smoke
дымянка, лекарственная common fumitory, *Fumaria officinalis* L.
дыра hole, knot hole

дыхание respiration
дышло connecting rod, shaft, reach pole
дюйм inch
дюймовка material one inch thick
дюна dune
дягиль, лекарственный *Archangelica officinalis* Hoffm.
дятел woodpecker
дятел, белоспинный white-backed woodpecker, *Dendrocopos leucotos* Bech.
дятел, большой пестрый greater spotted woodpecker, *Dryobates major* Brehm
дятел, зеленый green woodpecker, *Picus viridis* L.
дятел, малый пестрый lesser spotted woodpecker, *Dryobates minor* L.
дятел, черный black woodpecker, *Dryobates martius* L.
дятловина woodpecker hole

Е

единица unit
единица, хозяйственная block, management unit
единичный single, solitary, of sporadic occurrence
еж, обыкновенный hedgehog, *Erinaceus europaeus* L.
ежа, сборная cock's foot, *Dactylus glomerata* L.
ежевика, сизая dewberry, *Rubus caesius* L.
ежегодник yearbook, annual
ежемуха tachinid
ежовник *Anabasis aphylla* L.
елка small spruce tree
елка, новогодная New Year tree
елка, рождественская Christmas tree
еловый spruce
елочка 1. small spruce tree 2. herringbone (design)
ель spruce, *Picea* Dietrich
ель, аянская Yezo spruce, *Picea jezoensis* Carr.
ель, восточная oriental spruce, *Picea orientalis* Carr.
ель Глена Sakhalin spruce, *Picea glehnii* Mast.
ель, европейская Norway spruce, *Picea abies* (L.) Karst.
ель, кавказская oriental spruce, *Picea orientalis* Carr.
ель, колючая blue spruce, *Picea pungens* Eng.
ель, обыкновенная Norway spruce, *Picea abies* (L.) Karst.
ель, сибирская Siberian spruce, *Picea obovata* Ldb.
ель, тянь-шанская Schrenk's spruce, *Picea schrenkiana* F. et M.
ель Шренка Schrenk's spruce, *Picea schrenkiana* F. et M.
ельник spruce forest (for various types, see adjectives under бор), *Piceetum*
емкость capacity
ерник 1. dwarf birch (several species) 2. dwarf birch vegetation type
естественный natural

Ж

жаба toad, *Bufo* L.
жабник, полевой *Filago arvensis* L.
жаворонок, лесной woodlark, *Lullula arborea* L.
жар heat
жара heat
жезл rod, staff
желвак tumour, nodule
желвак, смоляной resin gall
желвачный nodular, knotty
железа gland
железка gland
железняк stand of *Parrotia persica*
железо iron
железобетон reinforced concrete
желна black woodpecker, *Dryobates martius* L.
желоб groove, trough, channel
желоб, спускной chute
желобок gutter (e.g. in resin tapping)

желтинник smoke-tree, *Cotinus coggygria* Scop.
желтогузка *Euproctis similis* Fssl.
желтолозник purple osier, purple willow, *Salix purpurea* L.
желудь acorn
женский female
жердняк pole-stage stand, pole stand
жердь pole, post
жереп mountain pine, *Pinus mugo* Turra, (*Pinus montana* Mill.)
жерепняк thicket of mountain pine
жеруха watercress, *Nasturtium officinale* R. Br.
жерушник watercress, *Nasturtium officinale* R. Br.
жестер buckthorn, *Rhamnus* L.
жесткокрылые Coleoptera
жесткость rigidity
живица resin, oleoresin
живица, засохшая scrape
живица, твердая scrape
животное animal
живучка, ползучая bugle, *Ajuga reptans* L.
жидкость fluid, liquid
жидкость, бордосская Bordeaux mixture
жижа brew, liquid
жижа, навозная liquid manure
жизнеспособность viability
жилка nerve, rib, vein
жилкование venation
жимолость honeysuckle, *Lonicera* L.
жимолость, грузинская Georgian honeysuckle, *Lonicera iberica* M.B.
жимолость, кавказская eastern honeysuckle, *Lonicera orientalis* Lam.
жимолость-каприфоль goat-leaf honeysuckle, *Lonicera caprifolium* L.
жимолость, лесная fly honeysuckle, *Lonicera xylosteum* L.
жимолость, обыкновенная fly honeysuckle, *Lonicera xylosteum* L.
жимолость, синяя blue honeysuckle, *Lonicera coerulea* L.
жимолость, татарская Tatarian honeysuckle, *Lonicera tatarica* L.
жир fat, grease
жировать **1.** to lubricate **2.** to feed, to feed up
жолудь acorn
жостер buckthorn, *Rhamnus* L. (see also крушина)
жужелица ground beetle, carabid
жужелица-краснотел *Calosoma sycophanta* L.
жузгун *Calligonum* L. (see джузгун)
жук beetle
жуки Coleoptera
жук-носорог *Oryctes nasicornis* L.
жук-олень common stag-beetle, *Lucanus cervus* L.

З

забелина accidental cambial wound in tapping
заболачивание swamp formation, bog formation, waterlogging
заболонник scolytid
заболонник, березовый *Scolytus ratzeburgi* Jans.
заболонник, вязовый small elm bark-beetle, *Scolytus multistriatus* Marsh.
заболонник, дубовый oak bark-beetle, *Scolytus intricatus* Ratz.
заболонник, морщинистый *Scolytus rugulosus* Ratz.
заболонник-пигмей *Scolytus pygmaeus* F.
заболонник, струйчатый small elm bark-beetle, *Scolytus multistriatus* Marsh.
заболонь sapwood
заболонь, внутренняя included sapwood
заболонь, двойная included sapwood
заболоченный waterlogged, swampy, boggy
забор fence
заведение establishment, institution
заведующий manager, director, chief
завирушка, лесная hedge-sparrow, *Prunella modularis* L.

зависимость dependence, relation, function
завиток local irregularity of annual rings
завод factory
завод, лесопильный sawmill
завод, фанерный plywood factory, veneer factory
завязь ovary
завязываемость set (of seed or fruit)
завязывание set (of seed or fruit)
загиб bend
заглушение suppression, smothering
заглушенный suppressed, smothered
загниваемость susceptibility to rot
загнивание rotting, decay, rot
загнивший rotten
загонщик beater
загораемость combustibility
загораживание enclosure, fencing
загорание ignition
заготовить to prepare, to convert, to harvest
заготовка 1. procurement, harvesting, logging, felling and primary conversion 2. blank, half-finished product, dimension stock 3. store, stock
заготовка, черновая rough-sawn stock
загружать to load, to charge
загрузка load, loading
загрязнение pollution
задел, страховой buffer stock (of material)
заделка 1. covering (e.g. seeds) 2. closing, sealing, stopping, stopping up
заделывание covering (e.g. seeds)
задержание осадков interception of precipitation
задержка в росте growth check
задернелость turfiness
задернелый turfy, overgrown by turf
задернение turf formation
задерненный overgrown with turf
задичалый gone wild, abandoned
задорина splinter
задыхание 'suffocation', incipient deterioration (of felled beech)
зажигание ignition
зажим 1. clamp 2. pinching (of saw)
зажимать to clamp
зазор play
зазубренный notched, jagged, indented
заиление silting
зайчик leveret
заказник reserve, reserved forest, reserved seed stand
закаливание hardening, tempering
закалка hardening, tempering
закладка laying out, establishment
закладка культуры plantation establishment
закладка лесосеки coupe demarcation
заклеймить to mark
заклепка rivet, pin
заклинивание wedging
закомелистость swollen butt, butt swelling, root swelling
закомлеватость swollen butt, butt swelling, root swelling
закон, лесной forest law
закономерность regularity, pattern, rule
закрепление fixing, strengthening, stabilization
закрепление лесосырьевой базы timber concession
залежь 1. layer, bed, stratum 2. ley, fallow, grass sward
залежь, торфяная peat bed, peat deposit
залесение afforestation, reforestation
залесить to afforest, to reforest
заложение laying out, establishment
залом log jam (that does not completely block the stream), pile-up of debris
залысить to blaze, to mark
замазка putty
замаривание slow dying
замачивание steeping, soaking, wetting
замена substitution, replacement
замерзание freezing
замок 1. lock, catch 2. castle
замораживание freezing
заморозки frosts
заморозки, весенние late frosts, spring frosts
заморозки, осенние early frosts, autumn frosts
замшелый moss-covered

заноза splinter
заостренный pointed, sharp, acuminate
заочник person taking correspondence course
запанщик boom worker
запань terminal boom, trap boom (see also бон)
запань, поперечная transverse boom, trap boom
запань, приемная holding boom
запань, спицевая spoke boom, needle boom
запань, шатровая tent-shaped boom
запас 1. growing stock, standing volume 2. supply, stock
запас леса на корню standing volume
запас, ликвидный 'realizable' volume (the exploitable volume after deduction of reserved species and stands of less than 30 m^3/ha)
запас насаждения stand volume
запас, нормальный normal growing stock
запас, общий total standing volume
запас, целевой growing-stock target
запас, эксплуатационный exploitable volume (the volume of wood suitable by size and quality for harvesting)
запас, элементарный 'basic stocking' (standing vol. per unit of b.a.)
запасный reserve, stored
запашка tillage
запил saw damage (to wood)
запись entry, record, inventory
заплесневелый mildewed, mouldy
заповедник reserve, nature reserve
заподлицо flush, level (with)
запрет prohibition
запрещение prohibition
запруда dam
запруживание damming
заработная плата wages, pay
заработок earnings
заражение infection
зараженность degree of infection
зараженный infected
зарастание 1. colonization (by vegetation) 2. healing over, occlusion
зарастать 1. to colonize, to be overgrown 2. to heal over, to occlude
зародыш embryo
заросль thicket, brushwood, undergrowth
зарплата wages, pay
заруб 1. cut area, cutover area 2. axe damage (to wood)
зарубание notching, undercutting
зарубка notch
зарянка robin, *Erithacus rubecula* L.
засадка planting
засаживать to plant
засека abatis, abattis, forest used as military defence
заселение colonization, population
засинелый bluestained
заслон shield, screen
засмол resin streak, wood affected by resinosis
засмоление resinosis
засмоленный resin-soaked
засмолок resin streak, wood affected by resinosis
засоление salinization
засоленность salinity
засоленный salinified
засоренность weediness
застаиваться to stagnate
застой stagnation
застойный stagnant
застрявание stranding
заступ spade
засуха drought, dry period
засуха, зимняя winter drying, winter drought
засухоустойчивость drought resistance, drought tolerance
засушливый droughty
засъемка survey
засыпать to cover (with soil), to fill in
засыхание drying up, withering
затенение shading
затенивать to shade
затеска 1. blaze 2. 'nosing' of logs
затесывать 1. to blaze 2. to hew (roughly)
затишка windbreak
затопление inundation, flooding, flood

затор log jam
заточка sharpening, grinding
заточка, косая bevel
затрата expense, expenditure, consumption (see издержка, расходы)
затраты, приведенные costs per unit produced and transported (cost per unit at factory gate)
затупление blunting, dulling
затычка bung
захват grab, grapple
захламленность лесосеки debris left on the felling area
захламленный obstructed by felling debris
захлестывание beating (a fire)
зацветание efflorescence
зацепка catch, hook
зачаток rudiment
защита protection, shelter
защита, биологическая biological control
защищать to protect, to shelter
заяц hare, *Lepus* L.
заяц-беляк mountain hare, *Lepus timidus* L.
заяц-русак brown hare, *Lepus europaeus* L.
звездочка drive sprocket (chain saw)
звездчатка *Stellaria* L.
звездчатка, дубравная wood stitchwort, *Stellaria nemorum* L.
звездчатка, злаковидная *Stellaria graminea* L.
звездчатка, ланцетовидная *Stellaria holostea* L.
звено link, section, member
звено, лесорубочное cutting section (in Wagner's Blendersaumschlag system)
зверобой St John's wort, *Hypericum* L.
зверь wild animal
здание building
зеленеть to become green
зеленик danewort, *Sambucus ebulus* L.
зеленомошник area of *Hylocomium* (see also бор зеленомошник)
зеленчук, желтый yellow archangel, *Gabeobdolon luteum* Huds.
зелень foliage, green-stuff
зелень, техническая small foliage-bearing shoots up to 8 mm diam. (see лапка)
земледелие farming, agriculture
земледелие, лесное shifting cultivation
земледелие, огневое slash and burn agriculture, shifting cultivation
земледелие, паловое slash and burn agriculture, shifting cultivation
земледелие, подсечное shifting cultivation
землемер surveyor
землепользование land use
землеройка, лесная common shrew, *Sorex araneus*
землеустройство land management
земля earth, land
земляника, лесная wild strawberry, *Fragaria vesca* L.
земляничник strawberry tree, *Arbutus* L.
земноводный amphibious
зенкер countersink bit
зенковка countersink bit
зеркало mirror
зеркало грунтовых вод water table
зеркало карры actual face area
зернистость granularity, grain size
зерно grain
зерно, пыльцевое pollen grain
зигота zygote
зима winter
зимнестойкий winter-hardy
зимник winter trail
зимнозеленый evergreen
зимование wintering, hibernation
зимующий wintering, overwintering, hibernating
злак grass
злаки grasses, Gramineae
злаковые grasses, Gramineae
златка buprestid
златка, двупятнистая узкотелая *Agrilus biguttatus* Fabr.
златка, дубовая бронзовая *Chrysobothris affinis* Fabr.
златка, желтопятнистая *Buprestis haemorrhoidalis* Hbst.

златка, лиственничная *Melanophila guttulata* Gebl.
златка, синяя *Phaenops cyanea* Fabr.
златка, узкотелая зеленая *Agrilus viridis* L.
златоглазок chrysopid, green lacewing fly
златогузка brown-tail moth, *Euproctis chrysorrhoea* L.
змееяд short-toed eagle, *Circaetus gallicus* Gmel.
змея snake
знак sign, mark
знак, товарный trade mark
зола ash
золотарник golden-rod, *Solidago virgaurea* L.
золототысячник centaury, *Centaurium* Hill
зольник ash pit
зольность ash content
зона zone
зона, водоохранная water-conservation zone, catchment-protection zone
зонд probe, borer
зондирование probing, boring
зонт, древесный tree(s) giving shade to livestock, shade clump
зонт, зеленый tree(s) giving shade to livestock, shade clump
зонтик umbel
зонтичные Umbelliferae
зрелость ripeness, maturity (see also спелость)
зуб **1.** tooth **2.** tine, prong
зуб, вставной inserted tooth
зуб, Г-образный gouge-type tooth, chipper tooth
зуб, ограничительный depth gauge
зуб, очищающий raker tooth
зуб, плющенный swaged tooth, swage-set tooth
зуб, подрезающий raker tooth, cutter-raker
зуб, разведенный spring-set tooth
зуб, режкообразный gauge-type tooth, chipper tooth
зуб, режущий raker tooth
зуб, резающий cutter tooth
зуб, скоблящий raker tooth
зуб, треугольный triangular tooth
зуб, фуговочный gauge-type tooth, chipper tooth
зубец **1.** tooth **2.** tine, prong
зубило chisel
зубр European bison, *Bison bonasus* L.
зубровка, душистая holy-grass, *Hierochloe odorata* (L.) Beauv.
зубчатый toothed, serrated, dentate, serrate
зубянка, луковичная coral-wort, *Dentaria bulbifera* L.
зюзник, высокий *Lycopus exaltatus* L.
зюзник, европейский gipsy-wort, *Lycopus europaeus* L
зяблик chaffinch, *Fringilla coelebs* L.
зяблина frost crack
зябь autumn ploughing, land ploughed in autumn

И

ива willow, *Salix* L.
ива, бабилонская weeping willow, *Salix babylonica* L.
ива, безвершинная pollard willow
ива, белая white willow, *Salix alba* L.
ива, грушанколистная *Salix pyrolifolia* Ldb.
ива, двуцветная *Salix phylicifolia* L.
ива, деревцовидная *Salix arbuscula* L.
ива, каспийская Caspian willow, *Salix caspica* Pall.
ива, козья goat willow, sallow, *Salix caprea* L.
ива, копьевидная hastate willow, *Salix hastata* L.
ива, корзиночная common osier, *Salix viminalis* L.
ива, ломкая crack willow, *Salix fragilis* L.
ива, миндальнолистная almond willow, *Salix amygdalina* L.
ива, миртолистная *Salix myrsinites* L.

ива, мохнатая woolly willow, *Salix lanata* L.
ива, остролистная willow, *Salix acutifolia* Willd.
ива, пепельная grey willow, *Salix cinerea* L.
ива, плакучая weeping willow, *Salix babylonica* L.
ива, ползучая creeping willow, *Salix reptans* Rupr.
ива, полярная polar willow, *Salix polaris* Wg.
ива, прутовидная common osier, *Salix viminalis* L.
ива, пурпурная purple osier, purple willow, *Salix purpurea* L.
ива, пятитычинковая bay willow, *Salix pentandra* L.
ива, русская Russian willow, *Salix rossica* Nas.
ива, серая grey willow, *Salix cinerea* L.
ива, серебристая white willow, *Salix alba* L.
ива, сетчатая reticulate willow, *Salix reticulata* L.
ива, сибирская Siberian willow, *Salix sibirica* Pall.
ива, сизая blue willow, *Salix glauca* L.
ива, синеватосерая *Salix livida* Whlb.
ива, тонколистная *Salix tenuifolia* Turcz.
ива, травянистая dwarf willow, *Salix herbacea* L.
ива, трехтычинковая willow, *Salix triandra* L.
ива, ушастая *Salix aurita* L.
ива, чернеющая *Salix nigricans* Sm.
ива, черничная *Salix myrtilloides* L.
Иван-да-Марья *Melampyrum nemorosum* L.
Иван-побег Lammas shoot
Иван-чай rosebay willow-herb, *Chamaenerion angustifolium* (L.) Scop.
ивняк willow wood, *Salicetum*
игла needle
иглица butcher's broom, *Ruscus aculeatus* L.
игголистные Coniferae, conifers
игголистный needle-leaved, coniferous
игра play
избирательный selective
избранник final crop tree
избыток surplus, excess
известкование liming
известковый calcareous, lime, calcium
известняк limestone
известь lime
известь, белильная bleaching powder
известь, гашенная slaked lime
известь, едкая quicklime
известь, жженая quicklime
известь, хлорная bleaching powder
извилистый curved, crooked, tortuous, crooked-grown
извлечение extraction
изгиб bend, bending
изгиб, динамический impact bending
изгиб, продольный buckling
изгиб, статический static bending
изгиб, ударный impact bending
изгородь enclosure, hedge, fence
изделие article, product
издержка expense, cost, outlay (see затрата, расходы)
излом fracture, failure
излучение radiation
измельчение comminution
изменение запаса change in standing volume
измерение measuring, measurement
изморозь hoar frost
измочаливать to shred
износ wear, abrasion
изобонитет average site-quality class
изолирование 1. insulation 2. isolation, releasing
изоляция isolation, releasing
израстание 1. proliferation 2. growing rank
изреженность degree of openness due to thinning
изреженный 1. thinned 2. widely spaced
изреживание thinning (including natural thinning by suppression)
изреживание, естественное natural thinning

изреживать to thin
изрубить to cut, to chop
изюбрь Far Eastern red deer, *Cervus elaphus xanthopygos* M.Edw.
ил clay
илистый clayey
илмь elm, *Ulmus* L. (see also **вяз**)
илмь, горный wych elm, *Ulmus glabra* Huds. (*U. scabra* Mill.)
ильмняк elm forest
ильмовник **1.** elm forest **2.** Siberian elm, *Ulmus pumila*
инвентаризация making an inventory
инвентарь inventory
иней hoar-frost
инжектор, древесный tree-injector
инженер engineer
инженерия, генетическая genetic engineering
инжир fig, *Ficus carica* L.
инкрустация wedge grafting
иноземный exotic
инсектицид insecticide
инсоляция insolation
интина intine
инфраструктура infrastructure
ирга, обыкновенная service-berry, *Amelanchier ovalis* Med.
ирригация irrigation
ископаемый **1.** fossil **2.** mineral
искривление bend, twist, distortion
искривление побегов shoot distortion
искусственный artificial
испарение evaporation
использование utilization
использование, комплексное multiple use
использовать to use, to utilize
исправлять to repair, to beat up
испытание test, testing, experiment, trial
исследование investigation
истечение outflow
истирание abrasion, wear
истлевание rotting, decay, decomposition
истление rotting, decay, decomposition
источник source
истощение exhaustion, impoverishment
истребление destruction
итог sum, total, result
итсигек *Anabasis aphylla* L.
иудино дерево Judas tree, *Cercis siliquastrum* L.

К

кабан wild boar, *Sus scrofa* L.
кабарга musk deer, *Moschus moschiferus* L.
кабель-кран cable-way transporter, cable-crane
каботаж coastal navigation
кадастр cadastre, cadastral survey
кадка tub, vat
кадры personnel, specialists, skilled workers
каемчатый border, narrow-strip (felling)
казан boiler, retort
казенка **1.** cabin on a log raft **2.** log deck in a sawmill
кайма border, narrow-strip (felling)
каландр calender
каланча, пожарная fire tower
калевать to mould, to groove
калевка **1.** moulding **2.** moulding plane
калевочник moulding plane
калибр gauge
калибр, предельный non-adjustable gauge
калиброване **1.** calibration **2.** processing to a standard thickness (by sanding or planing)
калибровка milling a log (butt) to eliminate taper and sweep
калий potassium
калина guelder rose, *Viburnum opulus* L.
каллюс callus
калопанакс prickly castor-oil tree, *Kalopanax pictus* (Thunb.) Nakai
калужница, болотная marsh marigold, *Caltha palustris* L.
кальций calcium
камбий cambium
камбий, пробковий cork cambium, phellogen

камбий, ярусный storeyed cambium
камедетечение gum bleeding
камедь gum
каменистость stoniness
каменоломня quarry
камень stone
камень, лизальный salt lick
камень, точильный sharpening stone
камера 1. chamber, cell, compartment 2. kiln (see also сушилка)
камера, брачная mating chamber
камера непрерывного действия progressive kiln
камера периодического действия compartment kiln
камера поры pit chamber
камера, сушильная drying chamber, kiln
камнеломня quarry
камфора camphor
камыш 1. *Scirpus* L. (usually) 2. sometimes *Phragmites communis* Trin. (see тростник)
камыш, лесной *Scirpus sylvaticus* L.
канава ditch, trench, gutter
канава, водоотводная drainage ditch
канава, водосборная collector ditch
канава, ловчяя trap ditch
канава, нагорная upper side-drain
канава, оросительная irrigation ditch
канава, осушительная drainage ditch
канава, сточная drainage ditch
канавокопатель ditch digger
канал canal, channel
канал поры pit channel
канал, сплавной timber-floating channel
канат line, rope, cable
канат, возвратный return line
канат, грузовой the loaded part of the operating line
канат двойной свивки ordinary lay rope
канат крестовой свивки ordinary lay rope
канат, несущий skyline
канат, несущий опускающий slacking skyline
канат односторонней свивки Lang's lay rope
канат, проволочный wire rope
канат, тяговой operating line, haul line
канат, тяговонесущий running skyline
канат, холостой the unloaded part of the operating line
кандым *Calligonum* L. (see джузгун)
канифоль colophony, rosin
канифоль, талловая tall oil rosin
кант arris, edge
кап burr
кап, прикорневый basal burr
капиллярность capillarity
капкан trap
капокорень burr
каприфоль goat-leaf honeysuckle, *Lonicera caprifolium* L.
капуста заячья *Sedum maximum* Sut.
капюшонник bostrychid
кара-арча 1. black juniper, *Juniperus polycarpos* Koch (in Caucasus) 2. (sometimes) Seravshan juniper, *Juniperus seravschanica* Kom. (in Central Asia)
карагана *Caragana* Lam.
карагана, высокая pea tree, *Caragana arborescens* Lam.
карагана карликовая pygmy pea tree, *Caragana pygmaea* DC.
карагана, колючая spiny pea tree, *Caragana spinosa* DC.
карагана, кустарная dwarf pea tree, *Caragana frutex* Koch
карагач 1. (usually) smooth elm, *Ulmus carpinifolia* Gled. (*U. foliacea* Gilib.) 2. (sometimes) Siberian elm, *Ulmus pumila* L., var. *arborea* Litv. (*U. pinnatoramosa* Dieck.)
карандаш 1. pencil 2. core (of veneer bolt)
кара-черкез *Salsola paletzkiana* Litw.
карбонатный containing carbonate, calcareous
каретка carriage
каркас 1. hackberry, *Celtis* L. 2. frame(work), skeleton
каркас, голый *Celtis glabrata* Stev.

каркас, кавказский Caucasian nettle-tree, *Celtis caucasica* Willd.
каркас, южный southern nettle-tree, *Celtis australis* L.
карликовый dwarf
карман-накопитель bin (container for assortments on sorting line)
карман, смоляной resin pocket
кармашек, смоляной resin pocket
карниз cornice
карра face site, face
карра, безжелобочная staggered (herring-bone) face
карра, возходящая ascending face
карра, гладкая smooth (herring-bone) face
карра, нисходящая descending face
карра, обычная ordinary (smooth herring-bone) face
карра, ребристая close-ribbed (herring-bone) face
карра, рифленая corrugated (herring-bone) face
карра с перемычками spaced-rib (herring-bone) face
карроподновка freshening
карта map
картография cartography
картон cardboard, board
карточка card
карточка, перфорированная punched card
карча snag, floating debris, e.g. tree with roots
касательная tangent
касатик *Iris* L.
катер launch, motor boat
катер, буксирный tug, tug boat
катище riverside landing
каток roller, sheave
каток, бороздковый seed-drill roller
каток, снежный snow roller
катушка spool, bobbin, reel
каучук caoutchouc, rubber
качественный qualitative
качество quality
качим *Gypsophila* L.
каштан chestnut
каштан, благородный sweet chestnut, *Castanea sativa* Mill.
каштан, конский horse chestnut, *Aesculus hippocastanum* L.
каштан, посевной sweet chestnut, *Castanea sativa* Mill.
каштан, сьедобный sweet chestnut, *Castanea sativa* Mill.
квартал **1.** compartment **2.** quarter
квебрахо quebracho, *Aspidosperma quebracho-blanco* Schlecht.
кедр 'cedar', often Siberian pine, *Pinus sibirica* Rupr. (but see below)
кедр, атласский Atlas cedar, *Cedrus atlantica* Man.
кедр, гималайский deodar, *Cedrus deodara* (Roxb.) Loud.
кедр, европейский Arolla pine, *Pinus cembra* L.
кедр, корейский Korean pine, *Pinus koraiensis* Sieb. et Zucc.
кедр, ливанский cedar of Lebanon, *Cedrus libani* Loud.
кедр, маньчжурский Korean pine, *Pinus koraiensis* Sieb. et Zucc.
кедр, речной incense cedar, *Libocedrus decurrens* Torr.
кедр, сибирский Siberian pine, *Pinus sibirica* Rupr.
кедр, стлан(н)иковый dwarf Siberian pine, *Pinus pumila* Rgl.
кедрач **1.** *Cembretum* **2.** *Pinus sibirica* forest
кедровка nutcracker, *Nucifraga caryocatactes* L.
кедровник **1.** dwarf Siberian pine, *Pinus pumila* Rgl. **2.** *Cembretum* **3.** *Pinus sibirica* forest
кейреук *Salsola rigida* Pall.
керн, приростный increment core
кивсяк millipede
кизил Cornelian cherry, *Cornus mas* L.
кизильник cotoneaster, *Cotoneaster* Ehrh.
кизильник, обыкновенный *Cotoneaster integerrima* Med.
кизильник, чернопильный *Cotoneaster melanocarpa* Lodd.
кинжал, сажальный planting dagger

кипарис cypress, *Cupressus* Tourn.
кипарис, болотный swamp cypress, *Taxodium distichum* Rich.
кипарис, обыкновенный Mediterranean cypress, *Cupressus sempervirens* L.
кипарисовик cypress, *Chamaecyparis* Spach.
кипрей willow-herb, *Epilobium* L.
кипрей, болотный *Epilobium palustre* L.
кипрей, горный *Epilobium montanum* L.
кипрей, розовый *Epilobium roseum* Schreb.
кипрей, узколистный rosebay willow-herb, *Chamaenerion angustifolium* (L.) Scop.
кирка pick
кирказон *Aristolochia* L.
кислица wood sorrel, *Oxalis acetosella* L.
кисличник area of *Oxalis acetosella* L. (see also бор кисличник)
кислород oxygen
кислота acid
кислота, варочная cooking acid (in pulping)
кислота, гуминовая humic acid
кислота, серная sulphuric acid
кислота, смоляная resin acid
кислотность acidity
кислотность, обменная exchange acidity
кладка stacking
кладка, плотная tight stacking
кладка, рыхлая loose stacking
кларк threshold allowable concentration (see пдк)
класс class
класс бонитета site class
класс возраста age class (usually 20 years in European Russia, and 40 years in Siberia)
класс высоты height class
класс добротности quality class
класс Крафта Kraft (dominance) class
класс товарности merchantability class (determined by the % of timber in the standing volume)
класс толщины diameter class
класс хозяйства management class
классификация classification
классификация Крафта Kraft tree dominance classification
клевер clover, trefoil, *Trifolium* L.
клеенаноситель glue-spreader
клей 1. glue 2. size
клей горячего отверждения hot-setting glue
клей, казеиновый casein glue
клей, карбамидный urea glue
клей, костный bone glue
клей, мездровый hide glue
клей, мочевино-формальдегидный urea-formaldehyde glue
клей, пленочный film glue
клей, рыбий fish glue
клей, термопластичный thermo-setting glue, thermoplastic glue
клей, термореактивный thermo-setting glue, thermoplastic glue
клей, фенольно-формальдегидный phenol-formaldehyde glue
клей холодного отверждения cold-setting glue
клеймение marking
клеймо mark, marking iron
клекачка bladdernut, *Staphylea* L.
клекачка, колхидская Colchis bladder-nut, *Staphylea colchica* Stev.
клекачка, перистая pinnate bladdernut, *Staphylea pinnata* L.
клен maple, *Acer* L.
клен, американский box-elder, *Acer negundo* L.
клен, бархатный velvet maple, *Acer velutinum* Boiss.
клен, зеленокорый greenbark maple, *Acer tegmentosum* Maxim.
клен, маньчжурский Manchurian maple, *Acer mandshuricum* Maxim.
клен, мелколистный small-leafed maple, *Acer mono* Maxim.

клен, остролистный Norway maple, *Acer platanoides* L.
клен, полевой field maple, *Acer campestre* L.
клен, приречный Amur maple, *Acer ginnala* Maxim.
клен, татарский Tatarian maple, *Acer tataricum* L.
клен явор sycamore maple, *Acer pseudoplatanus* L.
клен, яркий Cappadocian maple, *Acer laetum* C.A.M.
клен, ясенелистный box-elder, *Acer negundo* L.
кленовица maple sap
кленовник maple forest
клепка stave
клепка, бочарная cooperage stave
клепка, заливная wet stave
клепка, сухотарная dry stave
клест-еловик crossbill, *Loxia curvirostra* L.
клест-сосновик parrot-crossbill, *Loxia pytyopsittacus* Bork.
клетка **1.** cell **2.** cage
клетка, инициальная initial
клетка, каменистая stone cell
клетка, кроющая sheath cell
клетка, лучевая лежачая procumbent ray cell
клетка, лучевая стоячая upright ray cell
клетка, начальная initial
клетка, паренхимная перегородчатая septate parenchyma cell
клетка, ситовидная sieve cell
клетка, черепитчатая tile cell
клеточка cell
клетчатка cellulose
клещ mite, tick
клещ, панцирный oribatid
клещи tongs, vise, dogs (of sawmill carriage)
климат climate
климатип climatic type, provenance
клин wedge
клиновидный wedge-shaped, cuneiform
клинтух stock dove, *Columba oenas* L.
клиренс ground clearance (of vehicle)
клон clone
клоп, сосновый подкорный *Aradus cinnamomeus* Pz.
КЛП integrated forest enterprise
клубень **1.** (root) nodule **2.** tuber
клубника hautbois strawberry, *Fragaria moschata* Duch.
клумба raised bed, flower bed
клюква, болотная cranberry, *Oxycoccus palustris* Pers.
клюква, мелкоплодная *Oxycoccus microcarpus* Turcz.
ключ **1.** key **2.** spring, fountain **3.** wrench
ключ, разводной saw set
кница knee, crook (used in boatbuilding)
кница, кокорная knee, crook (used in boatbuilding)
кноль set of sliced veneers
кобла pollard
кобчик red-footed falcon, *Falco vespertinus* L.
ковер, стружечный chip mat
ковыль feather-grass, *Stipa* L.
коготь claw
когти climbing irons
код, штриховый bar-code
кодекс, лесной forest code
кодировщик digitizer
кожа **1.** skin, hide **2.** peel
кожица thin skin, epidermis
кожура coat, rind
кожух case, housing, sheath
коза goat (female)
козел goat (male)
козлобородник, луговой goat's beard, *Tragopogon pratensis* L.
козлы sawing horse, saw block
козуля roe deer, *Capreolus capreolus* L.
козырек **1.** lip **2.** fender
козявка small beetle
кокора knee, crook (used in boatbuilding)
кокорвяга snag, sunken log, submerged stump
кокорник snag, sunken log, submerged stump

кокорье snag, sunken log, submerged stump
кокша sunken tree, stranded tree
кол **1.** stake, post **2.** large cutting, sett
кол, сажальный dibble
колено, надсеменодольное epicotyl
колено, подсеменодольное hypocotyl
коленчатый bent, angled, crooked
колесо wheel
колея **1.** track, rut **2.** (railway) line, gauge
колин kolin (see фитолин)
количественный quantitative
количество quantity, amount, number
колка splitting, cleaving
колки **1.** pegs **2.** forest outliers
колкий cleavable
коллективный collective
коллектор main drain
колленхима collenchyma
колода **1.** fallen stem **2.** log, block
колодезь well
колодец well
колодка last (for shoes)
колодка, тормозная brake shoe
колок **1.** peg **2.** forest outlier
колокольчик *Campanula* L.
колонок Siberian weasel, *Mustela sibirica* Pall.
колонопряды Lasiocampidae
колосок, душистый *Anthoxanthum odoratum* L.
колотушка mallet
колотый cleft, split
колоть to cleave, to split
колпак cap, cowl, cover
колпак, трелевочный skidding cap
колун cleaver
колхоз kolkhoz, collective farm
кольматаж warping (of land), silt deposition
кольцевание ringing, girdling
кольцевать to ring, to girdle
кольцеобразный annular, ring-shaped
кольцепоровый ring-porous
кольцесосудистый ring-porous
кольцо ring, girdle, collar
кольцо, годичное annual ring, growth ring
кольцо, ловчее trap ring, grease band
колючка prickle, spine, barb
ком clod, ball (of earth), lump
команда command, squad, party
комбайн combine, combine harvester
комбайн, лесной tree-harvesting machine
комбинат industrial combine, integrated mill
комель butt
комлеватость swollen butt, butt swelling, root swelling
комлевой butt (adj.)
комок clod, ball (of earth), lump
компас compass
комплекс complex
комплексное лесное предприятие integrated forest enterprise
комплексный complex, integrated
комплект set
комплект, ящичный box set
компост compost
конвейер conveyor
конверт quincunx
кондак cant hook, peavy
конек, лесной tree-pipit, *Anthus trivialis* L.
коник bunk, cradle, frame
конкурент competitor
конкуренция competition
конопля hemp, *Cannabis sativa* L.
коносамент bill of lading
консервирование preservation, conservation
консоль console, cantilever, bracket, arm
конструкция construction, design
контора bureau, board, office
контрклин counter-wedge
контрнож pressure bar
контроль control, check, inspection
контрфорс buttress
контур contour, outline
конус cone
конус нарастания growing point

концеравнитель end-trimmer, saw (or other device) for levelling up the ends of logs or lumber
кончик tip, end, point
кооператив co-operative
копа measure of small assortments: 60 staves
копал copal
копать to dig
копулировка apical grafting, splice grafting
копулировка, простая splice grafting, whip grafting
копулировка с язычком whip-and-tongue grafting
копулировка, улучшенная whip-and-tongue grafting
копытень, европейский asarabacca, *Asarum europaeum* L.
копытный hoofed, ungulate
кора **1.** bark **2.** cortex
кора, дубильная tanbark
кора, зеркальная smooth bark
кора, трещиноватый fissured bark
корабельщик lymexylonid
кордон cordon
корень root
корень, адвентивный adventitious root
корень, воздушный aerial root
корень, всасывающий absorbing root
корень, главный main root
корень, досковидный buttress root
корень, зародышевый radicle
корень, опорный prop root, supporting root
корень, поверхностный surface root, horizontal root
корень, придаточный adventitious root
корень, проводящий conducting root
корень, ростовый extension root
корень, скелетный skeleton root, main root
корень, сосательный absorbing root
корень, сосущий absorbing root
корень, стержневой tap root
корень, ходульный stilt root, prop root
корень, якорный anchor root
корешок rootlet
корзина basket
корзиночник basket willow, osier
коридор **1.** corridor, strip, ride **2.** channel
коринка service-berry, *Amelanchier ovalis* Med.
корка **1.** crust **2.** rough outer bark
корм fodder, feed, forage
кормушка feed trough, feed rack
корневина root channel
корневище rhizome, rootstock
корнедушитель, дубовый *Rosellinia quercina* Hart.
корнежил, еловый *Hylastes cunicularius* Er.
корнежил, малый еловый *Hylastes opacus* Er.
корнежил, черный *Hylastes ater* Payk.
корненаселенность root occupation (of soil)
корненасыщенность root occupation (of soil)
корнеотпрыск sucker
корнеплод root crop
корнерез root pruner
коробиться to warp
коробка **1.** chest, box, case **2.** housing
коробление warping (see also покоробленность)
коробочка **1.** small box **2.** pod, capsule
коровка coccinellid
коровяк mullein, *Verbascum* L.
короед ipid, barkbeetle
короед-автограф *Dryocoetes autographus* Oliv.
короед, валежниковый *Orthotomicus proximus* Eich.
короед, вершинный *Ips acuminatus* Gyll.
короед-гектограф *Dryocoetes hectographus* Reitt.
короед-двойник *Ips duplicatus* Sahlb.
короед, двузубчатый *Pityogenes bidentatus* Fabr.
короед, двухцветный *Taphrorychus bicolor* Hbst.
короед, еловый большой *Ips typographus* L.

короед-крошка *Crypturgus* Erich.

короед, лиственничный большой *Ips subelongatus* Motsch.

короед, непарный западный *Anisandrus dispar* Fabr.

короед, непарный сосновый *Xyleborus eurygraphus* Ratz.

короед-пожарищ *Orthotomicus suturalis* Gyll.

короед, четырехзубчатый *Pityogenes quadridens* Hart.

короед, шестизубчатый *Ips sexdentatus* Boern.

короедина barkbeetle damage

королек, желтоголовый goldcrest, *Regulus regulus* L.

королек, красноголовый firecrest, *Regulus ignicapillus* Temm.

коромысло yoke, balance beam

корообдирка barker

кирорубка bark chopper

короснимaтель bark remover, barking blade

короставник, полевой *Knautia arvensis* Coult.

коротконожка, лесная *Brachypodium sylvaticum* Beauv.

коротыш short shoot

коротье short log, short timber

коррозия corrosion

кортаме short boards (for Mediterranean market)

кортик, охотничий hunting knife

корчевание rooting out, grubbing up, uprooting

корчевание леса forest clearance, deforestation

корчевание пней stump clearance

корчеватель stump-grubber

корчеватель, треножный tripod stump-puller

корчевка rooting out, grubbing up, uprooting

корыто bath, trough, vat

корье tanbark

корявый curved, crooked, tortuous, crooked-grown

коряга snag, sunken log, submerged stump

коряжник scrub, brush

коса **1.** scythe **2.** spit, tongue (of land) **3.** braid, plait

косарь chopper

косилка mower, mowing machine

косогор slope

косослой spiral grain

косослойность spiral grain

костенец spleenwort, *Asplenium* L.

костер **1.** stack, pile, heap (esp. for burning) **2.** bromegrass, *Bromus* L.

кострища slash bonfire site

костяника *Rubus saxatilis* L.

костянка drupe, stone fruit

косуля roe deer, *Capreolus capreolus* L.

косяк jamb

косяк, дверной door jamb

косяк, колесный felloe

косяк, оконный window jamb

котел **1.** boiler **2.** hollow, depression

котел, варочный digester

котел, паровой steaming tank

котел, пропиточный impregnating tank

котлован log pond, foundation pit

котловина **1.** deep hollow **2.** gap, group (in group system)

коуш eye, eye ring

кочедыжник *Athyrium* L.

кочедыжник, женский lady fern, *Athyrium filix-femina* Roth.

кочка hummock, hillock, tussock

кошель bag boom, towing boom

кошка **1.** cat **2.** drag anchor

кошка, дикая wildcat, *Felix silvestris* Schreb.

кошки climbing irons

кошма raft

коэффициент coefficient, quotient, factor

коэффициент, переводной converting factor

коэффициент полезного действия efficiency

коэффициент полнодревесности solid-volume/stacked-volume ratio

коэффициент формы form quotient

кравенир peeled veneer faced with waxed kraft paper
крампон lip (in resin tapping)
кран 1. crane 2. tap
кран, башенный tower crane
кран, кабельный cable-way transporter, cable-crane
кран, козловый gantry crane
кран, консольный cantilever crane
кран, мостовой overhead travelling crane
кран, поворотный jib crane
кран, стреловой jib crane
крапивник wren, *Troglodytes troglodytes* L.
крапиво nettle, *Urtica* L.
крапиво, глухая white dead-nettle, *Lamium album* L.
крапиво, двудомная stinging nettle, *Urtica dioica* L.
крапиво, жгучая small nettle, *Urtica urens* L.
краситель dye, dyestuff, pigment
краска paint, colour, dye
краскораспылитель paint sprayer, spray gun
краснина red stain
краснозем red earth, krasnozem
краснолесье conifer forest (pine or larch)
краснополосица red streak, red stain
краснотал willow, *Salix acutifolia* Willd.
краснохвост *Dasychira pudibunda* L.
Крафт Kraft tree dominance class (see классификация)
крахмал starch
крашение painting, dyeing, colouring
кремень 1. compression wood 2. flint
кремнезем silica
кремний silicon
кремнина compression wood
креноватость compression wood
кренометр clinometer
крень 1. compression wood 2. conifer scrub
кренюжник conifer scrub
креозот creosote
крепость strength
крест петров, чешуйчатый toothwort, *Lathraea squamaria* L.
крестовник, обыкновенный groundsel, *Senecio vulgaris* L.
крестоцветные Cruciferae
крестьянин peasant, farmer
кривая curve
кривая высот height curve
кривая хода роста growth curve
кривая частоты frequency curve
кривизна sweep, crook
криволесье elfin woodland, scrub
кривошип crank
крифал *Cryphalus* Erich.
кровля roof, covering
кровохлебка, обыкновенная *Sanguisorba officinalis* L.
кролик rabbit, *Oryctolagus cuniculus* L.
кромка border, edge
кромка, режущая cutting edge
крона crown
крона, развесистая branchy crown, spreading crown
крономер instrument for measuring tree height and crown length
кронциркуль calipers
кронштейн bracket
крот mole, *Talpa europaea* L.
кротовина molehill, mole run
круг circumference, circle, disk
кругляк roundwood, billet(s)
круговорот cycle, rotation, circulation
кружок от ствола stem disk
крупнозем coarse soil
крупномер 1. large timber (usually over 7 cm diam.) 2. large planting stock
крупнослойный wide-ringed, coarse-ringed
крупнотравье large herbs
крупный large, coarse
крутик *Amorpha fruticosa* L.
кручение torsion, twisting
крушина buckthorn, *Rhamnus* L.
крушина, вечнозеленая evergreen buckthorn, *Rhamnus alaternus* L.
крушина, ломкая alder buckthorn, *Rhamnus frangula* L. (*Frangula alnus* Mill.)

крушина, слабительная common buckthorn, *Rhamnus cathartica* L.
крыжовник gooseberry, *Ribes grossularia* L.
крылатка samara
крылатый winged, alate
крыло wing
крыловка board (1.3 cm thick, 6 m long, 22-27 cm wide)
крыса rat
крыса, водяная water vole, *Arvicola terrestris* L.
крыса, серая brown rat, *Rattus norvegicus* Berk.
крыса, черная black rat, *Rattus rattus* L.
крыша roof, covering
крыша, гонтовая shingle roof
крюк hook
крюк, багровый cant hook, peavy
крюк, поворотный cant hook, peavy
кряж **1.** log, block **2.** ridge
кряж, клепочный stave bolt
кряж, лыжный ski log assortment
кряж, пиловочный sawlog
кряж, спичечный match stock
кряж, тарный box wood
кряж, фанерный veneer log, peeler log
кряж, шпальный sleeper log
кряж, ящичный box wood
ксерофит xerophyte
ксилан xylan
ксилема xylem
ксилит xylitol
ксилоза xylose
ксилол xylene
ксилометр xylometer
кубатура cubic content, volume
кубометр cubic metre
кубометр, плотный solid cubic metre
кубометр, складочный stacked cubic metre
кувалда sledge-hammer
кудрявый curly
кузов basket (made of bark)
кузьмичева трава *Ephedra vulgaris* Rich.
куколка pupa
кукуруза maize, *Zea mays* L.
кукушка cuckoo, *Cuculus canoris* L.
кукушкин-лен *Polytrichum* Dill.
кукушкин-лен, малый *Polytrichum piliferum* Schreb.
кукушкин-лен, обыкновенный hair moss, *Polytrichum commune* L. Hedw.
кукушкин-лен, прямой *Polytrichum strictum* Banks
кукушкин-лен, стройный *Polytrichum gracile* Menz.
кукушкин-лен, утонченный *Polytrichum attenuatum* Menz.
кукушкин-цвет ragged robin, *Lychnis flos-cuculi* L.
кукушкины слезки spotted orchis, *Orchis maculata* L.
кулиса 'coulisse' (the broader unfelled strip between strip fellings)
кульбаба, осенняя *Leontodon autumnalis* L.
кульбаба, шершаволистная *Leontodon hispidus* L.
культивар cultivar
культиватор grubber, cultivator
культура (see also лес; насаждение) **1.** plantation (in general), young plantation **2.** culture **3.** crop **4.** cultivation
культура, географическая provenance trial plantation
культура, корнесобственная cultivation of a plant 'on its own root' (i.e. not grafted)
культура, кулисная strip plantation
культура, предварительная underplanting, planting in advance
культура, сидерационная green manure crop
кумяника *Rubus nessensis* W. Hall
куница marten, *Martes* L.
куница, каменная stone marten, *Martes foina* Erx.
куница, лесная pine marten, *Martes martes* L.
куньи Mustelidae
купальница, европейская *Trollius europaeus* L.
купена, лекарственная *Polygonatum officinale* All.

купена, многоцветная *Polygonatum multiflorum* All.
купена, мутовчатая *Polygonatum verticillatum* All.
купырь, лесной cow parsley, *Anthriscus sylvestris* (L.) Bernh.
куриные Galliformes
'курить в лесу запрешено' 'no smoking in the forest'
куропатка, белая willow grouse, *Lagopus lagopus* L.
куропатка, серая partridge, *Perdix perdix* L.
куропатка, тундровая ptarmigan, *Lagopus mutus* L.
куртина group, clump
курчавость листьев leaf curl
куст 1. shrub, bush 2. group
кустарник 1. shrub, bush 2. shrubbery, scrub
кустарь home craftsman, handicraftsman
кустовой bushy, shrubby
кусторез bush cutter, brush cutter
кутикула cuticle
куча pile, heap
кучина mahaleb cherry, *Prunus mahaleb* L.
кущение 1. bush formation 2. tillering
КФС urea-formaldehyde resin
кювет side drain, drain

Л

лабаз 1. raised platform (for shooting game) 2. store
лабазник, вязолистный meadow-sweet, *Filipendula ulmaria* (L.) Maxim.
лаборатория laboratory
лавина avalanche
лавр, благородный laurel, *Laurus nobilis* L.
лавровишня cherry-laurel, *Prunus laurocerasus* L.
лага joist, bolster
лад stave wood
ладьян coral-root, *Corallorhiza trifida* Chatel
лазоревка blue tit, *Parus caeruleus* L.
лазы climbing irons
лак varnish, lacquer
лакировка varnishing, lacquering
ламнеломка, бедреная *Pimpinella saxifraga* L.
ландшафт 'landscape' (i.e. soil, topography plus vegėtation)
ландыш lily-of-the-valley, *Convallaria majalis* L.
ланцетовидный lanceolate
лань fallow deer, *Dama dama* L.
лапа 1. foot, paw 2. root spur
лапа, корневая root spur
лапа-сковородень dovetail
ЛАПБ chipping headrig conversion line
лапина, кавказская Caucasian wingnut, *Pterocarya caucasica* C.A.M.
лапка needle-bearing twig
лапки кошачьи cat's foot, *Antennaria dioica* (L.) Gaertn.
лапник thin twig with needles
лапчатка *Potentilla* L.
лапчатка, белая *Potentilla alba* L.
лапчатка, кустарная *Potentilla fruticosa* L.
лапчатка, лесная common tormentil, *Potentilla erecta* (L.) Räusch.
лапчатка, песчаная *Potentilla arenaria* Borkh.
лапчатка, прямостоящая common tormentil, *Potentilla erecta* (L.) Räusch.
ласка weasel, *Mustela nivalis* L.
лата slat, lath, batten
латук, стенной *Mycelis muralis* (L.) Rchb.
лебедка winch, windlass
лебедка, подъемная hoisting winch
лебедка, тракторная tractor winch
ледник 1. refrigerator 2. glacier
лежень 1. log sawn (or hewn) on two opposite sides, cant 2. log used in a fore-and-aft road
лезвее edge, blade
лезвие edge, blade
лекарственный medicinal
лемех ploughshare

лемминг, лесной wood lemming, *Myopus schisticolor* Lill.

лен-кукушкин *Polytrichum* Dill. (see кукушкин-лен)

лента **1.** band, tape, strip **2.** blade (of band saw)

лента, мерная measuring tape

лепесток petal

лес (see also древесина; насаждение) **1.** wood, timber, lumber **2.** forest, wood(land)

лес, байрачный ravine forest

лес, берегозащитный bank protection forest

лес, бореальный boreal forest

лес, боровой pine forest, infertile pine site type, *Pinetum* (see бор)

лес, бочарный cooperage wood

лес, водоохранный water-conservation forest

лес, водорегулирующий water-regulating forest

лес, вторичный secondary forest

лес, выборочный selection forest

лес, высокоствольный high forest (i.e. of seedling origin)

лес, высушенный dry wood, seasoned wood

лес, галлерейный gallery forest

лес, горелый fire-damaged forest

лес, горный montane forest

лес, городской urban forest, city forest

лес государственного значения centrally managed forest (as opposed to kolkhoz forests)

лес, государственный state forest

лес, девственный virgin forest, primary forest

лес, деловой timber (all assortments except fuelwood)

лес, длиномерный long timber

лес, дождевой rain forest

лес, доступный accessible forest

лес, дровяной fuelwood, fuelwood forest

лес, заливной floodplain forest

лес, заповедный forest reserve

лес, защитный protection forest

лес, инструментальный wood for musical instruments

лес, кантовый squared timber

лес, климаксный climax forest

лес, кокорный root-spur wood (for shipbuilding)

лес, колотый cleft timber

лес, колочный forest outlier

лес, колхозный kolkhoz forest

лес, кондовый high-quality softwood

лес, корабельный shipbuilding timber

лес, корьевой (coppice) forest managed for tanbark

лес, крепежный mining timber, pitprops

лес, крестьянский farm forest

лес, кривой knee, crook (used in boatbuilding)

лес, круглый roundwood, billet(s)

лес, крупный large timber (usually over 7 cm diam.)

лес, курортный amenity forest in health resorts

лес, кустарный scrub forest, bush

лес, лесосечный uniform high forest

лес, летнезеленый deciduous forest

лес, лиственный broadleaved forest

лес, мачтовый mast timber

лес, мелколиственный forest of small-leaved deciduous species (e.g. birch)

лес, мелкотоварный small wood

лес, мертвый dead wood

лес, молевой loose-floated timber

лес, мягколиственный forest of soft-wooded deciduous species (e.g. aspen)

лес, мяндовый wide-ringed pine timber with high sapwood content

лес на корню standing timber

лес, недоступный inaccessible forest

лес, неосвоенный unexploited forest

лес, непрерывно-производительный sustained-yield forest

лес, низкоствольный coppice forest

лес, нормальный normal forest

лес, обтесанный hewn timber

лес, общинный communal forest, community forest

лес, освоенный exploited forest, forest in use
лес, островный forest outlier
лес, охранный protection forest
лес, первобытный virgin forest, primary forest
лес, перестройиый overmature forest
лес, пиленый sawn timber, lumber
лес, пильный sawtimber
лес, плодовый forest of (wild) fruit trees
лес, поделочный timber (all assortments except fuelwood)
лес, пойменный floodplain forest
лес, порослевый forest originating from coppice (or suckers)
лес, пригородный suburban forest
лес, приспевающий forest approaching maturity
лес, простой single-storeyed forest
лес, пьяной 'drunken' (leaning) forest in permafrost regions
лес, редкий sparse forest, open woodland, open forest
лес, резервный unexploited forest
лес, резонансовый resonance wood
лес, рудничный mining timber, pit-props
лес, рудовой high-quality softwood
лес, рукотворный artificial forest, plantation
лес, светлохвойный 'light' conifer forest (e.g. pine, larch)
лес, светлый open forest
лес, семенной high forest (i.e. of seedling origin)
лес, сложный multi-storeyed forest
лес, смешанный mixed forest
лес, совхозный sovkhoz forest
лес, сомкнутый closed forest
лес, спелый mature forest
лес, сплавляемый россыпью loose-floated timber
лес, сплавной floated timber
лес, средний coppice with standards
лес, срубленный felled timber
лес, стихийный virgin forest, primary forest
лес, столярный joinery wood
лес, строевой building timber
лес, строительный building timber
лес, сухостойный dead standing forest
лес, сучковатый knotty timber
лес, твердолиственный forest of hard-wooded broadleaved species
лес, темнохвойный 'dark' conifer forest (e.g. spruce, fir)
лес, тесанный hewn timber
лес, токарный turnery wood
лес, толстомерный large timber (usually over 7 cm diam.)
лес, тонкомерный small timber
лес, тугайный riparian woodland
лес, фаутный defective timber
лес, хвойный conifer forest
лес, хозяйственный **1.** forest under management **2.** commercial forest
лес, частновладельческий private forest
лес, частный private forest
лес, черневый taiga
лес, шпальный sleeper timber
лес, эксплуатируемый exploited forest, forest in use
лесистость area under forest, forest cover percent
лесистый forested, wooded
Лескомцентр Forest Commercial Centre
Лесная Биржа Wood Exchange
лесная опытная станция forest experiment station
лесник forester, forest warden, ranger
лесничество forest district (subdivision of a лесхоз), beat
лесничество, опытное experimental forest district
лесничий forester, forest warden, ranger
лесное промышленное хозяйство lespromkhoz (an administrative forestry enterprise, chiefly concerned with logging)
лесное хозяйство **1.** forestry, forest management **2.** leskhoz (an administrative forestry enterprise)
лесное энергетическое средство forestry power source
лесной **1.** wood, forest **2.** timber
лесобиржа timber yard

лесоботаника forest botany
лесоведение silvics, forest science
лесовик *Dryocoetes* Eich.
лесовик, пихтовый *Dryocoetes striatus* Egg.
лесовик, таежный *Dryocoetes hectographus* Reitt.
лесовик, хвойный *Dryocoetes autographus* Ratz.
лесовладелец forest owner
лесовод forester, silviculturist
лесоводственный silvicultural
лесоводство silviculture
лесовоз timber transport machine, timber lorry
лесовозобновление regeneration, renewal
лесовозобновление, естественное natural regeneration
лесовозобновление, искусственное artificial regeneration
лесовосстановление regeneration, renewal
лесозаготовитель logger
лесозаготовка procurement, harvesting, logging, felling and primary conversion
лесозащита forest protection
лесоистощение exhaustion of forest, depletion of forest
лесоистребление forest destruction
лесок small forest, wood
лесокатка log elevator
лесокомбинат forest combine (large integrated processing plant)
лесоматериал(ы) wood, timber, forest product(s)
лесомелиорация improving land by forestry measures (e.g. planting shelterbelts, sand fixation)
лесонакопитель bin (pocket) for logs on a sorting conveyor
лесонасаждение **1.** forest stand **2.** plantation
лесообразование formation of forest
лесообразователь forest-forming species
лесоохрана forest protection
лесоохранение forest protection
лесоочистка forest clearance, deforestation
лесопарк forest park
лесоперевалка transfer of wood from one kind of transport to another
лесоперевалочный appertaining to the transfer of wood from one kind of transport to another
лесопиление sawmilling
лесопилка sawmill
лесопильня sawmill
лесопользование forest exploitation, cut
лесопродукция forest products
лесопромышленный комплекс forest-industry complex
лесопункт **1.** production unit of lespromkhoz **2.** local timber conversion point
лесоразведение afforestation, reforestation
лесоразведение, богарное afforestation without irrigation
лесоразведение, защитное protective afforestation
лесоразведение, полезащитное shelterbelt establishment
лесоразработка forest exploitation, logging and conversion
лесорама frame saw
лесорез tree cutter, tree-cutting machine
лесоруб feller
лесосад forest orchard (of wild fruit and nut trees)
лесосводка removal of merchantable timber
лесосека coupe, felling area, cut-over area
лесосека, годичная annual coupe
лесосека, каемчатая strip felling
лесосека, обсеменительная seeding felling
лесосека, расчетная calculated cut, planned (authorized) cut
лесосека, сметная planned felling area
лесосека, сплошная clear felling, clear-felled area
лесосеменная плантация seed orchard
лесосеменный заказник seed reserve

лесосечный appertaining to felling (see работа, лесосечная and рубка, лесосечная)
лесоскат slide, chute (see лесоспуск)
лесосклад log dump, log storage place
лесосплав timber floating
лесосплав, первоначальный timber floating on waters not navigable or only seasonally navigable
лесосплав россыпью loose floating
лесоспуск slide, chute
лесоспуск, водяной flume
лесоспуск, деревянный wooden slide
лесоспуск, канатный cable slide, gravity cableway
лесоспуск, проволочный wire slide
лесостепь forest steppe
лесостой stand
лесосушилка wood drying plant, kiln, timber drier
лесотаска log conveyor
лесотехническая академия forest academy
лесотрест timber trust
лесотундра forest tundra
лесоуправление forest administration
лесоустроитель forest manager, working-plan officer
лесоустройство forest management (in the strict sense), forest regulation
лесоучасток forest district, forest plot
лесохимия 'forest chemistry', wood chemistry (including chemical utilization, resin tapping etc.)
лесохозяйственник forester, silviculturist
лесохозяйственный forestry
лесохранилище wood storage place, log pond
лесоэкономический район forest economic region
лесоэксплуатация forest exploitation, cut
леспедеца, двуцветная *Lespedeza bicolor* Turcz.
леспромхоз lespromkhoz (an administrative forestry enterprise, chiefly concerned with logging)
лесс loess
Лестехимпорт Agency for Forest Machinery Imports
лестница ladder, steps
лестничный scalariform
лестрансхоз transportable forestry enterprise, migratory forestry enterprise
лесхоз leskhoz (an administrative forestry enterprise)
лесхоз, опытный experimental leskhoz
лесхоззаг leskhozzag (enterprise combining the features of a leskhoz and a lespromkhoz)
лет пыльцы pollen shedding
летающий volatile, flying
летнаб-таксатор airborne mensuration officer
лето summer
летучий volatile, flying
лещина hazel, *Corylus* L.
лещина, древовидная Turkish hazel, *Corylus colurna* L.
лещина Зибольда Japanese hazel, *Corylus sieboldiana* Bl.
лещина, крупная filbert, *Corylus maxima* Mill.
лещина, маньчжурская Manchurian hazel, *Corylus mandshurica* Max.
лещина, обыкновенная common hazel, *Corylus avellana* L.
лещина, разнолистная hazel, *Corylus heterophylla* Fisch.
лещинник hazel wood
лжеакация black locust, *Robinia pseudoacacia* L.
лжетсуга, зеленая green Douglas fir, *Pseudotsuga menziesii* var. *viridis*
лжетсуга, серая grey Douglas fir, *Pseudotsuga menziesii* var. *caesia*
лжетсуга, сизая blue Douglas fir, *Pseudotsuga menziesii* var. *glauca*
лжетсуга, тиссолистная Douglas fir, *Pseudotsuga menziesii* Mirb.
лиана liane
ливень heavy shower, downpour
лигнамон densified wood treated with ammonia
лигнин lignin

лигноцеллюлоза lignocellulose
лизалка salt lick
ликвид merchantable material
ликвидность merchantability, realizability
ликвидный merchantable, realizable
лилия, лесная *Lilium martagon* L.
лим *Erythrophloeum fordii* Oliv.
лиман estuary, plot irrigated by spring snow melt
лимонник *Schizandra* Michx.
линейка **1.** ruler, rule **2.** a line of log bundles ready for inclusion in a raft
линейка, прижимая nose bar, pressure bar
линейка, счетная slide rule
линейный linear
линейчатый способ ribbon spreading (glue)
линия line
линия, визирная sight line, sighting line, survey line
линия, водораздельная watershed, divide
линия, квартальная compartment boundary line
линия реза line of cut
линия скольжения slip line
линия, теневая shadow line (in sawmilling)
линнеа *Linnaea borealis* L.
линька moult, moulting
линяние moulting
липа linden, lime, *Tilia* L.
липа, амурская Amur lime, *Tilia amurensis* Rupr.
липа, войлочная silver linden, *Tilia tomentosa* Moench
липа, кавказская Caucasian lime, *Tilia caucasica* Rupr.
липа, красная red lime, *Tilia rubra* DC.
липа, крупнолистная large-leafed linden, *Tilia platyphyllos* Scop.
липа, крымская Crimean linden, *Tilia euchlora* K. Koch
липа, маньчжурская Manchurian linden, *Tilia mandshurica* Rupr.
липа, мелколистная small-leafed linden, *Tilia cordata* Mill.
липа, серебристая silver linden, *Tilia tomentosa* Moench
липа, сибирская Siberian linden, *Tilia sibirica* Bayer
липняк lime wood, *Tilietum*
лисий хвост reciprocating log cross-cut saw
лисица fox, *Vulpes vulpes* L.
лисичка, настоящая chanterelle, *Cantharellus cibaris* Fr.
лисохвост fox-tail, *Alopecurus* L.
лист **1.** leaf, blade **2.** sheet
листва foliage
лиственница larch, *Larix* Mill.
лиственница, даурская Dahurian larch, *Larix gmelinii* Litv.
лиственница, европейская European larch, *Larix decidua* Mill.
лиственница, курильская Kurile larch, *Larix gmelinii* var. *japonica* Pilg.
лиственница, сибирская Siberian larch, *Larix sibirica* Led.
лиственница Сукачева Sukachev's larch, *Larix sukaczewii* Dyl.
лиственница, тонкочешуйчатая Japanese larch, *Larix leptolepis* Murr.
лиственница, чекановская *Larix czekanowskii* Szaf. (hybrid of *L. gmelinii* and *L. sibirica*)
лиственница японская Japanese larch, *Larix leptolepis* Murr.
лиственничный larch
лиственный leafy, broadleaved
листвяг larch forest, *Laricetum*
листоблошка psyllid
листовертка tortricid
листовертка-верхоед pine-shoot moth, *Evetria buoliana* Schiff.
листовертка, дубовая зеленая green oak-roller moth, *Tortrix viridana* L.
листовертка, лиственничная *Semasia diniana* Gn.
листовертка, пихтовая черноголовая *Cacoecia murinana* Hb.
листовертка-толстушка, пихтовая *Cacoecia murinana* Hb.

листовертка, шишковая *Laspeyresia strobilella* L.
листоед chrysomelid
листоед, сосновый *Luperus pinicola* Duft.
листоед, тополевый *Melasoma populi* L.
листопад leaf fall
листорасположение leaf arrangement, phyllotaxis
листораспускание flushing
листочек leaflet
литера **1.** letter **2.** area (designated by a letter)
лихтер lighter
личинка larva
лишай lichen
лишайник lichen
лоад load, 50 ft^3 (= *c.* 1.4 m^3) of sawn timber
лобзик fret saw
лов catching, hunting
ловля catching, hunting
ловушка trap
лог **1.** small ravine **2.** catchment
логовище lair
ЛОД forest experiment estate
лодка boat
ложбина ravine, hollow
ложе bed
ложе реки river bed
ложногусеница false larva (of sawflies)
ложнокороед bostrychid
лоза willow, *Salix triandra* L. or *Salix viminalis* L.
лоза корзинная common osier, *Salix viminalis* L.
лозина willow, *Salix triandra* L.
лом crowbar
ломкий brittle, fragile
ломкость brittleness
ломонос *Clematis vitalba* L.
лопастной lobed
лопасть lobe
лопата spade
лопата, клиновидная notching spade
лопата, полукруглая semi-circular (planting) spade
лопата, сажальная planting spade
лопатка small spade
лопатка для посадки в щель notching spade
лопатка, окорочная barking iron
лопаточка swaged saw tooth
ЛОС forest experiment station
лось elk, *Alces alces* L.
лоток flume, chute
лоток, водный flume
лоток, мокрый flume
лоток, сухий dry chute
лох oleaster, Russian olive, *Elaeagnus* L.
лох, узколистный oleaster, *Elaeagnus angustifolia* L.
лощадь horse
лощина ravine, hollow
ЛП-18 grapple skidder
ЛП-18А grapple skidder
ЛП-19 feller-buncher
ЛП-2 feller-buncher
ЛПК forest-industry complex
ЛСЗ seed reserve
ЛСП forest seed orchard
ЛТ forwarder
ЛТА forest academy
луб **1.** bast (in general) **2.** phloem
луб, включенный included phloem
луб, внутренный internal phloem
луб, вторичный secondary phloem
луб, первичный primary phloem
луб, проводящий phloem
лубоед bast beetle
лубоед, вязовый *Pteleobius vittatus* Fabr.
лубоед, еловый (большой) *Dendroctonus micans* Kug.
лубоед, сосновый (большой) *Blastophagus piniperda* L.
лубоед, сосновый (малый) *Blastophagus minor* Htg.
лубоед, ясеновый (большой) *Hylesinus crenatus* Fabr.
лубоед, ясеновый (пестрый) *Hylesinus fraxini* Panz.
луг meadow
луговик, дернистый *Deschampsia caespitosa* (L.) Beauv.

луговик, извилистый *Deschampsia flexuosa* (L.) Trin.
лук 1. bow 2. *Allium* L.
лука bend (of river)
луковица 1. bulb 2. onion
лукошко bast basket
лунка hole, seed-spot
лунка, посадочная planting hole
лунка, серебристая *Phalera bucephala* L.
луночка ring shake
лупин lupin, *Lupinus* L. (see люпин)
лупить to peel, to strip
луч ray
луч, древесный wood ray, xylem ray
луч, лубяный phloem ray
луч, сердцевинный primary ray
луч, сложный aggregate ray
лучина splinter
лущение 1. extraction (of seeds), shelling 2. skimming, shallow ploughing 3. peeling (of veneer)
лущильня seed extraction plant
лыжа ski
лыко bast (in general)
лыкодер bast stripper
лысение patch barking
лысина bald patch, bare patch
лысить to bark partially, to strip bark, to patch bark
лысый bald, bare
льнянка, обыкновенная toadflax, *Linaria vulgaris* Mill.
ЛЭР forest economic region
ЛЭС forestry power source, prime mover
любка, двулистная lesser butterfly orchid, *Platanthera bifolia* (Cust.) Rich.
люк hatch, hatchway
люлька cradle
люпин lupin, *Lupinus* L.
люпин, многолетний perennial lupin, *Lupinus polyphyllus* Lindl.
лютик *Ranunculus* L.
люцерна 1. lucerne 2. *Medicago* L.
лягушка frog, *Rana* L.
ляда land cleared by shifting cultivation
лядина 1. system of burning to clear ground for agriculture 2. land cleared by shifting cultivation

М

магистраль main line of communication (road, rail or water)
магнолия magnolia, *Magnolia* L.
мадаке, японское *Phyllostachys reticulata*
мадриери sawn timber 3-8 m long, 70 x 210-220 mm in cross-section
мазут masut (fuel oil)
мазь paste, grease
майдан tar pit
майна 1. hole (in ice) 2. mynah
майник, двулистный May lily, *Maianthemum bifolium* (L.) F.W. Schmidt
макет model
маклер broker
маклюра osage orange, *Maclura aurantiaca* Nutt.
макулатура waste paper
макушка tip, crown
малина raspberry, *Rubus idaeus* L.
малорослый small, small-growing
манипулятор grapple
марал Siberian red deer, *Cervus elaphus sibiricus* Sev.
марка sign, mark
марка, грузовая load line, Plimsoll line
маркировка marking
мартук *Emeropyrum orientale*
маршрут route, route survey
марь 1. mist, haze 2. shrubby sphagnum bog 3. *Chenopodium* L.
марьянник cow-wheat, *Melampyrum* L.
марьянник, дубравный *Melampyrum nemorosum* L.
марьянник, лесной *Melampyrum sylvaticum* L.
марьянник, луговой *Melampyrum pratense* L.
марьянник, полевой *Melampyrum arvense* L.

маслина, европейская olive, *Olea europaea* L.
масло 1. oil 2. butter
масло, всплывное light (flotation) oil
масло, дегтярное tar oil
масло, древесносмоляное wood-tar oil
масло, пихтовое fir oil
масло, смоляное 1. tar oil 2. resin oil
масло, флотационное flotation oil
масло, эфирное essential oil
масса 1. mass, volume 2. pulp
масса, древесная 1. ground-wood pulp, mechanical pulp 2. wood volume
масса, древесная, белая white groundwood
масса, древесная, химическая chemigroundwood
масса, ликвидная 'realizable' volume (the exploitable volume after deduction of reserved species and stands of less than 30 m^3/ha)
масса, плотная solid volume
массиф, лесной forest massif
мастер skilled workman, foreman, ganger
мастер лесозаготовок logging foreman
мастерская workshop
мастика mastic
масштаб scale
материал material
материал, горючий fuel
материал, древесный wood-based materials
материал, культурный planting stock
материал, лесной wood, timber, forest product(s)
материал, пиленый sawn timber, lumber
материал, пневой stumpwood
материал, посадочный planting stock
материал, посевной seed
материал, сырой raw material
материнский maternal, mother, parent(al)
маточник plant or group of plants used as a source of seed, cuttings, etc. (e.g. a stool or stool-bed)
мать-и-мачеха coltsfoot, *Tussilago farfara* L.
махровый double (of flowers)
мачта mast, spar
мачта, головая head spar
мачта, тыловая tail spar
машина machine (see also станок)
машина, валочная felling machine
машина, валочно-пакетирующая feller-buncher
машина, валочно-трелевочная feller-skidder
машина, дробильная chipper, crusher
машина, калевочная moulder
машина, корчевальная stump grubber
машина, окорочная debarker
машина, погрузочно-транспортная forwarder
машина, рубительная chipper, hog
машина, сажальная planting machine
машина, сплоточная bundling installation, bundle binder (for floated timber)
машина, сучкорезная branch trimmer, de-limber
машина, трелевочная skidder
машина, щепорубительная flaker
маяк 1. standard, reserve 2. signal tower, lighthouse
мгла mist, haze
МДФ MDF, medium-density fibreboard
мебель furniture
мебель, гнутая bent-wood furniture
медведица arctiid
медведка, обыкновенная *Gryllotalpa gryllotalpa* L.
медведь, бурый brown bear, *Ursus arctos* L.
медвежья ягода bearberry, *Arctostaphylos uva-ursi* (L.) Spreng.
медонос nectar-bearing plant
медоносный nectar-bearing
медуница lung-wort, *Pulmonaria* L.
медуница, лекарственная *Pulmonaria officinalis* L.
межа limit, boundary
межвидовой interspecific
междоузлие internode
междузубие gullet (of saw tooth)
междурядие space between the rows
межевание 1. surveying 2. demarcation

межевик surveyor
межень normal (summer) water level
межклетник intercellular space
межклеточный intercellular
межлесье opening in forest
межродовой intergeneric
мезотрофный mesotrophic
мезофилл mesophyll
мезофилловый mesophyllous
мезофильность mesophily
мезофильный mesophilous, mesophile
мезофит mesophyte
мел chalk
мелиорация improvement, amelioration (often drainage), reclamation
мелиорация леса forest improvement
мелиорация, лесная improvement of land by forestry measures (chiefly to control drought and erosion)
мелкий shallow, fine, small
мелководье shallow water, shoal
мелковолокнистый fine-grained, fine-fibred
мелкозем fine earth
мелколесье scrub
мелколиственный small-leaved
мелкослойный narrow-ringed, narrow-layered
мелкосопочник region of low hillocks and declivities (in Kazakhstan)
мелочь fines
мель sandbank
мельница mill
мендач tree (usually pine) with wide sapwood and wide rings
мендовый with wide sapwood, wide-ringed
мера measure, dimension
мера борбы control measure
мера плотнои массы solid measure
мера, складочная stacked measure
мергель marl
мерзлота frozenness, frozen ground
мерзлота, вечная permafrost
мерзлый frozen
меристема meristem
мерить to measure
мероприятие measure
мертвопокровный with no live ground cover
мертвяк deadman
местность locality, district, terrain
местный local
место place
место поворота turning place
место, пустое gap
место, узкое bottleneck
местонахождение **1.** station, locality **2.** site
местообитание **1.** habitat **2.** site
местоположение situation
местопроизрастание **1.** habitat, station **2.** site
месторождение **1.** deposit, layer **2.** site **3.** occurrence
метаморфоз metamorphosis
метелица snow storm
метелка panicle
метель snow storm
метеослужба weather service
метик heart shake (extending the length of the material)
метик, звездчатый star shake
метик, крестовый two shakes at an angle to each other
метик, лучистый star shake
метик, несогласный spiral heart shake
метик, согласный heart shake in a single vertical plane
метить to mark
метка mark
метла broom, brush
метла ведьмина witches' broom
метод method
метод горяче-холодной ванны hot-and-cold bath treatment
метод, комбинированно-периодный yield regulation by combined area and volume
метод, контрольный control method, méthode du contrôle
метод контроля control method, méthode du contrôle
метод, массовый yield regulation by volume

метод ограниченного поглощения (по Рюпингу) Rueping empty-cell process
метод, периодно-массовый yield regulation by volume
метод, периодно-площадный yield regulation by area
метод, площадный yield regulation by area
метод полного поглощения full-cell process
метод полуограниченного поглощения (по Лауру) Lowry empty-cell process
метод, поточный flow method (of working)
метод ухода, верховой crown thinning
метод ухода, низовой low thinning
метод, цикличный alternating pressure process
метр metre
метр, квадратный square metre
метр, кубический cubic metre
метр, объемный stacked cubic metre
метр, плотный solid cubic metre
метр, погонный running metre
метр, складной folding rule
метр, складочный stacked cubic metre
метр, тесьмовый tape measure
мех fur
мехлесхоз mechanized leskhoz
меч sword
меч Колесова Kolesov's planting iron
меч, сажальный planting iron
меченый marked, labelled, tagged
мешалка mixer
мешок bag, sack, sac
мешок, пыльцевой pollen sac
микология mycology
микориза mycorrhiza
микотрофия mycotrophy
микотрофность mycotrophy
микрограф *Pityophthorus micrographus* L.
микроклимат microclimate
микрорельеф micro-relief
микрофлора microflora
микрошпон very thin veneer (0.09-0.1 mm)
микроэлемент trace element
мимоза, серебристая silver wattle, mimosa, *Acacia dealbata* Link.
миндаль almond
миндаль, колючейший spiny almond, *Prunus spinosissima* Franch.
миндаль, низкий dwarf Russian almond, *Prunus tenella* Batsch
миндаль, обыкновенный almond, *Prunus amygdalus* Batsch
миндаль, степной dwarf Russian almond, *Prunus tenella* Batsch
минер leaf-miner
минерализация mineralization
Минлесхоз Ministry of Forestry
мирт, обыкновенный common myrtle, *Myrtus communis* L.
мицелий mycelium
млекопитающее mammal
млечник latex tube
многовершинность multiple top
многолетний many years old, of many years' duration, perennial
многоножка **1.** myriapod **2.** polypody, *Polypodium vulgare* L.
многорядник shield fern, *Polystichum* Roth.
многослойный multi-ply, multi-layered, laminated
многоярусный multi-storeyed, multi-layered
могильница *Vinca* L.
модальный modal, most frequent (applied to stands)
модель **1.** sample tree, sample **2.** model
модер moder
модуль modulus
модуль упругости modulus of elasticity
можжевелеягодник *Arceuthobium oxycedri* M.B.
можжевельник juniper, *Juniperus* L.
можжевельник, вонючий stinking juniper, *Juniperus foetidissima* Willd.
можжевельник, высокий Grecian juniper, *Juniperus excelsa* Bieb.

можжевельник, даурский Dahurian juniper, *Juniperus dahurica* Pall.
можжевельник, зеравшанский Seravshan juniper, *Juniperus seravschanica* Kom.
можжевельник, казацкий savin, *Juniperus sabina* L.
можжевельник, красный prickly juniper, *Juniperus oxycedrus* L.
можжевельник, многоплодный black juniper, *Juniperus polycarpos* C. Koch
можжевельник, обыкновенный common juniper, *Juniperus communis* L.
можжевельник, полушаровидный semiglobose juniper, *Juniperus semiglobosa* Rgl.
можжевельник, сибирский Siberian juniper, *Juniperus sibirica* Burgsd.
можжевельник, туркестанский Turkestan juniper, *Juniperus turkestanica* Kom.
можжевельник, туркменский Turcoman juniper, *Juniperus turcomanica* B.Fed.
можжевельник, тяжелопахучий stinking juniper, *Juniperus foetidissima* Willd.
мозаика листьев leaf mosaic
мозоль callus
мокрица 1. chickweed, *Stellaria media* (L.) Vill. 2. woodlouse, *Oniscus* L.
мокрослой wetwood, watercore
мокрый wet
молиния, голубая *Molinia caerulea* Moench
молния lightning
молодняк young stand (before thicket stage)
молодняк, густой dense young stand
молодняк, порослевой young stand originating from coppice (or suckers)
молодняк семенного происхождения young stand of seedling origin
молоко, известковое milk of lime
молот hammer
молот, деревянный mallet
молот, маркировочный marking hammer
молот, номеровальный die hammer
молот, разметочный marking hammer
молот, расколочный cleaving hammer
молот, штамповальный die hammer
молоток hammer (see молот)
молочай spurge, *Euphorbia L.*
молочай, миндалевидный wood spurge, *Euphorbia amygdaloides* L.
моль 1. loose floating 2. moth
моль, дубовая одноцветная *Tischeria complanella* Hb.
моль, лиственничная побеговая larch shoot-borer, *Argyresthia laevigatella* H.S.
моль, лиственничная чехликовая larch leaf-miner, *Coleophora laricella* Hb.
моль-пестрянка, платановая *Lithocolletis platani* Str.
моль, сосновая минирующая *Ocnerostoma piniariella* Zell.
моль, тополевая *Lithocolletis populifoliella* Z.
мольбище sacred grove (in Mari region)
монашенка nun moth, *Lymantria monacha* L.
мониторинг monitoring
монокультура pure plantation, pure crop, monoculture
монтаж assembly, installation, fitting
монтаж, накидной making a mosaic (air survey)
мор mor, raw humus
морали squared timber 4 m long or over (for Mediterranean markets)
морена moraine
морение staining, fuming
мороз frost
морозобой(на) frost crack
морозостойкий frost-resistant, frost-hardy
морозостойкость frost resistance, frost hardiness
морозоупорность frost resistance, frost hardiness
морошка cloudberry, *Rubus chamaemorus* L.
морфология morphology
московка coal tit, *Parus ater* L.

мосо *Phyllostachys edulis*
мост bridge
мостовая pavement, roadway
мостовая, деревянная wood pavement
мостовая, торцовая wood-block pavement
мотоблок 2-wheel (walking) tractor
мотобур powered hole-borer
мотоветверезка, древолазающая tree-climbing branch-pruner
мотокусторез bush cutter, brush cutter
мотопила power saw
мотыга hoe, mattock
мотылковые Papilionaceae
мох 1. bryophyte 2. moss
мох, болотный bog moss, *Sphagnum* L.
мох, боровой *Pleurozium schreberi* Mitt.
мох, исландский *Cetraria islandica* Ach.
мох, олений reindeer moss, *Cladonia rangiferina* Hoffm./L.
мох, перистый *Ptilium crista-castrensis* de Not.
мох, торфяной bog moss, *Sphagnum* L. (see also сфагнум)
моховик, зеленый *Boletus subtomentosus* L.
моховка *Ribes procumbens* Pall.
моховой mossy
мочага hollow, pool (in bog)
мочажина hollow, pool (in bog)
мочало bast fibre (from lime)
мочалыга bast cord
мочевина urea
мочежина hollow, pool (in bog)
мочение steeping, soaking, wetting
мочка 1. fibre 2. steeping, soaking, wetting
мочковатый fibrous
мощность 1. power, capacity 2. thickness
мрамор 1. mottled rot 2. marble
мраморный marbled, mottled
МСЛИО IUFRO
МСЛНИО IUFRO
мужской male
мука flour, meal
мука, буровая frass
мука, витаминная vitamin flour (from conifer needles)
мука, древесная wood flour
мука, костяная bone meal
мука, хвойно-витаминная vitamin flour (from conifer needles)
мулль mull
мульча mulch
мульчирование mulching
мурава sward, grass
муравей ant
муравей, большой черный black carpenter ant, *Camponotus herculeanus* L.
муравей, лесной common black ant, *Lasius niger* L.
муравей, малый лесной *Formica polyctena* Först.
муравей, рыжий лесной wood ant, *Formica rufa* L.
муравейник ant hill
муравьежук ant beetle, *Thanasimus formicarius* L.
мускатница moschatel, *Adoxa moschatellina* L.
мускусница moschatel, *Adoxa moschatellina* L.
мутация mutation
мутовка whorl
мутовчатый whorled
муха fly
муха, лиственничная *Chortophila laricicola* Karl.
мухоловка flycatcher
мухоловка-белошейка white-collared flycatcher, *Muscicapa albicollis* Temm.
мухоловка-пеструшка pied flycatcher, *Muscicapa hypoleuca* Pall.
мухоловка, серая spotted flycatcher, *Muscicapa striata* Pall.
мухомор, красный fly agaric, *Amanita muscaria* L.
мушмула medlar, *Mespilus germanica* L.
мхи 1. bryophytes 2. mosses
мхи, листостебельные Musci, mosses
мхи, печеночные Hepaticae, liverworts
мшара wet mossy site type

мшара, сосновая wet mossy pine forest type
мшистость 1. mossiness 2. roughness (of wood surface)
мшистый mossy
мытник, болотный red-rattle, *Pedicularis palustris* L.
мычка mat-grass, *Nardus stricta* L.
мышевидный mouse-like rodent (any small species of the Muridae) (see грызун)
мышиные Muridae
мышь mouse
мышь, домовая house mouse, *Mus musculus* L.
мышь, желтогорлая yellow-necked field mouse, *Apodemus flavicollis* Melch.
мышь, лесная long-tailed field mouse, *Apodemus silvaticus* L.
мышь, летучая bat
мышь -малютка harvest mouse, *Micromys minutus* Pall.
мышь, полевая striped field-mouse, *Apodemus agrarius* Pall.
мышьяк arsenic
мягколиственный soft-wooded broadleaved
мягкотелые molluscs
мякоть flesh (of fruit), pulp
мяндач tree (usually pine) with wide sapwood and wide rings
мяндовый with wide sapwood, wide-ringed
мята mint, *Mentha* L.
мята, водяная *Mentha aquatica* L.
мята, длиннолистная horse-mint, *Mentha longifolia* (L.) Huds.
мята, лесная horse-mint, *Mentha longifolia* (L.) Huds.
мятлик meadow-grass, *Poa* L.
мятлик, болотный *Poa palustris* L.
мятлик, боровой *Poa nemoralis* L.
мятлик, дубравный *Poa nemoralis* L.
мятлик, луговой meadow grass, *Poa pratensis* L.
мятлик, однолетний *Poa annua* L.

Н

набор assembly, set
набухание swelling
навес 1. shed 2. awning
навеска weighed portion (or sample)
навигация navigation, floating season
наводнение inundation, flooding, flood
навожение manuring
навоз manure
нагар scorch, scorch mark
нагель dowel, pin, peg
нагорный mountainous, upland
нагружаемость load capacity
нагружать to load
нагрузка 1. load, loading 2. face/girth ratio, face/diameter ratio (in resin tapping)
нагрузка, рекреационная recreational loading, recreational impact
надбавка на усушку extra allowance for shrinkage
надел portion, share
надел, земельный plot
надземный aerial, above ground
надзиратель inspector, overseer
надкожица cuticle
надрез incision, cut
надруб notch, undercut
надстройка superstructure
наездник, пихтовый шишковый *Megastigmus strobilobius* Ratz.
наждак emery
наземный terrestrial
назревание maturing, ripening
накалывание incising
накат 1. subfloor 2. layer of logs
накладка facing, cover
наклевываться to sprout (e.g. of acorns)
наклеп hammer hardening (saw blades)
наклон inclination, gradient, slope
наклон волокон slope of grain
наклонение inclination
наклон пилы hang, overhang (of vertical frame saw)
наклюнуться to sprout (e.g. of acorns)
накол incision
накопитель hopper, container, bin

налед impounding water by freezing
налет 1. natural regeneration (from seed) colonizing open areas 2. bloom (on fruit)
налет, грибный surface fungal disease (spots, mildew, etc.)
налог tax
намачивание wetting, soaking
наметельник broomhandle, broomhandle square (1-1½ in (= *c.* 2.5–4 cm) square, 48–52 in (= *c.* 122-132 cm) long)
намечание marking
намечать to mark
намораживание impounding water by freezing
нанос 1. alluvium 2. detritus
наносный alluvial
нападение infestation
напенный stump, stool
наперстянка *Digitalis* L.
напилок file
напильник file
наплыв callus, burr
наплыв, корневый swollen butt, butt swelling, root swelling
направка setting, adjustment
направление direction
направляющая 1. guide (e.g. of bandsaw) 2. fairlead (e.g. of winch drum)
направщик пил saw doctor
напряжение 1. stress 2. voltage
напряжение, внутреннее internal stress
напряжение, допускаемое working stress, allowable stress
напряженность stress
нарастание growth, increase, accumulation
наращивание growth, increase, accumulation
нарезка 1. incision 2. thread (of screw)
нарост callus, outgrowth, burr
нарост, морозобойный frost rib
нарубать to cut
наружный exterior, external
нарушение breaking, disturbance
нарушение сомкнутости opening of the canopy
насаждение (see also лес) 1. stand 2. plantation 3. planting
насаждение, главное main stand
насаждение, двухярусное two-storeyed stand
насаждение, защитное protective stand
насаждение, зеленое amenity stand
насаждение, изреженное thinned stand
насаждение леса planting of forest
насаждение, лесное forest stand
насаждение, материнское parent stand
насаждение, многоярусное multi-storeyed stand
насаждение, модальное modal stand
насаждение, молодое young stand
насаждение, одновозрастное even-aged stand
насаждение, одноярусное single-storeyed stand
насаждение, полезащитное shelterbelt, stand for protection of fields
насаждение, полное fully stocked stand
насаждение, порослевое stand originating from coppice (or suckers)
насаждение, приспевающее stand approaching maturity
насаждение, пробное trial stand
насаждение, простое single-storeyed stand
насаждение, разновозрастное uneven-aged stand
насаждение, разреженное thinned stand
насаждение, редкое sparse stand
насаждение, семенное high forest (i.e. of seedling origin)
насаждение, сложное multi-storeyed stand
насаждение, смешанное mixed stand
насаждение, сомкнутое closed stand
насаждение, спелое mature stand
насаждение с просветами gappy stand
насаждение, указательное indicator stand
насаждение, чистое pure stand
насекомое insect
насекомое, вредное harmful insect, insect pest

насекомое, листогрызущее leaf-eating insect
насекомое, минирующее leaf-miner
насекомое, полезное useful insect
насекомое, хвоегрызущее needle-eating insect
насекомоядный insectivorous
население **1.** colonization **2.** population
наследственность heredity
наследственный hereditary, inherited
наследство inheritance
наследуемость heritability
наст frozen snow crust
настил flooring, decking
насыпь fill, filling, embankment
насыщение saturation
насыщенный saturated
натачивать to sharpen
натрий sodium
натура nature
натура семян seed weight (in g/l)
натяжение tension, tensioning (of saw blades)
наука science
научно-техническое общество scientific and technical association
начальник head, chief
начерно roughly
начисто cleanly, smoothly
НВ field capacity
недефицитный sufficient, abundant
недозрелость unripeness, immaturity
недозрелый unripe, immature
недомер short measure
недопил key, crest (in tree felling)
недоруб **1.** felling less than the prescribed quantity, undercut **2.** key, crest (in tree felling)
недостаток **1.** shortage, deficiency **2.** defect
недостаточность питательных веществ nutrient deficiency
недотрога, желтая touch-me-not, *Impatiens nolitangere* L.
незабудка forget-me-not, *Myosotis* L.
незабудка, болотная *Myosotis palustris* L.
незабудка, полевая *Myosotis arvensis* Hill
незрелый unripe, immature
нейтральный neutral
неклен Tatarian maple, *Acer tataricum* L.
некроз necrosis
необработанный untreated, unfinished, rough
необрезный unedged
неопадающий persistent, evergreen
неполноценный of reduced value
неправильный incorrect, irregular
непрерывный uninterrupted, continuous
непродуваемый impermeable (of shelterbelts)
непроизводительный unproductive
непроклей glue-line defect
непроницаемый impermeable
неравномерный irregular
нерассортированный unsorted
нерв листа nerve of leaf
неспелый unripe, immature
несплавный unfloatable
несучковатый **1.** unbranched **2.** knot-free, clear
нетоварный unmarketable, unmerchantable
нетребовательный undemanding
неурожай poor crop, crop failure
неустойчивый unstable, non-resistant
неформатный of inexact size, not cut to size
нехрущ summer chafer, *Amphimallon solstitialis* L.
нива land cleared by shifting cultivation
нивелир level, levelling instrument
нивелировка levelling
низина low place, lowland
низкорослый of low growth
низкоствольник coppice, coppice stand, coppice tree
низменность lowland
низовой low thinning (see прореживание, низовое)
низовье lower part of a river
НИИЛГиС Research Institute for Forest Genetics and Breeding

нитрагин nitragin (a preparation of *Rhizobium* used as a bacterial fertilizer)
нитрификация nitrification
нитроаммофоска compound NPK fertilizer
нить, клеевая glueline
ниш niche
НЛУ continuous forest management, sustainable forestry
новатор innovator
новина new land (cleared by shifting cultivation)
нож knife, blade
нож -корнерез root-pruning knife
нож, расклинивающий riving knife
нож, садовый grafting knife
нож, сажальный planting iron
нож -скобель barking iron
нож, строгальный planer blade, slicing blade
ножевка hack-saw
ножницы shears
ножовка hack-saw
нонуис vernier
нора burrow, hole
норичник, шишковатый figwort, *Scrophularia nodosa* L.
норка European mink, *Mustela lutreola* L.
норм norm, standard
норм времени standard time
норм выработки standard time
нормирование standardization, setting of standards (esp. standard times)
нормировка standardization, setting of standards (esp. standard times)
ночница noctuid
ночница, сосновая pine noctuid, *Panolis flammea* Schiff.
НРВ growth substance derived from petroleum
НТО scientific and technical association
нумератор numbering hammer

О

обапол slab
обвал landslide, landslip
обвал, снежный avalanche
обветвление branching
обветвленный branched, branchy
обводнение irrigation
обвязка bundling, binding
обгородить to hedge, to fence, to enclose
обдир коры mechanical damage to bark, fraying
обезвершинивание topping
обезвершинивать to top, to decapitate, to pollard
обезвоженный dehydrated
обезлесение deforestation
обезлесенный deforested
обезлесить to deforest
обезличенный undressed (of lumber)
обсейвание sowing
обсеменение sowing, seeding, seeding up
обескорить to bark
обескрыливание de-winging (of seeds)
обескрыливать to de-wing (seeds)
обесплодить to sterilize
обесцвечивание discoloration
обжим squeezing, compression
обжинка mowing (weeds around plants)
обзол wane
обзольный waney
обзор survey, review
обилие abundance
обильный abundant
облава circular drive (of game into an enclosure)
облагораживание improvement
область region, district
облачность cloudiness
обледенение icing
облепиха sea buckthorn, *Hippophae rhamnoides* L.
облесение reforestation, regeneration, afforestation
облесенность area under forest
облесенный forested, wooded, regenerated
облесить to afforest, to reforest, to regenerate
обливина wane
облиствение foliation

облицовка cladding, facing, covering
облом broken end
облуп ring shake
облупить to peel, to strip
облысение becoming bald or bare
обмен exchange
обмен веществ metabolism
обмен опытом exchange of experience
обменный exchangeable
обмыватель бревен log washer
обнажать to bare
обод rim, felloe
ободок, перфорационный perforation rim
оболонь outer sapwood, sapwood
оболочка 1. envelope, coat 2. wall
оболочка, вторичная secondary wall
оболочка клетки cell wall
оболочка, первичная primary wall
оболочка, семенная seed coat
оболочка, третичная tertiary wall
обоновка installing a boom or booms, e.g. in preparing a river for timber floating
оборот 1. rotation, revolution 2. turnover
оборот, годовой annual turnover
оборот рубки rotation, rotation age
оборот хозяйства interval between fellings (in selective logging)
оборудование equipment, plant
обрабатывание (see обработка) 1. processing, conversion 2. cultivation
обработанный 1. processed, worked 2. dressed 3. cultivated
обработать 1. to cultivate 2. to dress 3. to convert, to process
обработка 1. processing, conversion 2. cultivation
обработка, довосходовая pre-emergence treatment
обработка, контурная (щитов) panel sizing
обработка, междурядная cultivation between the rows
обработка, первичная primary conversion
обработка почвы soil cultivation, tillage
обработка, сплошная complete cultivation
образец sample
образец, пусковой prototype
образец, чистый clear specimen
образование 1. formation 2. training
образование насаждения stand formation, stand development
обратимый reversible
обратный reverse, reciprocal
обращение circulation, revolution
обрез the first board cut after the slab
обрезание 1. trimming, cutting off, snedding 2. pruning 3. edging
обрезать to cut, to cut off
обрезка 1. trimming, cutting off, snedding 2. pruning 3. edging
обрезка ветвей 1. pruning 2. trimming
обрезка досок edging
обрезка засохших ветвей dry pruning
обрезка зеленых ветвей green pruning
обрезка сухих ветвей dry pruning
обрезка сучьев 1. pruning 2. trimming
обрезной edged
обрезок off-cut, trimming
обрешетка lathing
обрубать to prune, to trim, to cut off
обрубка сучьев pruning, trimming (see also обрезка)
обрубок block
обруч hoop
обсадка planting (around or along)
обсадка дорог street planting
обсеменитель seed tree
обсеменительный seeding (see рубка, обсеменительная)
обслуживание, массовое queuing
обстрогать to plane
обтачивать to turn
обтесанный hewn
обтеска hewing
обтесывать to hew
обточка turning
обугливание дерева charring wood (for preservation)
обух топора head of axe
обхват girth, circumference
обход beat, round, stint

обшивка cladding, sheathing
обшмыг коры mechanical damage to bark
общество society
объезд **1.** beat (of a mounted ranger) **2.** detour
объездчик mounted ranger
объем volume
объем крупной древесины volume of large timber (usually over 7 cm diam.)
объем, складочный stacked volume
объем ствола stem volume
оводнение irrigation
овраг gully, ravine
овсяница fescue, *Festuca* L.
овсяница, лесная *Festuca sylvatica* (Poll.) Vill.
овсяница, луговая *Festuca pratensis* Huds.
овсяница, овечья sheep's fescue, *Festuca ovina* L.
овсянка bunting, *Emberiza* L.
оглеение gleying
оглинивание clay deposition
оглобельник shaft wood, pole wood of hardwood species
оглобля shaft
огневка **1.** pyralid **2.** tinder fungus
огневка, еловая *Dioryctria abietella* Schiff.
огневка, сосновая *Dioryctria pinitella* Ratz.
огневка, шишковая *Dioryctria abietella* Schiff.
огнеопасный inflammable
огнестойкий fire-resistant
огнетушитель fire extinguisher
огнеупорный fire-resistant
огнище burnt site
оголение почвы baring the soil
огонь fire (see also пожар)
огонь, встречный counter fire, back fire
огонь, управляемый controlled burn, prescribed burn
огораживание fencing in, enclosing
ограничитель depth gauge (chainsaw teeth)
одежда clothes, clothing
одеревенение lignification
одерновение turf formation
одичалый wild, running wild, naturalized
одичание naturalization, running wild
одновозрастный even-aged
однодольные monocotyledons
однодомный monoecious
однолетний **1.** one-year-old **2.** annual
одноэтажный single-storeyed
одноярусный single-storeyed
одревесневший lignified
одубина tan waste
одуванчик, лекарственный common dandelion, *Taraxacum officinale* agg.
ожика wood rush, *Luzula* DC.
ожика, бледная *Luzula pallescens* (Wahl.) Bess.
ожика, волосистая *Luzula pilosa* (L.) Willd.
ожика, обыкновенная *Luzula campestris* (L.) DC.
ожина dewberry, *Rubus caesius* L.
ожог **1.** scorch, scald **2.** blight
ожог коры bark scorch, sunscald
ожог, солнечный bark scorch, sunscald
озеленение planting of greenery (for amenity and decoration)
озеленить to plant greenery (for amenity and decoration)
озимь **1.** winter crop **2.** immature cone(s)
окаймление поры pit border
окаменение petrifaction
окантовка profiling [logs] into cants
окаривание barking
оклад **1.** tax **2.** salary
окно **1.** window **2.** small gap (in canopy)
околоплодник pericarp
околоцветник perianth
окольцовывание ringing, girdling
оконтуровка contouring, face delimitation (in resin tapping)
окоренение rooting
окорение barking
окоренный barked
окоренок butt, butt end
окоривание barking
окорить to bark
окорка barking

окорка, групповая batch barking
окорка набело clean barking
окраина margin
окраска colour, colouring, paint, painting, stain, staining
округ district
округ, лесной forest district
округление rounding, rounding off
окружность circumference, girth
окукление pupation
окулировать to bud, to bud-graft
окулировка budding, bud-grafting
окулировка щитком shield budding
окуляр eyepiece
окупаемость pay-back period
окуривание fuming, fumigation
окучивание earthing up, heaping up
окучивать to earth up, to heap up
оледенение freezing, glaciation
олень deer
олень, благородный red deer, *Cervus elaphus* L.
олень, северный reindeer, *Rangifer tarandus* L.
олень, пятнистый sika deer, *Cervus nippon* Temm.
олесить to afforest, to reforest
ольс alder stand, *Alnetum*
ольха alder, *Alnus* Mill.
ольха, белая speckled alder, *Alnus incana* Moench
ольха, бородатая bearded alder, *Alnus barbata* C.A.M.
ольха, камчатская Kamchatka alder, *Alnus kamtschatica* Kom.
ольха, клейкая common alder, *Alnus glutinosa* Gaertn.
ольха, кустарникова dwarf alder, *Alnus fruticosa* Rupr.
ольха, серая speckled alder, *Alnus incana* Moench
ольха, сердцелистная *Alnus subcordata* C.A.M.
ольха, черная common alder, *Alnus glutinosa* Gaertn.
ольховник alder stand, *Alnetum*
ольшаник alder stand, *Alnetum*
ольшатник alder stand, *Alnetum*
ольшинник alder stand, *Alnetum*
омела mistletoe (usually *Viscum album* L.)
омела, белая mistletoe, *Viscum album* L.
омела, европейская *Loranthus europaeus* L.
омела, можжевельниковая *Arceuthobium oxycedri* M.B.
омела, обыкновенная mistletoe, *Viscum album* L.
омолаживание rejuvenation
омолодить to rejuvenate
омоложение rejuvenation
омшара wet mossy site type (see мшара)
ондатра muskrat, *Ondatra zibethicus* L.
опад litter, litter fall
опадающий deciduous
опадение fall, falling (off)
опадение хвои needle cast
опад семян seed fall
опаленный burned, scorched
опал шейки root-collar scorch
опасность danger
опасность, пожарная fire danger, fire risk
опахать to plough round
опашка ploughing
опенок honey fungus, *Armillaria mellea* Quél.
опережение leading, outstripping
опиленный **1.** sawn **2.** filed
опиливать **1.** to saw **2.** to file
опилка сучьев sawing off branches, pruning
опилки **1.** sawdust **2.** filings
опиловка **1.** sawing, sawing off **2.** filing
описание description
описание, таксационное mensurational description
опись list, inventory
оплодотворение fertilization
оплотник flexible boom round a raft
оподзоленный podzolized
опока sedimentary siliceous rock
оползень landslide
опора support
опорный bearing, supporting
определение determination

определение возраста age determination
определение запаса determination of standing volume
определение прироста increment determination
определение размера пользования yield determination
опробкование suberization
опрокидыватель tipper, dumper
опрыскивание spraying
опрыскиватель sprayer
опрыскиватель, ранцевый knapsack sprayer, backpack sprayer
опрыскиватель, тракторный tractor sprayer
опрятывание reburning (in shifting cultivation)
опудривание powdering, dusting
опустошение devastation, destruction
опустынивание desertification
опухоль swelling, tumour
опушка margin, edge, border
опушка, ветроупорная wind-break
опыление pollination
опыление, перекрестное cross-pollination
опыливание dusting
опыливатель duster
опыт **1.** experiment, test, trial **2.** experience
опыт, вегетационный pot trial
опыт, передовой advanced (progressive) work
орех **1.** walnut, *Juglans* L. **2.** nut
орех, грецкий walnut, *Juglans regia* L.
орех Зибольда Japanese walnut, *Juglans sieboldiana* Maxim.
орех, кедровый pine 'nut' (kernel) (e.g. of *Pinus sibirica*)
орех, маньчжурский Manchurian walnut, *Juglans mandshurica* Maxim.
орех, масляный butternut, *Juglans cinerea* L.
орех, медвежий Turkish hazel, *Corylus colurna* L.
орех, серый butternut, *Juglans cinerea* L.
орех, чернильный gall
орех, черный black walnut, *Juglans nigra* L.
орешник **1.** hazel, *Corylus* L. **2.** hazel grove (see also лещина)
орешник, медвежий Turkish hazel, *Corylus colurna* L.
орешок nutlet
орешок, буковый beech nut
орешок, чернильный gall
орляк bracken, *Pteridium aquilinum* Kuhn.
оросительный irrigation
орошать to irrigate
орошение irrigation
ортштейн ortstein, iron pan
орудие tool, implement
оса wasp
осадка **1.** draught **2.** settling
осадки, атмосферные atmospheric precipitations
осадки, кислотные acid rain
осадок deposit, sediment, precipitate
осветительный (see рубка, осветительная)
осветление first cleaning/weeding (usually up to age 10 years), admitting light
осветлитель weeding unit, weeder
осветлитель, цепной chain-flail weeder
осветлять to clean/weed, to admit light
освещение illumination
освобождать to release, to liberate
освобождение release
освоение utilization, development, opening up, roading
освоить to utilize, to develop, to open up
осеверение растений acclimatization of plants to more northerly regions
оселок whetstone, oilstone
осина aspen, *Populus tremula* L.
осина Давида David's aspen, *Populus tremula* var. *davidiana*
осинник aspen forest
осколок splinter
оскудение impoverishment
осмол resinous wood (usually Scots pine)
осмол, пневой resinous stumpwood
осмол, стволовой resinous stemwood
осмоление **1.** resinosis **2.** tarring

осмолоподсочка tapping to induce the formation of resinous wood
оснастка rig, rigging
основа base, foundation, principle
основание base, foundation, principle
осока sedge, *Carex L.*
осока, бутыльчатая *Carex rostrata* Stokes
осока, волосистая *Carex pilosa* Scop.
осока, двудомная *Carex dioica* L.
осока, дернистая *Carex caespitosa* L.
осока, лесная *Carex sylvatica* Huds.
осока, пальчатая *Carex digitata* L.
осока, пузырчатая *Carex vesicaria* L.
осока, топяная *Carex limosa* L.
осокорник stand of black poplar
осокорь black poplar, *Populus nigra* L.
осолонение salinization
остатки, лесорубочные slash, logging residue, felling debris
остатки, лесосечные slash, logging residue, felling debris
остатки, порубочные slash, logging residue, felling debris
остаток residue
остов skeleton, framework
остолоп stub, snag (high stump)
острие edge, cutting point
остров **1.** forest outlier **2.** island
осушать to drain
осушение **1.** draining, drainage **2.** drying up
осушитель drain
осушка **1.** draining, drainage **2.** drying up
осыпь talus
ось **1.** axis **2.** axle
ось соцветия rhachis
отбелка bleaching
отбор selection, elimination
отборные selects
отброс waste, refuse
отвал mouldboard
отвал, бульдозерный bulldozer blade
отвал, винтовый digger body
отвал, полувинтовый semi-digger body
отведение marking out (see отвод)
отверстие opening, aperture
отверстие поры pit aperture
ответвление branching
отвод лесосеки marking out a coupe, coupe demarcation
отвод места рубки marking out a coupe
отводок layer
отводок, воздушный air layer
отгружать to unload
отдел section (see отделение)
отделение **1.** section, division **2.** emission
отделение, маточное section for production of propagating material (seeds, cuttings, etc.)
отделение, посевное seedling section (of nursery)
отделение, школьное transplant section (of nursery)
отделка finishing
отдых recreation
отенение shade, shading
отеска rough hewing
отесывать to hew roughly, to hew
отжиг counter-fire, backfire
отжим thrust
откать, канатный cable slide
отклонение deviation
откомлевка **1.** piece cut off from butt end of log **2.** cutting off the butt end
откос slope
открытый open
отлог slope
отлуп ring shake
отмель sandbank, shallow
отмерять to measure off
отметить to mark
отметка mark, marking
отмечать to mark
отмирание **1.** dying off **2.** die-back **3.** decline
отмирать **1.** to die off **2.** to die back
отношение ratio, relationship
отпад **1.** litter, debris **2.** mortality
отпиливание sawing off
отправка dispatch
отпрыск **1.** sucker, root sucker **2.** offshoot
отпрыск корневой sucker, root sucker

отпугивать to frighten off, to repel
отпуск леса sale of timber, outturn of timber, removal or delivery of timber
отравление poisoning
отрасль branch
отрезок length, section, segment
отрог spur
отросток 1. shoot, offshoot 2. extension
отруб 1. end, transverse section 2. diameter (used in loose sense)
отруб, верхний 1. top end 2. top diameter
отруб, нижний 1. butt diameter 2. butt end
отрубать to cut off
отрубок log, section
отряд order
отряхивание (шишек) shaking off (cones)
отсекатель log sorter, log kicker
отсекать to cut off, to cut back, to stump
отслойка ring shake
отставать to fall behind, to lag behind
отсталь ring shake
отстрел shooting off, culling
отструг off-cut
оттаивание thawing
отходы waste
отходы, лесосечные slash, logging debris
отцепщик choker-release man
отчет report, record
охвоение degree of foliage (needle) development
охлестыватель whip
охлестывать to whip, to lash
охота hunting
охотник hunter
охрана 1. protection 2. guard
охрана леса forest protection
охрана, пожарная fire protection
охрана природа nature conservation
охрана труда labour protection, work safety, accident prevention
охранять to protect, to guard
оценивать to estimate, to evaluate, to appraise
оценка estimation, evaluation, valuation, appraisal
оценка, глазомерная ocular estimation
оценка, стоимостная valuation
очаг focus (of infection or infestation)
очередь рубки sequence of felling
очес tirr, undecomposed layer of peat
очистка clearing, cleaning
очистка лесосеки clearing the felling site
очистка, огневая slash burning
очистка разбрасыванием clearing by scattering slash
очистка ствола 1. pruning, natural pruning 2. trimming
очищаемость ствола natural pruning, self-pruning
очищать to clear, to clean
очищение cleaning, clearing (see очистка)
очко eye, bud
очкование budding, bud-grafting
ошибка error
ошибка, среднеквадратическая standard deviation
ошкуривание barking
ошкурить to bark
ошлаговка fastening a raft together with tackle
ошмыг коры mechanical damage to bark, fraying
ошпаривание boiling (to assist germination)

П

паводок flood, high water
падалица fallen fruit, windfall
падина hollow
падуб holly, *Ilex* L. (usually *Ilex aquifolium* L.)
падуб, морщинистый wrinkled holly, *Ilex rugosa* Fr. Schmidt
падуб, остролистный holly, *Ilex aquifolium* L.
паз groove, slot
пазуха 1. axil 2. gullet (of saw tooth)

пазушный axillary
пакет **1.** package, set **2.** bundle
пакетирование bunching, packaging
пакет лесоматериалов log bundle, log bunch
пакетование **1.** packaging (lumber) **2.** bunching, bundling (logs)
пакет пиломатериалов lumber package, packaged lumber
пакет шпона set of veneers or plies
паклен field maple, *Acer campestre* L.
пал burn, land cleared by shifting cultivation
палетка 'panel' (e.g. of bark removed for counting insect populations)
палить to burn
палка stick
палубник softwood sawlog assortment for decking
пальма palm
пальма, кавказская box, *Buxus sempervirens* L.
пальма, финиковая date, *Phoenix* L.
пальчатый palmate, digitate
памятник природы natural monument
папоротник fern
папоротник, женский lady fern, *Athyrium filix-femina* Roth. (see кочедыжник, женский)
папоротник, мужской male fern, *Dryopteris filix mas* L. (see щитовник, мужской)
папоротник-орляк bracken, *Pteridium aquilinum* Kuhn. (see орляк)
папоротниковые ferns, Filicales
папоротникообразные Pteridophyta
пар **1.** steam **2.** fallow
пар, черный bare fallow
пара pair
пара пор pit-pair
паразит parasite
парашютист, пожарный smoke-jumper
паренхима parenchyma
паренхима, вазицентрическая vasicentric parenchyma
паренхима, диффузная diffuse parenchyma
паренхима, крыловидная aliform parenchyma
паренхима, лубяная phloem parenchyma
паренхима, лучевая ray parenchyma
паренхима, разъединенная disjunctive parenchyma
паренхима, сомкнутокрыловидная confluent parenchyma
паренхима, тяжевая parenchyma strand
парк **1.** park **2.** fleet (e.g. of vehicles)
парк, национальный national park
паркет parquet
паркет, мозаичный mosaic parquet
паркет, наборный mosaic parquet
паркет, штучный parquet block
паркет, щитовый panel parquet
паркетина small board for making parquet
парник glasshouse, frame, hotbed
паровоз locomotive, steam engine
пароход steamer
партеногенез parthenogenesis
партия lot, parcel
парцелла 'parcel' (horizontal synusia)
парша scab
пасека **1.** section of a felling area **2.** apiary
пасечник bee-keeper
паслень, сладко-горький bittersweet, woody nightshade, *Solanum dulcamara* L.
пасок juice, sap
паспорт certificate, rating plate (of a machine)
паспорт пакета tally
паста, хлорофилло-каротиновая chlorophyll-carotene paste
пастбище pasture
пастьба grazing, pasturage
пастьба, лесная forest grazing
пасынкование disbudding
пасынок large splay knot
пасюк brown rat, *Rattus norvegicus* Berk.
патока, древесная wood molasses
патрон **1.** chuck **2.** cartridge
пауза freshening interval
паук spider

пахать to plough
пахота 1. plough-land, arable land 2. tillage
пахучка wild basil, *Clinopodium vulgare* L.
пачка bunch (of logs), packet, parcel
пашня plough-land, arable land
пдк threshold allowable concentration
пек pitch
пек, древесный wood pitch
пек, лесохимический wood pitch
пекан pecan, *Carya pecan* E. et G.
пенек 1. stub, stump 2. stump plant
пенек от сучка branch stub
пенек, посадной stump plant
пеночка warbler, *Phylloscopus* L.
пень 1. stub, stump 2. stool
пепел ash
первобытный primeval, original
первоцвет primula, *Primula* L.
перевалка transfer (from one method of transport to another)
перевешивание пилы hang, overhang (of vertical frame saw)
перевод 1. transfer, shift 2. translation
перевозка transport, conveying
перегной humus (see also гумус)
перегонка distillation
перегонка дерева wood distillation
перегонка смолы tar distillation
перегонка, сухая dry distillation
перегородка partition, wall
перегородчатый septate
перегретый superheated
перегружать to overload
перегрузка overloading
передача 1. transmission, transfer 2. gear, drive
передвижение translocation
передвижной movable
передержать to hold over
передовик leader, front-rank worker
перезимовка overwintering
перекат sandbank
перекати-поле *Gypsophila paniculata* L.
перекатывать to roll over, to turn over
перекладина cross-bar, cross-piece
переклейка plywood
перекоп cross-ditch
перекошенный crooked, twisted, distorted
перекрест cross-field
пролеска, многолетняя dog's mercury, *Mercurialis perennis* L.
перелесок small wood, copse
перелесье glade, clearing
перелог fallow, ley
перемычка 1. crosspiece, connector 2. dike, dam 3. strip of living bark left intact between freshenings or faces
перемычка, межкарровая vertical interface (in resin-tapping)
перенаселенность over-population
перепел, обыкновенный quail, *Coturnix coturnix* L.
перепелятник sparrow-hawk, *Accipiter nisus* L.
переплет transom
перепончатокрылые Hymenoptera
переработка conversion, processing, treatment
перерастание growing over, colonizing
перерезание 1. cross-cutting 2. shear perpendicular to grain
перерезывание 1. cross-cutting 2. shear perpendicular to grain
перерождение 1. regeneration 2. degeneration
переросший overgrown
переруб 1. over-cutting 2. over-cut area
перерыв break, interruption
перерыв покоя breaking dormancy
перерыв сомкнутости opening the canopy
пересадка transplanting, lining out
пересаживание transplanting, lining out
пересаживать to transplant, to line out
пересечение, линейное line intersect
перестой overmature tree, overmature stand
перестойный overmature, declining
пересыхание drying up
перехватка осадков interception of precipitation
перечень list
перечет enumeration

перечет, выборочный partial enumeration
перечет, ленточный strip enumeration
перечет, подеревный complete enumeration
перечет, сплошной complete enumeration
перечет, частичный partial enumeration
перешеек isthmus
перешколивание transplanting, lining out
перешколивать to transplant, to line out
перештыковка deep digging
перидерма periderm
перила banisters, railing
периметр perimeter
период period
период, вегетационный growing season
период возобновления regeneration period, regeneration interval
период, межвегетационный period between the growing seasons, winter
период, навигационный floating season, navigation season
период оборота рубки rotation period
период, ревизионный revision period
период рубки леса felling season
перистый pinnate
перловка, одноцветная wood melick, *Melica uniflora* Retz.
перловка, пониклая *Melica nutans* L.
персик peach, *Prunus persica* (L.) Batsch
перфорация perforation
перфорация, множественная multiple perforation
перфорация, простая simple perforation
песец arctic fox, *Alopex lagopus* L.
песок sand
пестик **1.** pestle **2.** pistil
пестицид pesticide
пестицид, контактный contact pesticide
пестование tending (usually by felling)
пестование запаса Vorratspflege, tending of the growing stock
пестрый variegated
пестряк ant beetle, *Thanasimus formicarius* L.
песчаник sandstone
песчаный sandy
песчанка **1.** gerbil **2.** *Arenaria* L.
петля noose, loop, sling
петров крест, чешуйчатый toothwort, *Lathraea squamaria* L.
печь oven, furnace
пижма, обыкновенная tansy, *Tanacetum vulgare* L.
пикет picket
пикирование pricking out, transplanting
пикировка pricking out, transplanting
пикульник hemp-nettle, *Galeopsis* L.
пила **1.** file **2.** saw
пила, ажурная fret saw
пила, балансирная pendulum saw
пила, висячая pendulum saw
пила, головная head saw
пила, двухстороннеконическая swage saw, taper saw
пила, делительная re-saw
пила, дисковая circular saw
пила для обрезки сучьев pruning saw
пила, кольцевая cylinder saw
пила, коническая swage saw, taper saw
пила, круглая circular saw
пила, ленточная band saw
пила летучая 'flying' saw, travelling saw
пила лисий хвост reciprocating log cross-cut saw
пила, лучковая bow saw
пила, маятниковая pendulum saw (vertical)
пила, моторная power saw
пила, ножевая hack saw
пила, обрезная edger
пила, одностороннеконическая swage saw, taper saw
пила, поперечная cross-cut saw
пила, правоконическая swage saw, taper saw
пила, продольная rip saw
пила, прорезная fret saw
пила, пьяная drunken saw
пила, разрезная rip saw
пила, рамная frame saw (see also рама)
пила, ручная hand saw
пила, столярная joinery saw

пила, строгальная planer saw, hollow-ground saw
пила, торцовочная docking saw
пила, фанерная veneer saw
пила, форматная panel-sizing saw, saw for cutting to size
пила, цепная chain saw
пила, цилиндрическая cylinder saw
пила, шаровая concave saw
пила, штанговая reciprocating cross-cut saw
пиление 1. sawing 2. filing
пиление, поперечное cross-cutting
пиление, попутное rip-sawing
пиление, продольное rip-sawing
пиленый 1. filed 2. sawn
пилильщик tenthredinid, sawfly
пилильщик, лиственничный большой large larch sawfly, *Lygaeonematus erichsoni* Htg.
пилильщик, лиственничный обыкновенный *Lygaeonematus laricis* Htg.
пилильщик, пихтовый *Lygaeonematus abietinus* Christ.
пилильщик, сосновый обыкновенный pine sawfly, *Diprion pini* L.
пилильщик, сосновый рыжий fox-coloured sawfly, *Diprion sertifer* Geoffr.
пилильщик, сосновый черно-желтый *Diprion similis* Htg.
пилильщик-ткач pamphiliid
пилильщик-ткач, еловый *Cephaleia abietis* L.
пилильщик-ткач, сосновый *Acantholyda erythrocephala* L.
пилильщик-ткач, сосновый общественный *Acantholyda erythrocephala* L.
пилить 1. to saw 2. to file
пилка 1. sawing 2. small saw
пиловочник 1. sawlog 2. sawtimber
Пилолесоэкспорт Lumber Export Agency
пиломатериалы sawn timber, lumber
пилоправ saw doctor
пилопродукция sawn timber
пилостав saw doctor
пильчатый toothed, serrated, dentate, serrate
пильщик sawyer
пиния umbrella pine, *Pinus pinea* L.
пионер pioneer
пиролиз pyrolysis
питание 1. nutrition, feeding 2. supply
питательный nutritious, feeding, nutrient
питать 1. to nourish, to feed 2. to supply
питиограф, пихтовый *Pityophthorus pityographus* Ratz.
питомник nursery
питомник, временный temporary nursery
питомник, лесной forest nursery
питомник, постоянный permanent nursery
пихта fir, *Abies* Mill.
пихта, белая silver fir, *Abies alba* Mill.
пихта, белокорая East Siberian fir, *Abies nephrolepsis* Maxim.
пихта, грациозная Kamchatka fir, *Abies gracilis* Kom.
пихта, гребеньчатая silver fir, *Abies alba* Mill.
пихта, дугласовая Douglas fir, *Pseudotsuga menziesii* Mirb.
пихта, европейская silver fir, *Abies alba* Mill.
пихта, кавказская Caucasian fir, *Abies nordmanniana* Spach.
пихта, камчатская Kamchatka fir, *Abies gracilis* Kom.
пихта Майра *Abies mayriana* M. et K.
пихта Нордманна Caucasian fir, *Abies nordmanniana* Spach.
пихта, сахалинская Sakhalin fir, *Abies sachalinensis* Mast.
пихта, сибирская Siberian fir, *Abies sibirica* Ledeb.
пихта, тонкая Kamchatka fir, *Abies gracilis* Kom.
пихта, цельнолистная Manchurian fir, *Abies holophylla* Maxim.
пихтарник fir forest, *Abietum*
пихтач fir forest, *Abietum*

пихтоварение extraction of oil from fir foliage
пихтовый fir
пихтоельник fir/spruce forest
пищуха 1. pika, *Ochotona* 2. tree-creeper, *Certhia*
пищуха, обыкновенная tree-creeper, *Certhia familiaris* L.
плавня floodland
плавучесть buoyancy
плаккат low-lying site
плакор upland plateau between valleys
план plan, map
план лесонасаждений stock map
план, перспективный intermediate plan (5- or 7-year)
план рубок felling plan
план хозяйства working plan
планировка grading, levelling
планка 1. plank, board 2. lath
планомерно according to plan
плансон board, plank
плантаж deep ploughing
плантация artificial forest, plantation
плантация, архиво-маточная clone bank
плантация, древесосырьевая fast-growing plantation, biomass plantation
плантация, клоновая clonal seed orchard
плантация, маточная stool bed, plantation for producing cuttings, seed, etc.
плантация, семенная seed orchard
плантация, черенковая section of nursery where cuttings are rooted
планшет 1. plane-table 2. compartment map 3. flight map
пласт 1. layer, bed, stratum 2. furrow slice
пластик plastic, hot-pressed sheet, moulding
пластик, древеснослоистый laminated densified sheet, compreg laminate
пластик, древесный hot-pressed wood-based sheet (made of e.g. sawdust, chips or fibres)
пластик, лигноуглеводный 'lignocarbohydrate' sheet (made of lignocellulose material)
пластина 1. split, half-round bar 2. plate, bark plate
пластинка small plate, lamina
пластинка, клеточная cell plate
пластинка листа leaf blade
пластинка, межклетная intercellular layer
пластинка, перфорафионная perforation plate
пластинка, ситовидная sieve plate
пластинка, срединная middle lamella
пластинка твердого сплава hard-alloy tip
пластинчатоус scarabaeid
пластичность plasticity
пластмасса plastic
плата wages, pay
плата, корневая stumpage price
плата, поденная day wage
плата, попенная fixed stumpage price
плата, почасовая hourly pay
плата, сдельная piece-work pay
платан plane, *Platanus* L.
платан, восточный oriental plane, *Platanus orientalis* L.
платан, западный American plane, *Platanus occidentalis* L.
платан, кленолистный London plane, *Platanus acerifolia* Willd.
плато plateau
платформа 1. platform, ramp 2. open railway wagon, flat car
плаун clubmoss, *Lycopodium* L.
плаун -баранец fir clubmoss, *Lycopodium selago* L.
плаун, булавовидный common clubmoss, *Lycopodium clavatum* L.
плаун, годичный *Lycopodium annotinum* L.
плаун, плоский *Lycopodium complanatum* L.
плаха 1. block, log 2. short split, short half-round bar
плач bleeding (e.g. of sap)
плашка log, billet

плашкаут 1. barge, lighter 2. pontoon
плашкот 1. barge, lighter 2. pontoon
плашник cleft wood
плашняк cleft wood
плащак beam of rectangular cross-section (not square)
плевел rye-grass, *Lolium* L.
пленка film, layer
пленка поры pit membrane
плесень mould
плетеный wattled, wicker
плетень wattle fence
плинтус plinth, skirting board
плита board, slab, plate
плита, древесноволокнистая wood fibreboard
плита, древесностружечная wood particle board
плита, изоляционная insulation board
плита плоского прессования flat-pressed board
плита, полутвердая semi-hard board
плита пресса press platen
плита, сверхтвердая superhard board
плита, столярная blockboard
плита, твердая hardboard
плита, экструзионная extruded board
Плитимпэкс Panel Import/Export Agency
плитка, кровельная roof shingle
пловучесть buoyancy
плод fruit
плод, косточковый stone fruit
плод, ложный false fruit, pseudocarp
плод, растрескивающийся dehiscent fruit
плод, сборный aggregate fruit, syncarp
плод, сложный compound fruit
плодовитый fruitful, fertile
плодожил, дубовый *Balaninus glandium* Marsh.
плодожил, ореховый *Balaninus nucum* L.
плодожорка tortricid (see also листоверка)
плодожорка, буковая *Carpocapsa grossana* Hw.
плодожорка, желудевая *Carpocapsa splendana* Hb.
плодокорм mast
плодокорм, буковый beech mast
плодокорм, дубовый oak mast
плодолистик carpel
плодоножка fruit stalk, peduncle, pedicel
плодонос 1. carpophore 2. rhachis
плодоносный fruit-bearing
плодоношение fruiting
плодородие fertility
плодородный fruitful, fertile
плодосмен crop rotation (agricultural)
плоскогорье plateau
плоскость plane, surface
плоскость скольжения slip plane
плоскоход platypodid
плоскоход, цилиндрический *Platypus cylindrus* F.
плот raft
плот, зимний raft made in the winter
плот, морской sea-going raft
плот, озерный lake raft
плот, пучковый bundle raft
плот, речной river raft
плот, секционный sectional raft
плот-сигара cigar-shaped raft
плот, шлюзуемый raft capable of passing through a lock
плотбище raft-construction point
плотина dam, weir
плотина, лесосплавная floating dam
плотина, ряжевая cribwork dam
плотник carpenter
плотность density, compactness
плотность древесины wood density
плотность загрузки bulk density (e.g. of chips)
плотность насаждения stand density, crop density, stocking
плотность населения population density
плотность полога canopy density
плотность почвы soil density
плотность упаковки 'packing density'
плотность, условная nominal density
плотовщик floating operator
плотоспуск dam sluiceway for rafts

площадка patch, small plot, landing (see also площадь)
площадка, погрузочная loading platform, loading ramp
площадка, разделочная conversion point
площадь 1. area 2. plot
площадь вырубки felling area
площадь, вырубленная felled area
площадь, лесная forest area (see Гослесфонд)
площадь, лесопокрытая forested area
площадь лесосеки felling area
площадь, летучая temporary plot
площадь, нелесная non-forest area
площадь, неосвоенная area not opened up (not utilized)
площадь, непокрытая лесом non-forested area
площадь, непроизводительная unproductive area
площадь, опытная experimental area, experimental plot
площадь, периодная periodic block
площадь питания feeding area (i.e. growing space)
площадь пользования felling area
площадь поперечного сечения cross-sectional area
площадь, пробная sample plot
площадь, пробная (временная) temporary sample plot
площадь, пробная (постоянная) permanent sample plot
площадь сечения basal area, sectional area
площадь, эксплуатационная area to be felled
ПЛСБ permanent seed collection base
ПЛСУ permanent seed collection area
плуг plough
плуг, безотвальный plough without a mouldboard
плуг, винтовый digger plough body
плуг, двухотвальный double-mouldboard plough, ridging plough
плуг, лесной forest plough
плуг, плантажный trenching plough, deep plough
плуг, полувинтовый semi-digger plough body
плюс-дерево plus-tree
плюска cupule, acorn cup
плющ ivy, *Hedera helix* L.
плющение swaging
плющенный swaged
пневмоокорка pneumatic barking
пневой stump
пни stumps
побег shoot (usually one year's growth)
побег, боковой side shoot, lateral shoot
побег, верхушечный leader, terminal shoot
побег, вершинный leader, terminal shoot
побег, водяной epicormic shoot
побег, годовой annual shoot
побег, замещающий replacement shoot
побег Иванов Lammas shoot
побег, конечный leader, terminal shoot
побег, корневой sucker, root sucker
побег, летний Lammas shoot
побег от пня coppice shoot
побег, пролептический proleptic shoot
побег, силлептический sylleptic shoot
побег, стволовой epicormic shoot
побег, укороченный short shoot, spur
побеговьюн *Evetria* Hb., (*Rhyacionia*) Hb.
побеговьюн, бурый *Evetria pinivorana* Z.
побеговьюн, зимующий pine-shoot moth, *Evetria buoliana* Schiff.
побеговьюн, летний *Evetria duplana* Hb.
побеговьюн, почковый *Evetria turionana* Hb.
побеговьюн-смолевщик pine resin-gall moth, *Evetria resinella* L.
побегопроизводительность shoot-forming capacity
побочный secondary, minor, subsidiary (see пользование, побочное)
повал felling, windthrow
поваленный felled, thrown down

поверхность surface, area
поверхность, боковая lateral (cambial) area
поверхность кроны crown cover, crown projection area
поверхность, торцовая end surface, end grain
повилика dodder, *Cuscuta* L.
повозка vehicle, wagon
поворот turn, bend, curve
поворотный turning, rotating, rotatable
повреждение damage, injury
повсходовый post-emergence
повторение repeating, repetition
повторение, сердцевинное pith fleck
повторение опыта replication
повторяемость рубок ухода thinning cycle
поганка **1.** toadstool, *Amanita* Pers. **2.** grebe, *Podiceps* L.
поглощение absorption, uptake
погода weather
поголовье number of head of livestock
погонаж, строганный planed stock
погонка planed stock
погрешность error
погружать **1.** to load **2.** to immerse
погрузка loading
погрыз коры bark stripping (by elk)
подать tax
подача feed, feeding, advance
подача древесины feeding wood
подача леса 'pre-skidding' wood, forwarding wood
подача, непрерывная continuous feed
подача, толчковая intermittent feed
подбел soil with a bleached horizon
подбел, многолистный *Andromeda polifolia* L.
подберезовик *Boletus scaber* Fr.
подбор selection
подборщик **1.** grader **2.** slash-collector
подборщик-сортиментовоз forwarder
подборщик сучьев slash-collector
подбур podzolized brown soil
подвид sub-species
подвижный **1.** mobile **2.** available (of nutrients)
подвоз transport
подвозка леса initial transport of wood, skidding
подвой rootstock, stock (for grafting)
подвяливание 'biological drying' of trees to reduce wood moisture (e.g. before floating)
подвяливание на корню girdling trees before felling
подвяливание после повала 'sour felling' (leaving the crown intact on felled trees)
подгон nurse, accessory species
подготовка preparation
поддон **1.** pallet **2.** caul
поддубица black soil under oak
поденщик day-labourer
поденщина day-labour
поджог arson
подземный underground, subterranean
подзол podzol
подзолистый podzolic
подзолообразование podzolization
подзона subzone
подкладка lining, backing, underlay
подкопка undercutting
подкормка feeding, top-dressing
подлежка sticker bluestain
подлесник, европейский sanicle, *Sanicula europaea* L.
подлесок underwood, undergrowth, layer of shrubs and small trees
подлесок, выполняющий auxiliary underwood, 'filler' underwood
подмаренник bedstraw, *Galium* L.
подмаренник, болотный *Galium palustre* L.
подмаренник, весенный *Galium vernum* Scop.
подмаренник, лесной *Galium schultesii* Vest.
подмаренник, мягкий *Galium mollugo* L.
подмаренник, настоящий lady's bedstraw, *Galium verum* L.
подмаренник, северный *Galium boreale* L.

подмаренник, топяной *Galium uliginosum* L.
подмаренник, цепкий goosegrass, *Galium aparine* L.
подмостки scaffold, scaffolding, trestle
подмыв undermining (of stream banks, etc.)
подновка freshening (general), streak (herringbone face)
подновка, первичная first streak
подновка, повторная repeat streak
подновка, предохранительная containing streak (to prevent spread of chemical stimulants)
подновка, сдвоенная double-cut streak
подоконник windowsill
подорожник plantain, *Plantago* L.
подорожник, большой *Plantago major* L.
подорожник, ланцетный ribwort, *Plantago lanceolata* L.
подорожник, средний *Plantago media* L.
подосинник *Boletus versipellis* Fr.
подосиновик *Boletus versipellis* Fr.
подошва sole, bottom
подошва, плужная plough sole
подошва пропила kerf bottom
подпар **1.** mottled butt rot, *Fomes annosus* (Fr.) Cke **2.** initial fungal development in freshly felled beech **3.** initial mottling
подпарина sticker bluestain
подпил undercut
подпилок file
подпильник file
подплав **1.** timber floated (e.g. to a mill) **2.** buoyancy aid
подполог beneath the canopy
подпор prop, support
подпорка prop, support
подпочва subsoil
подпочвенник subsoiler
подпрессовка pre-pressing
подпруда storage basin, storage pond
подразделение subdivision
подраменье spruce forest on wet site
подрастать to grow up
подрезать to cut (underneath), to trim, to prune
подрезка cutting, trimming, pruning
подрод subgenus
подрост advance growth, understorey, underwood
подруд undercut
подрумянивание removing coarse bark (to make a tapping face)
подрумянивать to remove coarse bark
подрядчик contractor
подсадка **1.** replanting, beating up **2.** underplanting
подсаживать **1.** to replant, to beat up **2.** to underplant
подсачивать to tap (resin)
подсев resowing, additional sowing
подсед natural regeneration under shelter (e.g. spruce under pine)
подсека cleared land
подсекать to cut down
подсерчивание bark removal (to kill trees in shifting cultivation)
подсечивание bark removal (to kill trees in shifting cultivation)
подсечка **1.** cleared land **2.** girdling (to kill trees in shifting cultivation)
подснежник, белый snowdrop, *Galanthus nivalis* L.
подсольнечник sunflower, *Helianthus annuus* L.
подсоченный tapped
подсочивание bark removal (to kill trees in shifting cultivation)
подсочка tapping, resin-tapping
подстилка litter plus surface humus
подстилка, лесная forest litter plus surface humus, forest floor
подстой dead standing tree
подстригать to cut, to clip, to trim
подсушина **1.** dead standing tree **2.** fire scar
подсушина, пожарная fire scar
подсушка pre-drying (e.g. of trees before felling)
подсчет calculation
подтаскивание dragging, skidding
подтоварник small roundwood

подтопление raised water table
подтрелевка skidding, skidding to (a particular point)
подтрелевщик skidder
подушечка, листовая pulvinus, leaf cushion
подушка pillow, cushion, bolster
подчерчивание bark removal (to kill trees in shifting cultivation)
подчиненный subordinate
подшипник bearing
подъельник, волосистый yellow bird's-nest, *Monotropa hypopithys* agg.
подъем lift, ascent, lifting, climbing
подъем пилы upstroke of saw
подъем, руководящий ruling gradient (road)
подъемник hoist, elevator, lift
поедание eating, eating up
поезд train
поезд, автомобильный lorry-plus-trailer, road train
пожар fire
пожар, беглый running fire, fast-moving but not intense fire
пожар, верховой crown fire
пожар, вершинный crown fire
пожар, встречный counter fire, back fire
пожар, лесной forest fire
пожар, наземный surface fire
пожар, низовой surface fire
пожар, повальный crown fire
пожар, подземный ground fire
пожар, почвенный ground fire
пожар, пятнистый spot fire
пожар, торфяной peat fire
пожар, устойчивый slow-moving intense fire
пожарище burn, burned area
пожелтение yellowing
пожог main burn (in shifting cultivation)
пойма floodplain
пойменный floodplain
показатель index
показатель, комплексный гидротермический Nesterov's fire-danger index
показатель местообитания site index
показатель пожарной опасности fire-danger index
покатый sloping
покой dormancy, rest
покой, лесосечный interval after felling (before establishing a new stand)
покой семян seed dormancy
поколение generation
покоробленность warping
покоробленность, винтовая twist
покоробленность, поперечная cupping
покоробленность, продольная bow
покоробленный warped
покров cover
покров, живой ground cover
покров, мертвый dead ground cover, litter
покров, напочвенный ground cover
покров, почвенный soil mantle
покров, растительный vegetation cover
покров, снежный snow cover, snow pack
покров, травяной herbaceous cover, grass cover
покровосдиратель scarifier
покровосдиратель, якорный anchor-type scarifier
покрывание covering
покрывать to cover
покрытосеменные angiosperms
покрытосеменный angiospermous
покрытосемянные angiosperms
покрытосемянный angiospermous
покрытый covered
покрытый лесом forested
покрышка cover, covering, lid
пол **1.** sex **2.** floor
пол- half
полдюймовка board ½ in (= *c.* 1.3 cm) thick
поле field
поле высокой частоты high-frequency field, radio-frequency field
поле зрения field of vision
поле, опытное trial field, experimental field
поле перекреста cross-field
поле под паром fallow field

поле, ситовидное sieve area, sieve field
полевица bent grass, *Agrostis* L.
полевица, белая *Agrostis alba* L.
полевица, обыкновенная *Agrostis vulgaris* With.
полевица, собачья *Agrostis canina* L.
полевичка, душистая *Eragrostis suaveolens* Beck.
полевка vole
полевка, красная northern red-backed vole, *Clethrionomys rutilus* Pall.
полевка, общественная social vole, *Microtus socialis* Pall.
полевка, обыкновенная common vole, *Microtus arvalis* Pall.
полевка, рыжая лесная bank vole, *Clethrionomys glareolus* Schr.
полевки Microtinae
полевой field
полегание 1. damping off 2. lodging
полезный useful
поленика *Rubus arcticus* L.
поленница wood pile, stack, cord
полено billet, bolt
полесье wooded district
полет flight, air dispersal
ползун slide block, slider, shoe
ползучесть creep
полив watering, irrigation
поливать to water, to irrigate
поливка watering, irrigation
полигон 1. polygon 2. testing area, graveyard
полиграф *Polygraphus polygraphus* L.
полировать to polish, to buff
полировка polishing, buffing
полиспаст pulley block
политика, лесная forest policy
политрихум *Polytrichum* Dill.
политрихум, обыкновенный hair moss, *Polytrichum commune* Hedw.
политура polish
полиуход multiple tending operation (e.g. thinning and fertilizing)
полка 1. weeding 2. shelf, rack
полнодревесность ratio of solid volume to stacked volume
полнодревесный approaching cylindrical, not tapering greatly
полнозернистость семян fullness of seeds
полнота леса 1. stand density, crop density, stocking (on a basal area basis) 2. canopy density (see сомкнутость)
полнотомер Биттерлиха Bitterlich's angle gauge, relascope
половинник split, half-round bar
половица floor board, board, plank
половодье high water, flood
полог canopy
полог насаждения stand canopy
полоз, санный sledge runner
полом broken tree
поломка breakage
поломка кроны crown breakage
полонина subalpine meadow
полоса strip, belt, shelterbelt, stripe, zone
полоса, ажурная open shelterbelt
полоса, железнодорожная railway shelterbelt
полоса, заградительная fire break, fire line
полоса, защитная shelterbelt, protective strip
полоса, лесная forest strip
полоса, лесозащитная protective forest strip
полоса, непродуваемая impermeable shelterbelt
полоса, охранная protective strip
полоса, полезащитная field shelterbelt
полоса, продуваемая permeable shelterbelt
полоса, противопожарная fire break, fire line
полоса, семянная strip of seed trees
полоса, снегозадерживающая snow-retaining strip
полоса, снегозащитная protective strip against snow
полоса, снегопоглощающая snow-absorbing strip
полосатость, красная red streak, red stain

полосопрокладыватель fire-break maker
полость hollow, cavity
полость клетки lumen
полость поры pit cavity
полотнище пилы saw blade
полотно 1. saw blade 2. road-bed 3. linen
полоть to weed
полу- half, hemi, semi
полужесткокрылые Hemiptera
полукокс semi-coke, charcoal produced by low-temperature pyrolysis
полукустарник sub-shrub, suffrutex
полумасса half-stuff (pulp)
полуница *Fragaria viridis* Duch.
полуокоренный half-barked
полуприцеп semi-trailer
полупустыня semi-desert
полусиб half-sib
полуфабрикат semi-manufactured product
полуцеллюлоза semi-chemical pulp
полый full, fully stocked
полынь *Artemisia* L.
полынь, веничная *Artemisia scoparia* L.
полынь, горькая wormwood, *Artemisia absinthium* L.
полынь, обыкновенная mugwort, *Artemisia vulgaris* L.
полынь, песчаная *Artemisia arenaria* DC.
полынь, полевая *Artemisia campestris* L.
полынь, понтийская *Artemisia pontica* L.
пользование 1. use, utilization 2. felling (see also рубка) 3. yield
пользование, главное principal felling, final yield
пользование лесом utilization, cut
пользование, побочное harvesting of minor forest products (products other than wood)
пользование, постоянное sustained use
пользование, предварительное intermediate felling, advance felling
пользование, промежуточное intermediate felling
поляна glade, clearing
помесь cross, hybrid
помет 1. dung, manure 2. offspring, brood
помост raised platform
понтон pontoon
понтонник softwood sawlog assortment for pontoon constructions
поперек across, transverse
поперек волокон across the grain, perpendicular to the grain
поперечина cross-bar, cross-piece
поперечный cross, transverse
поплавок float
поползень nuthatch, *Sitta europaea* L.
пополнение filling gaps, beating up
пополнять to fill gaps, to beat up
поправка correction, repair
популяция population
попуск, сплавной river drive, dam drive
пора 1. pit 2. pore 3. time
пора, зубчатая vestured pit
пора, окаймленная bordered pit
пора, простая simple pit
пора, разветвленная ramiform pit
пора, слепая blind pit
пора, украшенная vestured pit
поранение почвы soil wounding, scarification
пористость porosity
пористый porous, pitted
поровость pitting
поровость, лестничная scalariform pitting
поровость, односторонне-сложная unilaterally compound pitting
поровость, очередная alternate pitting
поровость, ситовидная sieve pitting
порог threshold
пороги rapids
порода 1. species 2. rock
порода, второстепенная secondary species
порода, главная principal species
порода, горная rock

порода, господствующая dominant species
порода, древесная tree species, woody species
порода, заболонная sapwood species, sapwood tree
порода, лиственная broadleaved species, hardwood
порода, материнская parent material, parent rock
порода, мягколиственная soft-wooded broadleaved species
порода, нежелательная undesirable species
порода, осадочная sedimentary rock
порода, основная principal species
порода, подчиненная subordinate species
порода, светолюбивая light-demanding species
порода, сопутствующая associate species
порода, твердая hard-wooded species
порода, твердолиственная hard-wooded broadleaved species
порода, теневыносливая shade-bearing species
порода, хвойная coniferous species, conifer, softwood
порода, ценная valuable species
порозаполнитель pore-filler
порозный porous
порок defect
порослевина coppice shoot
поросль shoots (usually coppice), coppice growth
поросль от корня suckers, root suckers
поросль от пня coppice shoots, stool shoots
поросль, пневая coppice shoots, stool shoots
порох powder, gunpowder
порошок powder
порт port, harbour
порубка illegal felling
порубщик timber thief
порыв ветра gust of wind
порядок order, sequence
посадить to plant
посадить на пень to cut back to stump
посадка 1. plantation 2. setting, fitting, fit 3. planting
посадка, бороздовая furrow planting
посадка в бугорки mound planting
посадка в пласт planting on the furrow slice
посадка в щель slit planting, notch planting
посадка в ямки hole planting, pit planting
посадка, географическая provenance trial
посадка глыбами ball planting
посадка, гнездовая nest planting
посадка, квадратная square planting
посадка, косая slant planting
посадка местами patch planting
посадка на пень 1. stump planting, planting out cut-back plants 2. coppicing
посадка площадками patch planting
посадка пучками bunch planting
посадка, рядовая row planting, line planting
посадка с глыбкой ball planting
посадка с защемлением slit planting, notch planting
посадка с комом ball planting
посадка хлыстами planting long cuttings
посадка черенками planting cuttings
посадка, щелевая slit planting, notch planting
посадочный planting
посев sowing
посев, бороздковый drill sowing
посев, бороздовый furrow sowing
посев вразброс broadcast sowing
посев в ямки hole sowing, pit sowing
посев, гнездовой nest sowing
посев, луночно-рядовой line seed-spot sowing
посев местами patch sowing
посев, озимый winter sowing
посев площадками patch sowing
посев полосами strip sowing

посев, рядовой line sowing, row sowing
посев, сплошной broadcast sowing
посев, строчно-луночный line seed-spot sowing (with gaps between seed-spot groups)
поселок settlement
посиневший bluestained
посинение bluestain development
посконник, коноляный hemp agrimony, *Eupatorium cannabinum* L.
последовательность лесосек sequence of fellings, order of fellings
послеледниковый post-glacial
постав the arrangement of blades in a frame saw, sawing schedule, sawing pattern
постепенный gradual (see also рубка, постепенный)
постоянный constant, sustained
постоянство constancy, continuity
постоянство пользования sustained yield
постройка building, construction
посылка feed per (double) stroke (frame saw)
поташ potash
потеря loss
потеря веса weight loss
потеря на пропил kerf loss
потеря при рубке loss in felling
поток stream, current, flow
поток, селевой mud and/or rock flow
потокообразователь current-maker (see also ускоритель)
потомство progeny
поточный flowing, continuous
потребление consumption, use
потребленость requirement, demand
початок **1.** spadix **2.** maize cob
почва soil, earth
почва, болотистая boggy soil
почва, болотная bog soil
почва, бурая brown soil
почва, влажная moist soil
почва, глинистая clayey soil
почва, горная montane soil
почва, дерновая soddy soil
почва, железистая ferruginous soil
почва, известковая calcareous soil
почва, илистая clayey soil
почва, карбонатная soil containing carbonate, calcareous soil
почва, каштановая chestnut soil
почва, кислая acid soil
почва, коричневая cinnamon-brown soil
почва, легкая light soil
почва, лесная forest soil
почва, лессовая loess soil
почва, луговая meadow soil
почва, меловая chalky soil
почва, мокрая wet soil
почва, наносная alluvial soil
почва, нелесная non-forest soil
почва, оглееная gleyed soil
почва, оподзоленная podzolized soil
почва, палевая yellowish soil, leached soil, lessivé
почва, песчаная sandy soil
почва, подзолистая podzolic soil
почва, пойменная floodplain soil
почва, рыхлая loose soil
почва, свежая fresh (i.e. fairly moist) soil
почва, серая grey soil
почва, суглинистая loam, clay-loam soil
почва, сырая wet soil
почва, торфяная peaty soil
почва, целинная virgin soil
почва, щелочная alkaline soil
почвенный soil
почвоведение soil science, pedology
почво-грунт soil and subsoil
почвозащитный soil-protective
почвообразовательный soil-forming
почвоуглубитель subsoiler
почвоулучшающий soil-improving
почвоухудшающий soil-degrading
починка mending, repairing
почка bud
почка, боковая lateral bud
почка, верхушечная terminal bud
почка, листовая leaf bud
почка, пазушная axillary bud
почка, придаточная adventitious bud
почка, спящая dormant bud

почка, цветковая flower bud
почка, цветочная flower bud
почка, центральная apical bud
пошлина duty, tax
появление appearance
пояс zone, belt
пояс, зеленый green belt
пояс, предохранительный safety belt
править to tension (saws)
правка tensioning (saws)
право right, law
право, лесное forest right, forest law
правонарушение infringement of law
практикант, лесной practising forester
пралес virgin forest, primary forest
превращение conversion, transformation
предгорье foothill, foothills
предел limit, boundary
предел насыщения волокна fibre saturation point
предел пропорциональности proportional limit
предохранение protection, prevention
предохранять to protect, to prevent
предпосевной pre-sowing, before sowing
предприятие enterprise, undertaking
предприятие, совместное joint venture
предсказание погоды weather forecast, weather forecasting
прелый rotten
преобладающий predominant (in numbers), prevailing
преобразование conversion, transformation
пресс press
пресс-крошки wood particles for making resin-bonded mouldings
пресс, многоэтажный multi-opening press, multi-daylight press
пресс, одноэтажный single-opening press, single-daylight press
прессование pressing, compressing
прессование, контурное contour compressing
прессование, одноосное uni-axial compressing
прессованный pressed, compressed, densified
прибавка 1. increase, addition 2. allowance
прибавка на усушку allowance for shrinkage
прибор apparatus, device
прибыль, генетическая genetic gain
прививальщик grafter
прививание 1. graft, grafting 2. inoculation
прививать to graft, to inoculate
прививка 1. graft, grafting 2. inoculation
прививка в вырез wedge graft
прививка в пенек apical grafting, splice grafting
прививка в приклад apical graft with scion pith to stock cambium
прививка в расщеп cleft graft
прививка за кору rind graft
прививка копулировкой apical grafting, splice grafting (see копулировка)
прививка мостиком bridge graft
прививка сближением approach graft
прививка сбоку side graft
прививка седлом saddle graft
прививка сердцевиной на камбий pith-to-cambium graft (a spliced or veneer side-graft)
прививок graft, scion
привод drive, gear
привой scion
пригонять to surface
приготовление preparation
придаточный adventitious
придорожный roadside
прием cut, felling (in shelterwood system)
приемник receptacle, cup (in resin tapping)
приживаемость survival
прижим pressure device (esp. a chipbreaker) in woodworking tools
прижимание pressing
приземистый low-growing
призма 1. prism 2. square-sawn timber, log with two opposite slabs removed, cant
признак sign, feature, characteristic

приказ order
приказ облесения afforestation order
прикапывать to heel in
приклад butt (of firearms)
прикол decrease in stacking density caused by cleaving the wood
прикорка heeling in
прилив flow, flood (tide)
прилистник stipule
приманка bait
примесь admixture
примшара mossy forest type
примыкание, кулисное 'coulisse' arrangement (coupes narrower than the intervening unfelled strips)
примыкание лесосек arrangement of the coupes in clear fellings, sequence of felling
примыкание, непосредственное progressive coupes on contiguous strips
примыкание, чересполосное coupes on alternate strips (equalling in width the intervening unfelled strips)
припуск allowance, overmeasure, overlength
прирастать **1.** to grow **2.** to adhere
приращение increase, increment
прирельсовый railside
приречный riverside, riparian
природа nature
природный natural
природоведение natural history
прирост increment, growth
прирост в высоту и т. д (see прирост по высоте и т. д.)
прирост, вторичный (в высоту) Lammas growth, Lammas shoot
прирост, ежегодный annual increment
прирост, массовый volume increment
прирост насаждения stand increment
прирост, нормальный normal increment
прирост, периодический periodic (annual) increment
прирост по высоте height increment
прирост по диаметру diameter increment
прирост по запасу growing stock increment
прирост по массе volume increment
прирост по объеме volume increment
прирост по тольщине diameter increment
прирост по ценности value increment
прирост, световой light increment
прирост, средний mean (annual) increment
прирост, текущий current (annual) increment
приспевать to approach maturity
приспевающий approaching maturity
приспособление adaptation
приспособляемость adaptability
приставка attachment (e.g. cross-arm on telegraph pole)
пристань waterside landing, wharf, quay
притенение shading
приток tributary
притык butt joint
прифуговка jointing
прицветник bract
прицел aim, sight
прицел, таксационный sighting bar (modification of Bitterlich's angle gauge)
прицеп trailer
прицеп, колесный wheeled trailer
прицеп, санный sledge trailer
причал hawser, mooring rope
прищипка pinching off, pinching out
прищипывать to pinch off, to pinch out
проба **1.** trial, test **2.** specimen, sample
пробитие (клея) bleed through (of glue)
пробка cork, phellem
пробный trial, test, sample
просверливать to drill through, to bore through
провешивание линии marking out a line (with stakes etc.)
проводимость conductivity (usually electrical or thermal, but sometimes used for the flow of fluids through wood)
проволока wire
проволока, колючая barbed wire

проволочник wireworm, elaterid larva
прогалина 1. gap, clearing (in forest) 2. place free from ice (on water)
прогиб deflection, sag
прогул absenteeism
продажа sale
продажа леса timber sale
продажа леса на корню sale of standing timber
продубина stain in floated wood of tannin-rich species
продукт product
продукт, побочный by-product, minor forest produce
продуктивность production, yield
продукция production, output
проект plan, project
проекция кроны crown projection
прозенхима prosenchyma
произведение 1. production 2. product
производитель producer, parent
производительность productivity, productiveness, output, yield
производить 1. to produce 2. to manufacture
производство 1. production 2. manufacture 3. industry
произрастание growing, growth, sprouting
произрастать to grow, to sprout
происхождение origin, provenance
происхождение, вегетативное vegetative origin
происхождение, естественное natural origin
происхождение, искусственное artificial origin
происхождение, семенное origin from seed
происхождение семян seed provenance
прокладка 1. laying, building 2. sticker (in stack) 3. packing piece, distance piece (in frame saws) 4. washer, packing
проковка hammering, tensioning (saws)
пролеска, сибирская *Scilla sibirica* Andr.
пролесник (see перелеска)
пролет span
пролысина bare patch, barked patch
пролысить to bark in patches (or strips)
пролыска bark blaze, barked patch, patch or strip barking
Проминлес Agency for Forest Sector Joint-Ventures
промоина scour, washed-out place, gully
промысел trade, business
промышленность industry
промышленность, бумажная paper industry
промышленность, деревообрабатывающая wood-working industry (see деревообработка)
промышленность, деревоперерабатывающая wood-converting industry (see деревообработка)
промышленность, лесная forest industry
промышленность, лесозаготовительная logging industry
промышленность, лесопильная sawmilling industry
промышленность, лесохимическая 'forest chemical industry' (resin tapping, chemical utilization of wood, etc.)
промышленность, целлюлозная pulp industry
промышленный industrial
проникать to penetrate, to permeate
проницаемость permeability
пропалывать to weed
пропаривание steaming
пропаривать to steam
пропарка steaming
пропахивать to plough (thoroughly)
пропил kerf
пропил, валочный felling cut, back cut
пропил, таранный boring cut (with a power saw)
пропитка impregnation, preservation
пропитка на корню impregnation of standing tree(s)
пропитывание impregnation, preservation
пропитывать to impregnate, to preserve

прополка weeding
пропсы pit props
пропуск gap, blank
пропускание passing, transmission (e.g. of light)
прорастание germination
прорастать to germinate
прореживание 1. thinning (in younger stands, usually of age 21-40 years) 2. thinning out
прореживание, верховое crown thinning, thinning from above
прореживание, низовое low thinning, thinning from below
прореживать to thin
прорез cut, slot, notch
прорось overgrown bark or dead wood
проросток 1. germinating seedling 2. one of the successive shoot extensions during the growing season
прорость overgrown bark or dead wood
прорубка 1. cutting through 2. ride
просвет 1. gap, opening 2. pore
просвет, одиночный solitary pore
просвет сосуда pore
просветы сомкнутые pore multiple
просвирник, лесной common mallow, *Malva sylvestris* L.
просека ride, cleared strip
просека, квартальная ride between compartments
просека, противопожарная firebreak
просинь bluestain
прослойка layer, interlayer
просмоление (древесины) resinosis, lightwood formation
просмолка resin pocket
просмолок resin streak, wood affected by resinosis
простой 1. standstill, idle time 2. simple, single-storeyed
простор late heavy thinning, increment felling (see рубка простора)
пространство space, expanse, area
пространство, жизненное growing space
пространство, занимаемое деревом growing space of a tree
пространство, межклеточное intercellular space
пространство, межполосное space between strips
просянка corn bunting, *Emberiza calandra* L.
протрава 1. disinfectant, fungicide, herbicide 2. mordant
протравитель 1. disinfectant, fungicide, herbicide 2. mordant
протравливание 1. disinfection 2. staining
протравливание почвы soil disinfection
протравливание семян seed disinfection
протяженность кроны crown length
протяжка pulling [a tree] through [a debranching unit]
профилировка profiling, shaping (e.g. of saw teeth)
профиль profile
профиль дороги road cross-section
профиль резца cutter profile
проходимость passability, practicability (of terrain), trafficability
проходимость трактора ability of a tractor to manoeuvre and operate
проходимый passable
процент percent, percentage
процент всхожести germination percent
процент лесистости area under forest, forest cover percent
процент прироста increment percent
процент, указательный indicating percent
процесс process
процесс, непрерывный continuous process
процесс, периодический batch process, intermittent process
прочистить to clean
прочистка cleaning/weeding (usually done to improve composition, in stands of age 11-20 years)
прочность strength, resistance, durability (see also сопротивление)
прочность на изгиб bending strength

прочность на растяжение tensile strength
прочность на сдвиг shear strength
прочность на сжатие compression strength, crushing strength
прочность на смятие compression strength perpendicular to grain
прочность на срез shear strength perpendicular to grain
прочность на срезывание shear strength perpendicular to grain
прочность на удар impact strength
пр. пл. sample plot
пруд pond
прут **1.** rod, withy, osier **2.** large cutting
прутняк osier stand
прямой straight
прямокрылые Orthoptera
прямослойный straight-grained
прямоствольный straight-stemmed
прятание reburning (in shifting cultivation)
психрометр psychrometer
птица bird
птица, хищная bird of prey
пузырек vesicle
пузырник bladder-senna, *Colutea* L.
пузырник, восточный eastern bladder-senna, *Colutea orientalis* Mill.
пузырник, древовидный bladder-senna, *Colutea arborescens* L.
пузырь bubble, blister
пункт point
пустельга, обыкновенная kestrel, *Falco tinnunculus* L.
пустой hollow, empty
пустоцвет barren flower
пустошь waste land
пустыня desert
пустырь waste land
путь way, path, track
путь резания distance travelled by each cutter or tooth
путь, рельсовый railway track
пух down, fluff
пучение, мерзлотное frost heave
пучок bundle
пучок бревен bundle of logs
пучок, сосудистый vascular bundle
пушистый pubescent
пушица cotton-grass, *Eriophorum* L.
пушица, влагалищная hare's-tail, *Eriophorum vaginatum* L.
пушица, узколистная *Eriophorum angustifolium* Honck.
пушица, широколистная *Eriophorum latifolium* Hoppe
пушнина fur
пуща dense virgin forest
пчела bee
пчела-листорез leaf-cutter bee, *Megachile*
пчела, медоносная honey-bee, *Apis mellifera* L.
пчеловод bee-keeper
пчеловодство bee-keeping
пчеловодство, бортевое forest bee-keeping
пчелояд bee-eater, *Pernis apivorus* L.
пчельник apiary
пыж log jam
пыжедержатель log-jam breaker, pier
пыжелом log-jam breaker, pier
пыл flame, blaze
пыль dust, silt
пыль, плодотворная pollen
пыль, торфяная peat dust, ground peat
пыльник anther
пыльца pollen
пырей, ползучий couch-grass, *Agropyron repens* Beauv.
пырей, собачий *Agropyron caninum* Beauv.
пьезотермопластик board or moulding made of particles hot-pressed without binders
пэн skidding pan
пяденица geometrid, looper
пяденица, березовая *Biston betularia* L.
пяденица, дубовая углокрылая *Ennomos quercinaria* L.
пяденица, зимняя winter moth, *Operophtera brumata* L.
пяденица-обдирало mottled umber moth, *Erannis defoliaria* L.

пяденица, сосновая pine looper, *Bupalus piniarius* L.
пяденица-шелкопряд, тополевая *Biston stratarius* Hfn.
пятилетка five-year plan
пятка heel (of cutting)
пятнистость spottiness, mottling
пятнистый spotted, mottled
пятно spot, stain
пятно, сердцевинное pith fleck

Р

работа work, labour
работа, лесосечная work at the felling site
работа на излом work to failure
работа, поденная day work
работа разрушения work to failure
работа, ручная hand work
работа, сдельная piece work
работа, ударная impact work
работа, удельная specific work
работать to work
работник worker
рабочий worker
равнина plain
равновесие equilibrium
равномерный uniform
радиальный radial
радиация, ультрафиолетовая ultraviolet radiation
радиус radius
разбивка лесосеки dividing up the felling area
разбивка по сортам sorting
разбирать to take apart, to dismantle
разборный collapsible
разборщик bundle-opener
разбрасывание broadcasting, spreading
разброс broadcasting, spreading
разбухание swelling
разбухать to swell
развал 1. (sawing) through and through without first slabbing 2. opening up (undoing) a bundle of logs
разведение 1. raising, cultivating, breeding 2. spring-setting
разведение леса afforestation, reforestation
разведение отводками raising by layers
разведение черенками raising by cuttings
разведенный spring-set (saw)
разветвление branching
разветвленный branched
развивать to develop
развилина fork, forking
развилистый forking
развилка forked tree
развитие development
развитие культуры plantation development
разводить 1. to raise, to cultivate, to breed 2. to spring-set
разводить зубья to spring-set teeth
разводить лес to afforest, to reforest
разводить пилу to spring-set a saw
разводка 1. spring-setting 2. saw set, setting pliers
развод пилы spring set of a saw
разворачивание бревна turning the log (in sawing)
разгружать to unload
разгрузка unloading
разделение division
разделение куста propagation by division of clumps
разделение леса division of a forest (into compartments, etc.)
разделка cross-cutting into lengths
раздробление chipping, hogging
разлагать 1. to decompose 2. to analyse
разлет семян seed dispersal
разложение 1. decomposition, breakdown 2. analysis
разлучка packing piece, distance piece (in frame saws)
размеживание demarcation
размер dimension, size
размер ежегодного отпуска size of the annual outturn
размер пользования yield, allowable cut

размер рубки yield, allowable cut
разметка marking
разметка карр face siting
разметчик marker
размещение arrangement, spacing
размещение листьев leaf arrangement
размещение растений plant spacing
размножать to multiply, to reproduce, to propagate
размножение multiplication, reproduction, propagation
размножение, вегетативное vegetative propagation
размножение, массовое mass reproduction
размножение, семенное propagation by seed
размолевание untying (log bundles)
размолевка untying (log bundles)
размыв gully erosion (see also эрозия)
размывание gully erosion
размыкание breaking, opening
размыкание насаждения opening stand canopy
разница, селекционная selection differential
разновидность variety
разновозрастный uneven-aged
разнообразие, биологическое biodiversity
разнообразие, видовое species diversity
разнообразие, генетическое genetic diversity
разнородный hcterogeneous
разность difference
разнотравный mixed herbs (see e.g. бор, разнотравный)
разнотравье mixed herb cover
разрабатывать to work, to process
разработка working, exploitation, conversion
разработка леса wood conversion
разработка лесосек layout of felling areas, logging organization
разрастание regrowth, spread, vigorous growth, proliferation
разреживание thinning, thinning out
разрез **1.** section **2.** cut
разрез, поперечный cross-section, transverse section
разрез, продольный longitudinal section
разрез, радиальный radial section
разрез, тангенциальный tangential section
разрез, торцовый cross-section, end grain
разрезывать to cut
разрушать to destroy, to rupture
разрушение destruction, rupture, failure
разрыв rupture, interruption, break
разрыв, противопожарный fire break, fire line
разрыхление почвы soil loosening
разрыхлять to loosen
разряд class, category (for survey work, forests are divided into 4 accuracy categories, based on % crop utilization)
разряд I category I (all wood sold)
разряд II category II (all wood from later thinnings and principal felling sold)
разряд III category III (75% of wood from principal felling utilized)
разряд IV category IV (25-75% of wood from principal felling utilized)
разряд высоты **1.** 'height series', height quality class (defined by the height/diameter curve) **2.** series of height curves
разъезд passing place (on road)
район region
рак canker
рак, опухолевидный erumpent swelling canker
рак, смоляной stem canker with resin flow (caused by *Cronartium flaccidum* Wint. or *Peridermium pini* Kleb.), 'resin top'
рак ствола stem canker
рак, ступенчатый target canker
рак, язвенный open or enlarging canker
ракита goat willow, sallow, *Salix caprea* L.
ракитник broom, *Cytisus* L.
ракитник, днепроский Dnieper broom, *Cytisus borysthenicus* Grun.

ракитник, русский Russian broom, *Cytisus ruthenicus* Fisch.

ракитник, чернеющий black broom, *Cytisus nigricans* L.

рама frame saw, frame

рама, вертикальная vertical frame saw

рама второго ряда deal frame, frame re-saw

рама, горизонтальная horizontal frame saw

рама, двухэтажная 'two-storey' frame saw (i.e. so tall that the logs must be lifted high above mill floor level)

рама, лесопильная frame saw

рама, одноэтажная 'single-storey' frame saw (i.e. working at abouι mill floor level)

рама первого ряда head frame saw, log frame

рама, пильная frame saw, saw frame

рама, распускная frame re-saw

рамень fertile spruce forest type

рамщик frame-saw operator

рана wound

РАН Russian Academy of Sciences

рана-подсочка freshening, streaking, streak (in resin tapping)

ранг class, rank

рангоут roundwood for masts, yards and spars

ранение wounding, injury

раса race

раскалываемость cleavability

раскалывание cleaving, splitting

раскалывать to cleave, to split

раскидистый sprawling, spreading

раскидник wood and brush debris left by spring floods

расклепка зубьев swaging saw teeth

раскол crack, split

расколимость cleavability

расколка cleaving, splitting

раскомлевка removing or trimming the butt end of a log

раскорчевка пней stump grubbing, stump removal

раскрой **1.** re-sawing (to produce dimension stock) **2.** breakdown (of stem or log) **3.** breaking down (a log)

раскрой вразвал thro' and thro' sawing

раскрой, бруского-круговой circular canting

раскрой с брусовкой slabbing

раскрой, сегментно-радиальной segment-radial canting

раскрой, сегментно-развальной segment thro' and thro' sawing

раскрой, сегментно-тангенциальной segment-tangential canting

раскрой, сегментный segment canting

раскряжевать to cross-cut (into logs)

раскряжевка cross-cutting (into logs)

распад decomposition, breakdown

распадение decomposition, breakdown

распаривание steaming

распахивание ploughing up

распахивать to plough up

распашка ploughing up

распил sawing, sawing up

распиливать to saw, to saw up

распиловка **1.** sawing **2.** re-sawing

распиловка, поперечная cross-cutting

распиловка, продольная rip-sawing

распиловка, радиальная quarter sawing

распиловка, ребровая edging

распиловка, тангенциальная flat sawing

распиловка, торцовая cross-cutting to length

расплющивание swaging

расположение arrangement, disposition

распоряжение arrangement, disposition

распределение distribution

распределение классов возраста age-class distribution

распределение пород species distribution

распределитель distributor

распространение spread, dispersal, spreading, range

распространение ветром wind dispersal

распространение семян seed dispersal

распускание flushing, opening

распутица season of bad roads, muddy period in autumn and spring
распыление **1.** spraying **2.** dusting
рассверловка boring
расселение settlement (in new places), colonization, migration
рассеяннопоровый diffuse-porous
рассеянососудистый diffuse-porous
расстояние distance, spacing
расстояние в ряду spacing in the row
расстояние между рядами distance between rows
расстройство роста growth disturbance
растаптывание trampling, crushing
растаскивание пачки breaking open a bunch of logs
раствор solution
раствор, питательный nutrient solution
растворять to dissolve
растение plant
растение, бобовое leguminous plant
растение, высшее higher plant
растение, вьющееся twining plant
растение, двулетнее **1.** two-year-old plant **2.** biennial
растение, декоративное ornamental plant
растение, древесное woody plant
растение, злаковое grass
растение-индикатор indicator plant
растение, кормовое fodder plant
растение, лекарственное medicinal plant
растение, лесное forest plant
растение, лечебное medicinal plant
растение, многолетнее perennial
растение, однолетнее **1.** annual **2.** one-year-old plant
растение, сорное weed
растение, сосудистое vascular plant
растение, споровое cryptogam
растение, тайнобрачное cryptogam
растение, травянистое herbaceous plant, herb
растение-хозяин host plant
растение, цветочное flowering plant, phanerogam
растение, чужеядное parasite
растение, явнобрачное flowering plant, phanerogam
растение, ядовитое poisonous plant
растениевод grower, plant breeder
растениеводство plant growing, plant breeding
расти to grow
расти в сомкнутости to close canopy
растительность vegetation
растительность, древесная tree vegetation, woody vegetation
растительный vegetable, plant
растрескивание cracking, splitting, checking
растяжение tension
растяжимость tensility, extensibility
растяжка tension member, guy
расходы expenses, costs, outlay
расхождение divergence, separation
расчет calculation
расчет, хозяйственный business accounting, profit-and-loss accountancy, cost accounting
расчистка clearing
расшифровка interpretation, deciphering
расшлаговка dismantling à raft
расщеп **1.** splintering **2.** sloven
расщепление splitting, fission, segregation
рашпиль rasp
реакция reaction
ребро **1.** rib **2.** edge
ребросклейка edge-gluing
ревизия **1.** revision **2.** inspection
ревизор, лесной forest inspector
регенерация regeneration
регулирование regulation
редина sparse stand, open stand
редкий rare
редколесье sparse forest, open woodland, open forest
редкослой abnormally wide-ringed wood
редкостой sparse stand, open stand
редкостойный open, sparse
редька, полевая wild radish, *Raphanus raphanistrum* L.
режим regime, system, schedule
режим сушки drying schedule

режущий cutting
резак bark-blazer
резание, прямолинейное orthogonal cutting
резание, силовое shearing (usually)
резать to cut
резец cutter, knife
резчик carver, engraver
резьба по дереву wood-carving
рей 1. yard 2. wing boom, fin boom
рейд roads, floating or water transport depot
рейд, сплавной floating or water transport depot
рейка narrow board, lath, batten
рейка, мерная measuring stick, measuring rod
рейс trip, journey
рейсмус 1. thicknesser 2. scribe
река river
река, сплавная floating river (for driving or rafting)
река, судоходная navigable river
рекогносцировка reconnaissance
реконструкция conversion, rehabilitation
рекоплав ice-free season (of rivers)
рекостав ice-bound season (of rivers)
рекультивация reclamation
реласкоп relascope
реласкоп, зеркальный mirror relascope
реликт relict
рельеф relief
рельс rail
ремез penduline tit, *Parus pendulinus* L.
ремень 1. belt, strap 2. vertical or horizontal interface (in resin tapping)
ремиза clump of broadleaves (planted in conifers)
ремнецветник, европейский mistletoe, *Loranthus europaeus* L.
ремонт repair, maintenance
ремонтантный ever-bearing
рендзина rendzina
рента financial return
рента, дифференциальная differential return attributable to land (in the Ricardian and Marxist theory of rent)
рента, лесная net return from forest land
рентабельный profitable
ресурсы resources
ресурсы, возобновляемые renewable resources
речка small river, stream
решетка 1. grill, grid 2. lattice
решето sieve, screen
рея 1. yard 2. wing boom, fin boom
ржавчина rust
ржавчина, пузырчатая blister rust
ржавчинник rust fungus
ржавчинник, елово-багульниковый *Chrysomyxa ledi* D.B.
ржавчинник, елово-грушанковый *Chrysomyxa pyrolae* Rostr.
ржавчинник, сосново-ласточниковый *Cronartium flaccidum* Wint.
ржавчинник, сосново-осиновый pine twisting rust, *Melampsora pinitorqua* Rost.
ржавчинник, сосново-смородинный *Cronartium ribicola* Fisch.
ризоид rhizoid
рисберм apron (of dam)
риски saw tooth marks on lumber surface
рисунок 1. illustration 2. figure, grain
рихтовка hammering, tensioning (saws)
ров ditch, trench, pit
рогожа bast matting
рогохвост siricid, woodwasp
рогохвост, березовый *Tremex fusicornis* Fbr.
рогохвост, большой хвойный *Sirex gigas* L.
рогохвост-гигант *Sirex gigas* L.
рогохвост, олховый alder woodwasp, *Xiphydria camelus* L.
рогохвост, синий сосновый *Paururus juvencus* L.
рогохвост, черный *Xeris spectrum* L.
род 1. genus 2. family, kind, sort, type
родина homeland, habitat
родник spring, source
родовой generic
рододендрон rhododendron, *Rhododendron* L.

рододендрон, даурский Dahurian rhododendron, *Rhododendron dauricum* L.
рододендрон, желтый yellow rhododendron, *Rhododendron luteum* Sweet
рододендрон, кавказский Caucasian rhododendron, *Rhododendron caucasicum* Pall.
рододендрон, камчатский Kamchatka rhododendron, *Rhododendron camtschaticum* Pall.
рододендрон, понтийский Pontine rhododendron, *Rhododendron ponticum* L.
родословие genealogy, pedigree
рожечник carob, *Ceratonia siliqua* L.
роза rose, *Rosa* L. (see also шиповник)
роза собачья dog rose, *Rosa canina* L.
розга, золотая golden-rod, *Solidago virgaurea* L.
розетка rosette
розетка сучьев whorl
розоцветные Rosaceae
ройка flute, fluting
рой пчел swarm of bees
рольганг roll-way
ромашка, непахучая *Matricaria inodora* L.
роса dew
роса, медвяная honeydew
роса, медовая honeydew
роса, мучнистая powdery mildew
роса, мучнистая (дуба) oak powdery mildew (caused by *Microsphaera alphitoides* Gr.)
Рослесбиржа Russian Wood Exchange
Рослесинвест Russian Forest/Wood Investment Company
роспуск trailer
россомаха glutton, *Gulo gulo* L.
россыпь 1. loose-floated wood 2. scattering
рост growth (see also прирост)
рост в высоту height growth
рост в толщину diameter growth
рост, вторичный secondary growth
рост, корявый crooked growth, stunted growth
росток germinated seedling, sprout
росянка, круглолистная sundew, *Drosera rotundifolia* L.
роща grove
рубанок plane
рубец scar
рубить 1. to fell 2. to cut, to chop, to chip
рубка 1. felling, cut 2. cutting
рубка, бухтообразная wavy-fronted felling
рубка, возможная possible cut
рубка, возобновительная regeneration felling
рубка, воспитательная tending felling, intermediate felling (weeding, cleaning or thinning)
рубка, вторичная secondary felling
рубка, выборочная 1. selective logging 2. selection felling
рубка, выборочная добровольная selection felling
рубка, выборочная на прииск selective logging, creaming (for particularly valuable assortments)
рубка, выборочная подневольная selective logging, creaming (to a diameter limit)
рубка, выборочная с вершка selective logging, creaming (to a diameter limit)
рубка, главная principal felling
рубка главного пользования principal felling
рубка, гнездовая group felling
рубка, группово-выборочная group selection felling
рубка, добровольно-выборочная selection felling
рубка, допустимая allowable cut, permissible cut
рубка, каемчатая narrow strip felling
рубка, комплексная combined principal felling plus thinning
рубка, комплексно-выборочная combined individual and group selection felling
рубка, концентрированная extensive ('concentrated') clear felling
рубка, котловинная group felling

рубка, кулисная alternate clear-strip felling (the unfelled strips 2-3 times wider than the felled)
рубка, ландшафтная landscape felling
рубка, лесосечная felling under systems involving coupes
рубка, линейная line thinning, row thinning
рубка на кобло pollarding
рубка на приск selective logging, creaming
рубка, нелесосечная felling under systems not involving coupes
рубка, несплошная fellings other than clear fellings (i.e. selection and shelterwood fellings)
рубка, обсеменительная seeding felling
рубка, окончательная final felling
рубка, осветительная secondary felling
рубка, очистительная final felling
рубка, очистная final felling
рубка переформирования conversion felling
рубка, подготовительная preparatory (seeding) felling
рубка по диаметру felling to a diameter limit
рубка, подневольная selective logging, creaming (to a diameter limit)
рубка, полосная strip felling
рубка, последовательная successive [regeneration] felling
рубка, постепенная shelterwood felling
рубка, присковая selective logging, creaming (for particularly valuable assortments)
рубка, промежуточная intermediate felling (any felling, e.g. cleaning or thinning, between stand establishment and the principal felling(s))
рубка промежуточного пользования intermediate felling (any felling, e.g. cleaning or thinning, between stand establishment and the principal felling(s))
рубка прореживания thinning (in younger stands, usually of age 21-40 years) (see прореживание)
рубка простора late heavy thinning, increment felling
рубка, проходная thinning (in older stands, usually over 40 years)
рубка, проходная верховая crown thinning
рубка, проходная низовая low thinning
рубка прочистки cleaning/weeding (usually done to improve composition, in stands of age 11-20 years)
рубка, санитарная **1.** sanitation felling **2.** salvage felling
рубка с вершка selective logging, creaming (to a diameter limit)
рубка, светлая any felling admitting light
рубка, семенная seeding felling
рубка, семенно-лесосечная uniform shelterwood felling
рубка, сметная planned cut
рубка, сплошная clear felling
рубка, сплошно-концентрированная extensive ('concentrated') clear felling
рубка, сплошно-лесосечная localized clear felling
рубка, схематическая systematic (i.e. non-selective) thinning
рубка, улучшительная improvement cutting
рубка, условно-сплошная 'clear' felling (that leaves small and inferior trees standing)
рубка ухода tending felling, intermediate felling (weeding, cleaning or thinning)
рубка, черезполосная alternate strip felling
рубка, шахматная felling in chequer arrangement
рубщик feller, woodcutter
руда ore
руддолготье long pitwood
рудник mine, pit
рудстойка pit-prop
рукава flute
руководитель leader
рукоятка handle
рулетка measuring tape
русак brown hare, *Lepus europaeus* L.

русло bed, channel
рухляк marl
ручей small stream
ручка handle
рыжик saffron milk-cap, *Lactarius deliciosus* Fr.
рыльце stigma
рынок market
рыночный merchantable
рысь lynx, *Lynx lynx* L.
рыхление loosening
рыхление почвы soil loosening
рыхлитель scarifier, ripper
рыхлый loose, friable
рыхляк marl
рычаг lever
рюм temporary riverside landing
рябина rowan, whitebeam, *Sorbus* L.
рябина глоговина wild service tree, *Sorbus torminalis* Crantz
рябина, круглолистная whitebeam, *Sorbus aria* Crantz
рябина, крупноплодная true service tree, *Sorbus domestica* L.
рябина, обыкновенная rowan, mountain ash, *Sorbus aucuparia* L.
рябина, садовая true service tree, *Sorbus domestica* L.
рябчик hazel-hen, *Tetrastes bonasia* L.
ряд **1.** row, line **2.** series
ряд, лесосечный felling series
ряжевый cribwork
рям raised bog (in W. Siberia)

С

сабель a type of stem burr
сабельник, болотный *Potentilla palustris* (L.) Scop.
саванна savanna
сад garden
сад, ботанический botanical garden
сад, лесной forest orchard (of wild fruit and nut trees)
сад, маточный plantation for producing grafting material and seeds, stool-bed
садовник **1.** pine-shoot beetle **2.** gardener
садовник, большой лесной larger pine-shoot beetle, *Blastophagus* (*Myelophilus*) *piniperda* L.
садовник, малый лесной lesser pine-shoot beetle, *Blastophagus* (*Myelophilus*) *minor* Hart.
сажалка planting machine
сажальщик planter
сажать to plant
сажать на пень to cut back to stump
саженец transplant, transplanted plant, sapling
саженец, перешколенный transplant, transplanted plant
саженец с закрытыми корнями containerized plant
саженец с обнаженными корнями bare-rooted plant
саженец, черенковый rooted cutting
саженец, штамбовый sapling pruned as a standard
сажень fathom (English, 6 ft (= *c.* 1.8 m); Russian, 7 ft (= *c.* 2.1 m)
сажень, кубическая piled fathom
сайга saiga, *Saiga tatarica* L.
сайгак saiga, *Saiga tatarica* L.
саксаул saxaul, *Haloxylon* Bge.
саксаул, белый white saxaul, *Haloxylon persicum* Bge.
саксаул, зайсанский saxaul, *Haloxylon ammodendron* Bge.
саксаул, песчаный white saxaul, *Haloxylon persicum* Bge.
саксаул, черный black saxaul, *Haloxylon aphyllum* Iljin
самец male
самка female
самовоспламенение self-ignition, spontaneous combustion
самоизреживание natural thinning
самолет aeroplane
самооблесение natural reforestation
самоопыление self-pollination
самоочищение ствола natural pruning, self-pruning

самосев natural regeneration (from seed), self-sowing
самотаска conveyor, hoist
самотаска, канатная cable slide
самшит box, *Buxus sempervirens* L.
сандарак *Tetraclinis articulata* Mast.
сани sledge
сани, лесовозные timber-transport sledge
сапрофит saprophyte
сапсан peregrine falcon, *Falco peregrinus* Tunst.
сарай shed, barn
саранчевые Acrididae
сарыч buzzard, *Buteo buteo* L.
саур-арча semiglobose juniper, *Juniperus semiglobosa* Rgl.
сахар, древесный wood sugar
сбег taper
сбежистость degree of taper, taper
сбежистый tapering, tapered
сближение, вегетативное vegetative rapprochement (in Michurinist theory)
сбор collection, harvesting, gathering
сборка assembling
сбор семян seed collection
сбрасывание **1.** ejecting, kicking (e.g. from a conveyor) **2.** launching, throwing (into water)
сбрасыватель log-kicker
сброска **1.** ejecting, kicking (e.g. from a conveyor) **2.** launching, throwing (into water)
сбруя harness
сбыт sale, market
сбыт леса timber sale
сваливать to fell, to cut out, to cut down
свая pile (for construction in marshy ground, etc.)
свежесрубленный freshly felled, green
свежий fresh, fairly moist
сверление drilling, boring
сверление, пазовое slot mortising
сверлило lymexylonid
сверлило, корабельное *Lymexylon navale* L.
сверлило, лиственное *Hylecoetus dermestoides* L.
сверлило, хвойное *Hylecoetus flabellicornis* Schn.
сверлильщик lymexylonid
сверлить to bore, to drill
сверло borer, drill, bit
сверло, винтовое twist bit
сверло, ложечное spoon bit
сверло, пазовое slot-mortising bit, grooving cutter
сверло, спиральное twist bit
сверло, центровое centre bit
сверхнагрузка overloading
сверчок warbler (Sylvidae)
свес **1.** hang, overhang (in frame saw) **2.** lip, projection
свет light
свет, верхний vertical light, overhead light
свет, рассеянный diffuse light, diffused light
световыносливый light-resistant, light-tolerant
светолюбивость love of light
светолюбивый light-loving, light-demanding
светолюбие love of light
светоустойчивый light-resistant, light-tolerant
свидина dogwood, *Cornus sanguinea* L.
свилеватость irregular grain, wavy grain, curly grain
свилеватый irregular, wavy, curly
свиль irregular grain, wavy grain, curly grain
свинья, дикая wild boar, *Sus scrofa* L.
свиристель waxwing, *Bombycilla garrulus* L.
свита suite, series, formation
свищ internal check, flaw, knothole
свободный free
сводка леса forest clearance
свойство property, character
свойство, лесоводственное silvicultural character
СВЧ-камера VHF kiln
связка **1.** bundling, binding **2.** bundle
связность почвы soil cohesion
связывание bundling, binding

связывать to bind, to bundle
сгиб bend
сгибать to bend
сгнивать to rot, to decay
сгон driving (of wood)
сгонщик wood floater
сгораемость combustibility
сгребание подстилки raking litter
сгущенный crowded
сдача delivery
сдвиг **1.** shear **2.** shift, displacement
сдельщина piecework
сдирание stripping, skimming, scalping, shredding
сдирание кожицы с рогов fraying, antler burnishing
сдирание коры bark stripping
сдирать кору to bark, to remove bark
себестоимость net cost, cost price, production cost, prime costs
себестоимость, корневая stumpage value
Севморпуть northern sea route
севооборот crop rotation
седач hemp agrimony, *Eupatorium cannabinum* L.
седмичник, европейский chickweed wintergreen, *Trientalis europaea* L.
сезон season
секатор secateurs
секач cutter, billhook, machete, barking iron
секвоя sequoia, *Sequoia* Endl.
секвоя, гигантская giant sequoia, *Sequoia gigantea* Decne.
сексуализация sexual specialization
секундомер stopwatch
секция, силовая section for testing case-hardening
секция, хозяйственная working circle
селевой (see сель; поток, селевой)
селезеночник, очереднолистный *Chrysosplenium alternifolium* L.
селекционер breeder
селекция breeding, selection, improvement
селение settlement, village
селитра, аммиачная ammonium nitrate
сель mud and/or rock flow
сельга esker
сельский village, rural
сельскохозяйственный agricultural
семеед, еловый *Megastigmus strobilobius* Ratz.
семейство family
семена seeds
семенник seed tree, seedbearer
семенной seed, of seedling origin
семеноводство seed production
семенодоля cotyledon
семенозаготовка seed harvesting, seed collection
семеномер seed trap, seed gauge
семеносушильня seed kiln, seed extraction plant
семеношение seeding, seed production
семечко seed
семья family
семя seed
семя, инбридная inbred seed
семя, полноценная sound seed, full seed
семядоля cotyledon
семяед, еловый *Megastigmus strobilobius* Ratz.
семяносушильня seed kiln, seed extraction plant
семяноуловитель seed trap
семяпочка ovule
семячковые pip fruits
сено hay
сенокос hay-making, mowing
сера sulphur
сердечник, горкий *Cardamine amara* L.
сердечник, извилистый *Cardamine flexuosa* With.
сердечник недотрога *Cardamine impatiens* L.
сердцевидный cordate
сердцевина pith
сердцевина, двойная double pith
середина middle
сережка catkin
серка hardened resin, scrape
серница resin pocket, resin streak
серозем s(i)erozem, grey desert soil
серп sickle

серпантин winding forest road on a slope
серпуха, красильная saw-wort, *Serratula tinctoria* L.
сертификация certification
сертификация семян seed certification
серянка stem canker with resin flow (caused by *Cronartium flaccidum* Wint. or *Peridermium pini* Kleb.), 'resin top'
сетка net, network (see also сеть)
сетчатокрылые Neuroptera
сеть system, net, network
сеть, дорожная road system, road network
сеть, дренажная drainage system
сеть, железнодорожная railway system
сеть, сортировочная sorting installation
сеча land cleared (by shifting cultivation)
сечение section, cross-section
сечь, лесорубочная felling series
сеялка seed drill, sowing machine
сеянец seedling (usually up to age 2)
сеять to sow
сжатие compression
сжатие вдоль волокон compression parallel to grain
сжатие поперек волокон compression perpendicular to grain
сжигание burning
сжигать to burn
сжимать to compress, to squeeze
сигара cigar-shaped raft
сидерат green manure crop
сидерация green manuring
сидячий sessile
сизоворонка roller, *Coracias garrulus* L.
сикомор sycamore maple, *Acer pseudoplatanus* L.
сила force, power
сила, двигательная driving force, motive power
сила, лошадиная horse power
сила растяжения tensile force
сила, срезывающая shearing force
сила тяжести gravity
силок noose, trap
симбиоз symbiosis
синева bluestain
синева, прокладная sticker bluestain
синица, большая great tit, *Parus major* L.
синица, длиннохвостая long-tailed tit, *Aegithalos caudatus* L.
синица, хохлатая crested tit, *Parus cristatus* L.
синь bluestain (see also синева)
сипуха barn owl, *Tyto alba* Scop.
сирень lilac, *Syringa vulgaris* L.
система system
система, географическая информационная (ГИС) geographic information system (GIS)
система, корневая root system
система позиционирования, глобальная (ГПС) global positioning system (GPS)
система, равномерная uniform (shelterwood) system
системный systemic
ситник rush, *Juncus* L.
ситник, развесистый soft rush, *Juncus effusus* L.
ситник, расходящийся soft rush, *Juncus effusus* L.
ситник, трехраздельный *Juncus trifidus* L.
сито sieve, screen
ситовина honeycomb rot
с/к over bark, o.b.
скала **1.** rock, cliff **2.** scale
скалывание shear parallel to grain
скантлинги scantlings
скарификация scarification
скат **1.** slope **2.** chute, slide
скат для саней sledge-way
скатка леса rolling wood down, launching wood (into water)
скатчик wood launcher
скатывать to roll, to slide
скашивание **1.** bevelling, chamfering **2.** mowing
скважность pore space, porosity
сквер square
сквозистость transparency (of canopy)
скворец, обыкновенный starling, *Sturnus vulgaris* L.

скелет skeletal content of soil
скерда, болотная *Crepis paludosa* (L.) Moench
скиддер skidder
скипидар turpentine
скипидар, древесный wood turpentine
скипидар, живичный gum turpentine
склад store, depot, landing, yard
склад, верхний forest log depot, upper landing
склад, лесной depot for primary conversion
склад, нижний industrial log depot (usually at railway or river), lower landing, logging terminal
складосвал landing
складывание stacking, storing
склеивание, линейчатое ribbon spreading of glue
склеивать to glue
склейка gluing
склеренхима sclerenchyma
склон slope
скоба clamp, cramp
скоба, мерная 1. diameter ruler 2. sighting bar (for measuring height)
скоба, проволочная staple
скобель barking iron, drawknife
сковородень dovetail
скол breaking off (of the stem in felling)
скольжение slip, slipping, sliding
скоп recovered stock, savings, fibre waste
с корой over bark, o.b.
скороспелый early, early-maturing
скорость speed, velocity
скорость витания suspension velocity
скорость резания cutting speed
скорпула ореха nutshell
скос bevel, chamfer
скосарь, черный *Otiorrhynchus niger* F.
скот livestock
скребок scraper
скрейбер scribe, scriber
скрещивание crossing, hybridization
скрещивание, внутривидное intraspecific crossing
скрещивание, возвратное back crossing
скрещивание, межвидное interspecific crossing
скрещивание, обратное reciprocal crossing
скрещивание, отдаленное distant crossing, remote crossing
скрещивать to cross
скрипун, большой осиновый large poplar borer, *Saperda carcharias* L.
скрипун, малый осиновый smaller poplar borer, *Saperda populnea* L.
скрытохоботник, ольховый poplar borer, *Cryptorrhynchus lapathi* L.
скумпия smoke-tree, *Cotinus coggygria* Scop.
славка warbler, *Sylvia* L.
сладкокорень common polypody, *Polypodium vulgare* L.
сланец shale, schist, slate
слань flooring, decking, planking
слесарь fitter
слетинги slats
слешер slasher
слива, домашная plum, *Prunus domestica* L.
слива, колючая blackthorn, sloe, *Prunus spinosa* L.
слизетечение slime flux
слизь slime, mucilage
слипер sleeper block (baulk of 10 x 10 in (= *c.* 25 x 25 cm) cross-section, sawn to give two sleepers - see шпала)
слоевище thallus
сложение constitution, structure
сложный complex, composite, multi-storeyed
слой layer, stratum, sheet, ply
слой, водоносный aquifer
слой, годичный annual ring, growth ring
слой, гумусовый humus layer
слой, пахотный ploughed layer, ploughed horizon
слой, приземной layer near the soil
слой, пробковый cork layer
слоник curculionid, weevil (see also долгоносик)
служба, лесная forest service
случай, несчастный accident

смазка lubricant
смачиваемость wettability
смежный adjacent, contiguous
смена change, shift, exchange
смена, восстановительная succession in regeneration (after felling)
смена, денная day shift
смена пород succession or alternation of species, replacement of species
смесь mixture
смета estimate
смешанный mixed
смешение mixing
смешение пород species mixture
смоковница fig, *Ficus carica* L.
смола **1.** resin **2.** tar
смола, генераторная generator tar
смола, древесная wood tar
смола, каменноугольная coal tar
смола, камерная retort tar
смола, карбамидная urea resin
смола, мочевино-формальдегидная urea-formaldehyde resin
смола, отстойная settled tar
смола, садовая grafting wax
смола, сланцевая shale tar
смола, сосновая **1.** pine tar **2.** pine resin
смолевка **1.** weevil, *Pissodes* Germ. **2.** catchfly, campion, *Silene* L.
смолевка, еловая *Pissodes harcyniae* Hbst.
смолевка, пихтовая *Pissodes piceae* Ill.
смолевка, сосновая banded pine weevil, *Pissodes pini* L.
смолевка, сосновая вершинная *Pissodes piniphilus* Hbst.
смолевка, сосновая жердняковая *Pissodes piniphilus* Hbst.
смолевка, точечная banded pine weevil, *Pissodes notatus* F.
смолевка, шишковая *Pissodes validirostris* Gyll.
смоление tarring
смолистость resinousness, resin content
смолистый **1.** resinous **2.** tarry
смолка, клейкая *Viscaria vulgaris* Bernh.
смоловместилище resiniferous cavity
смологрунт tar/earth (for stabilized earth roads)
смолокурение tar distillation
смолоносный resiniferous
смолопродуктивность resin productivity
смолотечение resin flow, resin bleeding
смольняк resinous wood
смородина currant, *Ribes* L.
смородина, альпийская alpine currant, *Ribes alpinum* L.
смородина дикуша *Ribes dikuscha* Fisch.
смородина, золотиситая golden currant, *Ribes aureum* Pursh.
смородина, красная redcurrant, *Ribes rubrum* L.
смородина, пушистая *Ribes pubescens* Hedl.
смородина, черная blackcurrant, *Ribes nigrum* L.
смыв sheet erosion (see also эрозия)
смывание sheet erosion
смыкание closure, closing
смыкание крон canopy closure
смыкание молодняка canopy closure of young stand
смыкаться to close canopy
смятие compression perpendicular to grain
снабжение provision, supplying
снасть **1.** tool, equipment **2.** rope, tackle
снег snow
снегирь bullfinch, *Pyrrhula pyrrhula* L.
снеговал lean or uprooting due to snow pressure
снегование storage in snow
снегозадержание snow retention
снеголом snow breakage
снегоочиститель snow clearer, snow plough
снегопад snowfall
снежеягодник snowberry, *Symphoricarpos* Duh.
снимок photograph
снытъ, обыкновенная goutweed, ground elder, *Aegopodium podagraria* L.

СО Siberian Section
собирание collection
собирание живицы resin collection
соболевание sable hunting
соболь sable, *Martes zibellina* L.
собственность, лесная forest property
сова owl
сова, белая snowy owl, *Nyctea scandiaca* L.
сова, болотная short-eared owl, *Asio flammeus* Pont.
сова, ушастая long-eared owl, *Asio otus* L.
сова, ястребиная hawk-owl, *Surnia ulula* L.
советник, лесной forestry adviser
Совет Экономической Взаимопомощи Comecon
совка noctuid
совка, сосновая pine noctuid, *Panolis flammea* Schiff.
совхоз sovkhoz, state farm
согнутый bent, curved
согосподствующий co-dominant
согра wet fairly fertile site type
СОД = структурно-ориентированная доска OSB = oriented strand board
содействие возобновлению assistance to regeneration
содержание content(s)
содержание влаги moisture content
соединение 1. compound 2. connection 3. joint
соединение вполдерева scarf joint
соединение впритык butt joint
соединение гребнем в паз tongue and groove joint
соединение, зубчатое finger joint
соединение, клеевое glue joint
соединение, клиновидное finger joint
соединение на зубчатый шип finger joint
соединение на ласточкин хвост dovetail joint
соединение на ус scarf joint
соединение, усовое scarf joint
соединение, шиповое tenon joint
соединение шипом в гнездо mortise and tenon
соединение, шкантовое dowelled joint
соединение, шпунтовое tongue and groove joint
соединение шпунтом в фальц tongue and groove joint
сожительство symbiosis
созревание ripening, maturing
созрелый ripe, mature
сойка jay, *Garrulus glandarius* L.
сок sap, juice
сок, клеточный cell sap
сокоистечение sap bleeding
сокол peregrine falcon, *Falco peregrinus* Tunst.
солестойкость salt resistance
солитер single tree, specimen tree
солнцепек sun scorch
солодь solod
солома straw
соломка, спичечная match stick, match splint
солонец solonetz
солончак solonchak
соль salt
солянка *Salsola* L. (see also черкез)
сомкнутость degree of closure, canopy density
сомкнутость крон canopy density
сомкнутость насаждения stand canopy density
сомкнутость полога canopy density
сомкнутый closed
сомкнуться to close canopy
соня, лесная forest dormouse, *Dyromys nitedula* Pall.
соня, орешниковая dormouse, *Muscardinus avellanarius* L.
соня-полчок fat dormouse, *Glis glis* L.
соня, садовая garden dormouse, *Eliomys quercinus* L.
сообщение communication, report
сообщественный associative
сообщество community
сооружение building, construction, installation

соотношение классов возраста ratio of age classes

соплодие infructescence

соприкосновение крон crown contact

сопротивление resistance, strength (see also прочность)

сопротивление изгибу bending strength

сопротивление износу wear resistance, abrasion resistance

сопротивление разрыву tensile strength

сопротивление раскалыванию cleaving resistance

сопротивление растяжению tensile strength

сопротивление резанию cutting resistance

сопротивление сдвигу shear strength

сопротивление сжатию compression strength

сопротивление скалыванию shear strength parallel to grain

сопротивление скручиванию torsional strength

сопротивление срезу shear strength perpendicular to grain

сопротивление трению friction resistance

сопротивление удару shock resistance, impact strength

сора sunken tree, snag

сорность specks (in pulp)

сорняк weed

сорока magpie, *Pica pica* L.

сорокопут shrike, *Lanius* L.

сорт **1.** assortment, grade **2.** variety

сортимент assortment

сортиментовоз vehicle for transport of assortments, forwarder

сортировать to sort, to grade, to screen

сортировка sorting, grading, screening

сортировка, прочностная strength grading

сортировка, силовая stress grading

сортоводство breeding, production of varieties

сортоиспытание variety testing

сортообразец variety, clone, hybrid, mutant, form or other category of breeding material

сортообъект area for testing varieties etc.

сосенка small pine tree

сосна **1.** pine, *Pinus* L. (see also кедр; стланик) **2.** Scots pine, *Pinus sylvestris* L.

сосна, алепская Aleppo pine, *Pinus halepensis* Mill.

сосна, веймутова white pine, Weymouth pine, *Pinus strobus* L.

сосна, горная mountain pine, *Pinus mugo* Turra, (*Pinus montana* Mill.)

сосна, желтая western yellow pine, *Pinus ponderosa* Dougl.

сосна, итальянская umbrella pine, *Pinus pinea* L.

сосна, калабрийская Corsican pine, *Pinus brutia* Ten.

сосна, кедровая сибирская Siberian pine, *Pinus sibirica* Mayr.

сосна, корейская Korean pine, *Pinus koraiensis* Sieb. et Zucc.

сосна, крымская Crimean pine, *Pinus nigra* var. *caramanica*

сосна, маньчшурская кедровая Korean pine, *Pinus koraiensis* Sieb. et Zucc.

сосна, мяндовая wide-ringed pine

сосна, обыкновенная Scots pine, *Pinus sylvestris* L.

сосна, пицундская Pitsunda pine, *Pinus pithyusa* Stev.

сосна, приморская maritime pine, *Pinus pinaster* Sol.

сосна, румелийская Macedonian pine, *Pinus peuce* Griseb.

сосна, сибирская Siberian pine (the Siberian subspecies of *Pinus sylvestris*)

сосна, черная black pine, *Pinus nigra* Arn.

сосна, эльдарская Eldar pine, *Pinus eldarica* Medwed.

сосновый pine

сосняк pine forest, infertile pine site type, *Pinetum* (see also бор)

состав composition

состав, возрастный age structure

состав насаждения stand composition
состояние condition, state
состояние насаждения stand condition
сострагивать to plane off
сосуд vessel, trachea
сосудистый vascular
сотрудник co-worker
соттомизура narrow boards (for Mediterranean market) (10-16 cm wide, 3-6 m long)
соха wooden plough
сохнуть to dry, to wither
сохранение preservation, conservation
сохраняемость древесины wood durability
соцветие inflorescence
сочный sappy, juicy, succulent
сошник plough-share, coulter
союз alliance (bot.)
спелость maturity, ripeness
спелость, возобновительная maturity for regeneration (giving best natural regeneration)
спелость, естественная natural maturity (age when decline sets in)
спелость, качественная qualitative maturity (yielding max. price per unit of wood volume)
спелость, количественная quantitative maturity (yielding max. volume production)
спелость насаждения stand maturity
спелость почвы mellowness of soil
спелость, техническая technical maturity (yielding the max. of the required assortment(s))
спелость, физическая physical maturity
спелость, финансовая financial maturity
спелость, хозяйственная economic maturity (yielding highest income)
спелость, экономическая economic maturity (yielding highest income)
спелый ripe, mature
спелый к рубке ripe for felling, mature
спеть to ripen, to mature
специалист specialist, expert
спецодежда working clothes
спил sawn-off piece or sample (e.g. a disc)
спиливать to saw off
спирея *Spiraea* L. (see also таволга)
спирт alcohol
спирт, древесный wood alcohol
список list
спица spoke
спичечный match
спичка match
сплав 1. floating 2. alloy
сплав в кошеляк floating in bag booms
сплав леса floating of wood
сплав, молевой loose floating
сплав, плотовой rafting
сплав россыпью loose floating
сплав, твердый hard alloy
сплавлять to float
сплавной floating, floatable
сплавщик timber floater
сплотить to raft
сплотка 1. raft assembly, bundling, binding 2. jointing
сплоточный rafting
сплотчик raft maker
сплошной continuous, clear, entire
сплющивание swaging
спора spore
способ method
способ посадки planting method
способ рубки felling system
способ хозяйства silvicultural system
способность capacity, ability, power
способность, производительная productivity, yield capacity
спуск slide, slope, lowering (of wood downhill)
спускать to slide down, to lower
спутник associate
спячка, зимняя hibernation
спящий dormant
срастание growing together, coalescing, uniting
срастание корней natural root-grafting
срастаться to grow together, to coalesce, to unite
сращивание growing together, coalescing, uniting

среда environment, medium
среда, внешняя external environment
среда, питательная nutrient medium
СредазНИИЛХ Central Asian Forest Research Institute
средиземноморский Mediterranean
срединный middle, mid
средневозрастный middle-aged
среднее mean, average
среднее, скользящее running mean
средний mean, average
средство, пропитывающее preservative
срез 1. section, slice 2. shear perpendicular to grain
срезание 1. shear perpendicular to grain 2. cutting off, severing
срезка на обратный рост cutting back
срезок offcut
срезывание 1. shear perpendicular to grain 2. cutting off, severing
срок time, date, period
срок окупаемости pay-back period
срок примыкания лесосек interval between successive adjacent fellings (see примыкание)
сруб frame, shell
срубать to fell
срубка felling
срубленный felled
срывка launching (loose wood into water)
ссыхаться to dry up, to shrink
ставень shutter
стадия stage (the following are stages in shelterwood systems)
стадия, обсеменительная seeding stage
стадия, окончательная final stage
стадия, осветительная secondary stage
стадия, подготовительная preparatory stage
стадо herd, flock
стамеска chisel
стандарт standard
стандарт, ленинградский Leningrad (Petrograd) standard
стандартизация standardization
становище landing
станок machine, machine tool (see also машина)
станок, гвоздезабивный nailing machine
станок, головной древнопильный head saw
станок, делительный re-saw machine
станок, долбежный mortising machine
станок, дровокольный firewood-cleaving machine
станок, калевочный moulding machine, moulder
станок, копировальный copying lathe
станок, кромкофуговальный edge-jointer
станок, круглопалочный rod-maker, turning machine
станок, круглопильный circular saw
станок, ленточнопильный band saw
станок лисий хвост reciprocating log cross-cut saw
станок, лобзиковый fret saw
станок, лущильный rotary veneer peeler, veneer lathe
станок, обрезной edger
станок, окорочный barking machine, debarker
станок, пазовальный groover, slotter, router
станок, пригонно-строгальный rotary planer
станок, пригоночный rotary planer, jointer
станок, прирезный re-saw, dimension saw
станок, проволокосшивный wire-stitching machine
станок, продольно-фрезерный peripheral milling machine, planer
станок, пропускной thicknesser
станок, ребровый re-saw, slab re-saw
станок, ребросклеивающий edge-gluer
станок, реймусовый thicknesser
станок, рубильный chipper, chopper
станок, сборочный set-maker
станок, сверлильно-пазовальный slot mortiser

станок, сверлильный boring machine, drill
станок, строгальный 1. planer 2. veneer slicer
станок, токарный turning lathe
станок, торцовочный docking saw, end-trimming saw
станок, универсальный universal wood-working machine
станок, фанерострогальный veneer slicer
станок, форматный panel-sizing machine
станок, фрезерно-брусующий chipper-canter headrig
станок, фрезерный spindle moulder, moulding machine
станок, фуговальный rotary planer, jointer
станок, цепнодолбежный chain mortiser
станок, четырехсторонниий four cutter
станок, шипорезный tenoner
станок, шкантовый dowel-maker
станок, шлифовальный sander
станция station
станция, лесная опытная forest experiment station
стаскивание pulling down (trees)
статистика, лесная forest statistics
стационар permanent study area
стация station, site
ствол stem, bole, trunk
стебель stem, stalk, culm
стеклянница sesiid, clearwing
стеклянница, осиновая *Sciapteron tabaniforme* Rott.
стеклянница, тополевая большая hornet clearwing, *Sesia apiformis* L.
стеклянница, тополевая малая *Sciapteron tabaniforme* Rott.
стеллит stellite
стена wall (see also стенка)
стена, водоподпорная quay
стена леса forest border, forest edge
стена, руслостеснительная stream-confining wall
стена, струенаправляющая stream-directing wall
стенка wall (see also стена)
стенка, вторичная secondary wall
стенка клетки cell wall
стенка, первичная primary wall
стенка, третичная tertiary wall
стенограф *Ips sexdentatus* Boern.
степень degree, class
степень полноты stand density
степень прореживания thinning grade
степень сомкнутости canopy density
степь steppe
стержень 1. core, pivot, rod, axis 2. tap-root
стерильный sterile
стерня stubble
стесывание hewing off
стивидор stevedore
стижка shearing, clipping
стижка, фигурная topiary
стимулировать to stimulate
стланец scrub, prostrate or shrubby tree
стланец, березовый dwarf birch, *Betula nana* L.
стланец, сосновый mountain pine, *Pinus mugo* Turra (*Pinus montana* Mill.)
стлан(н)ик scrub, prostrate or shrubby tree
стлань flooring, decking, planking
стлань, поперечная corduroy road
стоимость cost(s), price, value
стоимость вывозки extraction costs
стоимость, корневая stumpage value
стоимость охранения protection costs
стоимость, полная корневая stumpage price
стоимость производства production cost
стоимость, таксовая stumpage value
стойбище landing
стойка support, post
стойка, дверная door post
стойка, крепежная pit-prop
стойка, рудничная pit-prop
стойкость resistance, hardiness, durability
сток flow, run-off, effluent

сток, внутрипочвенный subsurface run-off
сток, поверхностный surface run-off
сток, подземный subsurface run-off
сток по стволам stem flow
сток, русловый stream flow
стол table
столб **1.** column, post, pole **2.** mine shaft
столб изгороди fence post
столб, телеграфный telegraph pole
столбик style (of flower)
столяр carpenter, joiner
стопа pile, stack
сторож guard, keeper
стояние standing, position
стояние деревьев position of the trees
стояние, свободное open position, isolation
страж guard, keeper
страна света point of the compass
стратификация stratification
страхование insurance
стрела **1.** arrow **2.** jib
стрела прогиба amount of deflection, sag
стрелевать to skid (away)
стрепет little bustard, *Otis tetrax* L.
стригун pine-shoot beetle (see also садовик)
стрипсы strips
строгальщик planer, slicer
строгание planing, slicing
строганный planed, sliced
строевой building, constructional
строение **1.** structure **2.** construction
строение древесины wood structure
строжка planing
строительство construction, building
строительство, дорожное road building
строительство, зеленое amenity planting
стройлес constructional timber, building timber
стройность 'slenderness' (height/d.b.h.), 'shapeliness' (height/d.b.h.)
строп sling
стропило rafter
струбцинка cramp
струг **1.** plane, freshening tool **2.** draw-knife
стружка particle, shaving, chip, wood-wool
стружка, витая shaving
стружколоматель chip-breaker
стружкообразование chip formation
структура structure
структура насаждения stand structure
структура почвы soil structure
стручок pod
стул chair
ступенчатый stepped, gradual
ступень **1.** step **2.** stage **3.** class, gradation
ступень возраста age class
ступень высоты height class
ступень толщины diameter class
ступень толщины, естественная natural diameter class (expressed in decimal fractions of mean stand diam.)
стык joint (see also соединение)
субальпийский subalpine
суболоть area starting to become waterlogged
суборь fairly infertile pine site type, pine forest type
субучина fairly fertile beech forest type
сувель-вал stem burr on birch
суглинистый loamy, clay-loam
суглинок loam, clay-loam
сугрудок fairly fertile (usually hardwood) site type
судно ship
судостройлес boatbuilding timber
судубрава fairly fertile oak forest type
сук **1.** branch **2.** knot (see also сучок)
сука land cleared by shifting cultivation
суковатка harrow made of a spruce top
суковатый **1.** branchy **2.** knotty
сумах sumac, *Rhus* L.
сумах, дубильный tanners' sumac, *Rhus coriaria* L.
сумах, оленерогий staghorn sumac, *Rhus typhina* L.
сумах, пушистый staghorn sumac, *Rhus typhina* L.

сумшара wet, mossy, fairly infertile pine site type
супесчаный sandy loam
супесь sandy loam
супка пастушья shepherd's purse, *Capsella bursa-pastoris* Med.
суппорт lathe carriage
супротивный opposite
сурамень fairly fertile spruce forest type
суровый rough, severe, coarse
сурок marmot, *Marmota* L.
суслик suslik, souslik, *Citellus* L.
сустав joint, articulation
сухобойна seasoning check
сухобокость dying of one side of stem due to damage
сухобочина dying of one side of stem due to damage
суховей dry wind
суховершинность stagheadedness
суховершинный stagheaded
суходол dry valley
сухоподстой dead standing tree(s)
сухостой dead standing tree(s)
сухостойный dead standing
сучкоблок building block made of baled twigs
сучковатость 1. branchiness 2. knottiness
сучковатый 1. branchy 2. knotty
сучкоподборщик slash collector
сучкорез 1. branch remover 2. trimmer, pruner
сучняк branchwood
сучок 1. branch 2. knot
сучок, вросший enclosed knot
сучок, выпадающий loose knot
сучок, гнилой unsound knot
сучок, живой live knot, intergrown knot
сучок, загнивший unsound knot
сучок, закрытый enclosed knot
сучок, заросший enclosed knot
сучок, здоровый sound knot
сучок, ивлевый knot with incipient decay
сучок, крапивный punk knot
сучок, круглый round knot
сучок, лапчатый branched knot, wing knot
сучок, несросшийся dead knot
сучок, открытый visible knot
сучок, плотный sound knot, tight knot
сучок, продолговатый splay knot
сучок, роговой resinous splay knot
сучок, сросшийся live knot, intergrown knot
сучок, сшивный splay knot
сучок, табачный punk knot
сучья 1. branches 2. knots
сушение drying, seasoning (see also сушка)
сушеница, лесная *Gnaphalium sylvaticum* L.
сушилка drier, kiln
сушилка, вентиляционная ventilating kiln
сушилка, высокочастотная high-frequency kiln
сушилка, ежекционная recirculating external-fan kiln
сушилка, конвекционная convection kiln
сушилка, конденсационная condenser kiln
сушилка, контактная contact drier
сушилка непрерывного действия progressive kiln
сушилка периодического действия compartment kiln
сушилка, роликовая roller drier
сушилка, туннельная progressive kiln, tunnel kiln
сушильня drier, kiln
сушка drying, seasoning
сушка, атмосферная air drying
сушка, биологическая 'biological drying' of trees (see подвялывание)
сушка, воздушная air drying
сушка, естественная air drying
сушка, искусственная artificial (usually kiln) drying
сушка, камерная kiln drying
сфагнум bog moss, *Sphagnum* L.
сфагнум, болотный *Sphagnum palustre* L.

счет account, bill, calculation
счетник, присяжный tallyman
счетовод book-keeper
счетчик meter, counter
съем листов шпона veneer take-off
съемка survey
съемщик surveyor
сырец raw material
сырой **1.** raw **2.** green **3.** wet
сырье raw material
сыч owl
СЭВ Comecon

Т

таблица table
таблица высот height table
таблица, массовая volume table
таблица, местная массовая local volume table
таблица, объемная volume table
таблица объемов бревен log volume table
таблица, опытная yield table
таблица производительности performance table
таблица сбега taper table
таблица сбежистости taper table
таблица с двумя входами double-entry table
таблица с одним входом single-entry table
таблица, сортиментная single-tree assortment table
таблица, товарная stand assortment table
таблица хода роста yield table
таволга *Spiraea* L.
таволга, вязолистная meadow-sweet, *Spiraea ulmaria* L.
таволга, зверобойнолистная *Spiraea hypericifolia* L.
таволга, иволистная *Spiraea salicifolia* L.
таволга, средняя *Spiraea media* Schm.
таежный taiga
тайга taiga
тайга, темнохвойная 'dark' coniferous taiga (spruce and/or fir)
тайга, черневая 'dark' coniferous taiga (spruce and/or fir)
тайник twayblade, *Listera* R.Br.
тайнобрачные cryptogams
такелаж tackle, cable
такса **1.** statutory price, fixed price **2.** dachshund
такса, лесная statutory (prescribed) stumpage price
таксатор enumeration officer, cruiser, working-plan officer
таксацировать to cruise, to appraise
таксация inventory, enumeration, cruising, appraisal, mensuration
таксация, лесная forest inventory, mensuration
таксация, перечислительная enumeration, enumeration survey
таксономия taxonomy
такыр takyr (flattish clayey area in arid Central Asia, with severe surface cracking in summer)
тал purple osier, purple willow, *Salix purpurea* L.
талина purple osier, purple willow, *Salix purpurea* L.
талреп turnbuckle
талый thawed, due to thawing
тальвег thalweg, valley line
тальник **1.** purple osier, purple willow, *Salix purpurea* L. **2.** scrub trees
тальреп turnbuckle
тамарикс tamarisk, *Tamarix* L. (see гребенщик)
таннид tannin
таннидоносность tannin content
таннин tannin
тара **1.** tare **2.** packing, container, box shook
таран ram, battering ram
таранный (see пропил, таранный)
тарбаган Siberian marmot, *Marmota sibirica* Radde
тарировка calibration
тариф tariff, tariff table
тарник boxwood assortment

тасбиюргуп *Nanophyton erinaceum* Pall.
тасва-арча Turkestan juniper, *Juniperus turkestanica* Kom.
тахина tachinid
тачка wheelbarrow
таяние thaw, thawing
таять to thaw, to melt
твердолиственный hard-wooded broadleaved
твердость hardness
текстура figure, grain, texture
текучесть flow, fluidity
телега cart, wagon
тележка small truck, trolley, carriage
тележка, впередирамная frame-saw log carriage
тележка, захватывающая saw carriage, log carriage
тело body
тело, плодовое fruiting body
темнина incipient decay, dark heart
темнохвойный 'dark' coniferous (spruce, fir)
температуропроводность diffusivity
теневой shady, shaded
теневыносливость shade tolerance
теневыносливый shade-tolerant, shade-bearing
тенелюбивый shade-loving
тень shade
теплица greenhouse, hothouse
тепло heat, warmth
тепловоз diesel-engined vehicle
теплоемкость heat capacity, specific heat
теплолюбивый heat-loving, thermophilous
теплопроводимость thermal conductivity
теплопроводность thermal conductivity
теплостойкость heat resistance
теплота heat, warmth
теплоход diesel-engined vehicle
теребление removal of useful assortments (from felled trees in shifting cultivation)
ТерКСОП territorial integrated schemes for nature conservation
термит termite
термолиз thermolysis, dry distillation
терн blackthorn, sloe, *Prunus spinosa* L.
тернистый thorny
терновик blackthorn, sloe, *Prunus spinosa* L.
терновлив bullace, *Prunus institia* L.
терпен terpene
терраса terrace
террасирование terracing
тес softwood boards
тесание hewing
тесать to hew
теска hewing
тесло adze
тесьма, мерная measuring tape
тетерев black grouse, *Lyrurus tetrix* L.
тетерев-косач black grouse, *Lyrurus tetrix* L.
тетеревиные Tetraonidae (grouse, ptarmigan, capercaillie, hazel hen)
техника technique, technology, equipment
техника, агрегатная multi-function logging machinery
техника безопасности accident prevention
техника валки felling technique
техникум, лесной forestry technical school
технологическая щепа industrial chips
технология technology
течение current, flow
течение, пластическое plastic flow
течка rut, rutting
тик teak, *Tectona grandis* L.
тилла tylosis
тиллоид tylosoid
тимофеевка timothy, *Phleum pratense* L.
тимьян thyme, *Thymus* L.
тимьян, обыкновенный wild thyme, *Thymus serpyllum* L.
тип type
тип леса forest type
тип почвы soil type
тип условий произрастания type of site conditions
типограф *Ips typographus* L.

типология, лесная forest typology, study of forest types
типчак sheep's fescue, *Festuca ovina* L.
тиски vice, vise, clamp
тисс yew, *Taxus* L.
тисс, восточный Japanese yew, *Taxus cuspidata* Sieb. et Zucc.
тисс, европейский common yew, *Taxus baccata* L.
тисс, ягодный common yew, *Taxus baccata* L.
ткань, клеточная cellular tissue
ткань, образовательная meristem
ткань, палисадная palisade tissue
ткань, проводящая conducting tissue
ткань, столбчатая palisade tissue
ткач pamphiliid (see also пилильщик-ткач)
тли Aphididae, aphids
тля aphid
ТМАУ peat/mineral fertilizer
ТММ thermo-mechanical pulp, TMP
товар merchandise, goods
товар, бессортный unsorted goods
товар, бракованный rejected goods
товар, пиленый sawn timber
товар, строганный planed timber
товарность древостоя merchantability of a stand (classified by the % timber in the total standing volume)
товарный merchantable, marketable
ток current, flow
токарь turner, lathe operator
токсичность toxicity
толокнянка bearberry, *Arctostaphylos uva-ursi* (L.) Spreng.
толстомер large timber, tree(s) of large diameter
толщина diameter, thickness (see also диаметр)
толщиномер thickness gauge
томасшлак Thomas slag, basic slag
томбуй buoy, float
тонкомер wood of small diameter, tree(s) of small diameter
тонконог *Koeleria* Pers.
тонкослойный narrow-ringed
тонна metric ton (1000 kg)
топка **1.** furnace **2.** heating, firing
топливо fuel
топляк sinker, sunken wood, sunken timber
топняк sinker, sunken wood, sunken timber
тополевник poplar forest
тополь poplar, *Populus* L.
тополь, бальзамический balsam poplar, *Populus balsamifera* L. (eastern balsam poplar, *P. tacamahaca* Mill.)
тополь, белый white poplar, *Populus alba* L.
тополь Болле Bolle's poplar, *Populus alba* var. *pyramidalis* L. (Bge.) (*P. bolleana* Lauche)
тополь, душистый fragrant poplar, *Populus suaveolens* Fisch.
тополь, канадский 'Canadian' poplar, *Populus deltoides* Marsh (but including also various hybrid black poplars)
тополь, китайский Chinese poplar, *Populus simonii* Carr.
тополь, лавролистный laurel-leafed poplar, *Populus laurifolia* Ldb.
тополь Максимовича Maximovich poplar, *Populus maximowiczii* Henry
тополь, "мощный" robusta poplar, *Populus 'Robusta'*
тополь, пирамидальный Lombardy poplar, *Populus nigra* var. *italica* L. (Duroi) (*P. pyramidalis* Roz.)
тополь, разнолистный *Populus diversifolia* Schrenk
тополь, серебристый white poplar, *Populus alba* L.
тополь, сереющий grey poplar, *Populus canescens* L.
тополь, серый grey poplar, *Populus canescens* L.
тополь, сизолистный blue poplar, *Populus pruinosa* Schrenk
тополь, сизый blue poplar, *Populus pruinosa* Schrenk
тополь, черный black poplar, *Populus nigra* L.
топор axe
топорик hatchet
топорище helve

топорник 1. axe-man 2. small roundwood (cut by axe)
топь fen, swamp
топяный fen, swampy, fenny
торговля trade
торговля, внешняя foreign trade
торговля, внутренняя home trade
торговля, лесная timber trade
торец 1. end, end grain 2. paving block
торможение 1. braking 2. inhibition
торус torus
торф peat
торф, верховой bog peat, sphagnum peat
торфование incorporation of peat (e.g. in sand)
торфо-минералное удобрение peat/mineral fertilizer
торфяник peat bog
торцевание 1. cross-cutting to length 2. truing up the ends of a log bundle
торцевать to cross-cut to length
торце-выравнивание truing up log ends, squaring off log ends
торцовка cross-cutting to length
торчок 1. stump plant 2. plant browsed or burned back to stump 3. branch stub
точение 1. turning 2. sharpening
точеный 1. turned 2. sharpened
точильщик anobiid, furniture beetle
точильщик, домовый *Anobium pertinax* L.
точильщик, мебельный common furniture beetle, *Anobium punctatum* De G.
точильщик, мягкий *Ernobius mollis* L.
точильщик, пестрый death-watch beetle, *Xestobium rufovillosum* De G.
точить 1. to turn 2. to sharpen
точка 1. point 2. sharpening
точка замерзания freezing point
точка насыщения волокна fibre saturation point
точка пил sharpening saws
точка роста growing point
точка росы dew point
точность accuracy, precision
точный accurate, precise
трава herb, grass
трава альфа *Stipa tenacissima* L.
трава кузьмичева *Ephedra distachya* L.
трава, сорная weed
траверс traverse, cross-piece
траверса traverse, cross-piece
травить 1. to hunt 2. to poison 3. to stain
травление древесиный staining of wood
травосеяние sowing herbaceous species
травоядный herbivorous
травянистый herbaceous, grassy
трактор tractor
трактор, гусеничный crawler tractor
трактор, колесный wheeled tractor
трактор, плавающий amphibious tractor
трактор, трелевочный skidding tractor
тракторист tractor driver
транспирация transpiration
транспорт transport
траншея trench, ditch
трасса route, route-line, power-line ride, right-of-way
трассирование route location, location survey
траурница *Vanessa antiopa* L.
трахеида tracheid
трахеида, вазицентрическая vasicentric tracheid
трахеида, волокнистая fibre tracheid
трахеида, лучевая ray tracheid
трахеида, разъединенная disjunctive tracheid
трахеида, сосудистая vascular tracheid
трахеида, тяжевая strand tracheid
трахей vessel, trachea
требовательный demanding, exacting
трелевать to skid
трелевка skidding
трелевка, аэростатная balloon logging
трелевка, лебедочная winch skidding, cable skidding
трелевка, наземная ground skidding
трелевка, подвесная aerial skidding
трелевка, тракторная tractor skidding
трелевщик skidding tractor, skidder

трение friction
тренога tripod
треножник tripod
трест trust
трехэтажный three-storeyed
трехярусный three-storeyed
трещина shake, split, crack, check
трещина, внутренняя internal shake, internal check
трещина, кольцевая ring shake
трещина, метиковая heart shake (extending the length of the material) (see also метик)
трещина, морозная frost crack
трещина, морозобойная frost crack
трещина, отлупная ring shake
трещина, поверхностная surface check
трещина, сердцевинная heart shake
трещина, солнечная sun crack
трещина, солнцепечная sun crack
трещина усушки check, shrinkage check
трещиноватый fissured (of bark)
трилистник clover, trefoil, *Trifolium* L.
тропа path, track
трос cable, rope, line (see also канат)
тростник, обыкновенный *Phragmites communis* Trin.
трость stick
трость, почвенная soil probe
трость, таксатора measuring stick (for diameter, height, plot demarcation, etc.)
трофность nutrient status, nutrient requirement
труба pipe, tube, duct
трубка pipe, tube, duct
трубка, млечная latex tube
трубка, ситовидная sieve tube
трубковерт attelabid, leaf-roller
трубковерт, дубовый *Attelabus nitens* Scop.
трубковерт, орешниковый *Apoderus coryli* L.
трубковерт, тополевый *Byctiscus populi* L.
трубочка small tube
трубочка, пыльцевая pollen tube
труд work, labour
трудодень work-day (as a unit of work)
трудоемкий laborious, labour-consuming
трут **1.** tinder **2.** fungus (see also гриб, губка)
трутовик **1.** fungus (see also гриб, губка) **2.** member of Polyporaceae
трутовик, дубовый корневой *Polyporus dryadeus* Fr.
трутовик, дубовый ложный *Fomes robustus* Karst.
трутовик, дуболюбивый *Polyporus dryophilus* Bark.
трутовик, ложный *Fomes igniarius* Gill.
трутовик, настоящий *Fomes fomentarius* Pill.
трутовик, окаймленный *Fomes pinicola* (Gill.) Fr.
трутовик, осиновый *Phellinus tremulae* B. et B.
трутовик-печеночник *Fistulina hepatica* Fr.
трутовик, северный *Polyporus borealis* Fr.
трутовик, серно-желтый *Polyporus sulphureus* Fr.
трутовик, чешуйчатый *Polyporus squamosus* Fr.
трутовик Швейница *Polyporus schweinitzii* Fr.
трутовик, щетинистый ясеневый *Polyporus hispidus* Fr.
труха rotten wood
трухлявый rotten
трюм ship's hold
трясина quaking bog, swamp
трясогузка wagtail, *Motacilla* L.
тугай riparian woodland
туземный native, indigenous
туман mist, fog
тумервал chaining (as a means of felling)
тунг tung, *Aleurites* Forst.
тундра tundra
тундра, кочковатая hummocky tundra
тундра, кустарниковая shrub tundra
тундра, моховая moss tundra
тундра, ягельная lichen tundra
тургор turgor

турча, болотная water violet, *Hottonia palustris* L.
тут mulberry, *Morus* L. (see also шелковица)
тутовик mulberry, *Morus* L. (see also шелковица)
тушение extinguishing
тушить пожар to extinguish a fire
тушканчик jerboa (Dipodidae)
туя arbor-vitae, *Thuja* L.
туя, восточная Oriental arbor-vitae, *Thuja orientalis* L.
тын paling
тысячелистник, обыкновенный yarrow, milfoil, *Achillea millefolium* L.
тычина stake, thin pole
тычинка stamen
тюлька sleeper length, sleeper timber
тяга **1.** pull, traction **2.** connecting rod **3.** roding
тягач tractor, truck-tractor, prime mover
тягач, колесный wheeled skidder
тяж strand
тяжесть gravity, weight, load
тяпка chopper, cleaver

У

ублюдок hybrid
уборка **1.** harvest, removal **2.** arranging
убыль decrease
убыток loss, damage
увлажнение moistening
увядание wilt, wilting
углевод carbohydrate
углежег charcoal burner
углежжение charcoal burning
углекислота carbon dioxide, carbonic acid
углерод carbon
угломер goniometer
угнетать to suppress
угнетение suppressing, suppression
угнетенность suppression
угнетенный suppressed
угодья lands
угодья, лесные forest lands
угодья, пахотные arable lands
угол angle
угол врастания (branch) insertion angle
угол встречи angle of incidence (between direction of tool travel and grain)
угол, задний clearance angle
угол заострения sharpness angle
угол заточки sharpness angle
угол карры face angle
угол косой заточки angle of bevel
угол обжима pressure-bar angle
угол, передний **1.** hook (of saw) **2.** cutting angle (of rotary cutter)
угол поворота sweep-back angle (of chipper knives)
угол подновки angle of streak (freshening)
угол резания complement of hook or cutting angle (i.e. clearance angle plus sharpness angle)
уголь coal
уголь, древесный charcoal
уголь-сырец lump charcoal
удар impact, blow, shock
удельный specific
удобрение **1.** fertilizer, manure **2.** fertilizing, manuring
удобрение, зеленое green manure
удобрение, минеральное mineral fertilizer
удобрять to manure, to fertilize
удод hoopoe, *Upupa epops* L.
удушье сеянцев smothering of seedlings (e.g. by *Thelephora terrestris* Ehr.)
ужовник, обыкновенный adder's tongue, *Ophioglossum vulgatum* L.
узел **1.** node **2.** junction, centre **3.** unit
узик common tormentil, *Potentilla erecta* (L.) Räusch.
узкоколейка narrow-gauge line
узкослойный narrow-ringed, narrow-layered
узор figure, grain
узорчатый figured
укатывать to roll
укладка **1.** stacking **2.** packing
укладка листов шпона veneer lay-up

укладчик stacker
укладывать to stack, to pile
уклон inclination, gradient, slope
уклономер clinometer
уклон пилы hang, overhang (in frame saw)
укоренение taking root, rooting
укореняться to take root, to root
укрепление fixing, strengthening, stabilization
укрепление берега bank stabilization
укрепление дюн dune fixing
уксусник staghorn sumac, *Rhus typhina* L.
улей beehive
улитка snail
УЛТИ Urals Forestry Institute
ультразвук ultrasonics, ultrasound
умеренный temperate, moderate
унаби common jujube, *Zizyphus jujuba* Mill.
унаваживание manuring
уничтожение destruction
упаковка packing
упил increase in stacking density caused by sawing the wood into shorter lengths
уплотнение compaction, condensation
упорядоченность (целлюлозы) state of order (of cellulose)
управление administration, management, control
управление, лесное forest administration
управление, устойчивое sustainable management
упрессовка degree of compression
упругость elasticity
упряг saw-shift, period of work between saw-sharpenings
упряжь harness
ураган storm, hurricane
урема riparian woodland, bottom-land forest
урман dense taiga forest (usually spruce or fir)
уровень level
уровень воды water level
уровень грунтовой воды groundwater level
уродливый deformed
урожай harvest, crop
урожай древесины timber harvest
урожай семян seed harvest
урожайность **1.** fruitfulness, yield **2.** seed yield
урожайный fruitful, productive
урочище area demarcated by natural boundaries
урюк dried apricot
урюк-арча Turkestan juniper, *Juniperus turkestanica* Kom.
ус **1.** spur **2.** antenna **3.** tendril **4.** torn or broken grain **5.** first freshening of the season
ус, лесовозный spur road, spur railway
усадка shrinkage
усадьба farm, estate
усач cerambycid, longhorn beetle
усач, домовый черный house longhorn, *Hylotrupes bajulus* L.
усач, дубовый (большой) *Cerambyx cerdo* L.
усач, еловый бурый *Tetropium castaneum* L.
усач, еловый черный (большой) *Monochamus urussovi* Fisch.
усач, еловый черный (малый) *Monochamus sutor* L.
усач, осиновый (see скрипун)
усач, тополевый large poplar borer, *Saperda carcharias* L.
усач, черный сосновый *Monochamus galloprovincialis* Ol.
усваивать to assimilate
усвоение assimilation
усвояемый assimilable, available
усик **1.** antenna **2.** tendril
усилие резания cutting force
ускоритель log or bundle mover, current-maker
ускоритель, барабанный drum-type mover
ускоритель, гладкотросовый bare-cable mover
условия conditions

условия, внешние environmental conditions
условия произрастания site conditions, growth conditions
условный conventional, provisional
устав regulations, rules
усталость fatigue
установка 1. installation, plant 2. setting, placing
установка, аэростатная трелевочная balloon logging installation
установка, воздушная трелевочная aerial cableway, skyline
установка, деревопропиточная timber impregnation plant
установка, подвесная aerial cableway, skyline
установка, энергохимическая installation for producing power and chemicals (from wood)
устой pier
устойчивость resistance, stability
устойчивый resistant, stable, sustainable
устройство organization, arrangement
устьице stoma
усушка shrinkage
усыхание 1. drying out, dying off, dieback 2. shrinking
усыхать 1. to dry out, to die off, to dieback, to wilt 2. to shrink
утка duck
утолщение thickening, swelling
утомление почвы soil exhaustion
утоп sinker, sunken wood, sunken timber
утренник morning frost
ухо медвнжье Aaron's rod, *Verbascum thapsus* L.
уход tending (usually by felling - see рубка ухода)
уход за запасом Vorratspflege, tending of the growing stock
уход за культурой tending a plantation
уход за лесом forest tending
уход за почвы soil cultivation, tillage
уход за приростом later thinnings (see рубка, проходная)
уход за составом cleaning (see прочистка)
уход за формы ствола и кроны thinning in younger stands (see прореживание)
участок section, plot, district
участок, комплексный subcompartment
участок, лесной site (any small area in forest)
участок, мастерский 1. section run by a logging foreman 2. logging camp
участок, мелкоконтурный small area
участок, семенный seed orchard, seed stand
участок, таксационный subcompartment
участок, элементарный subcompartment
учение instruction, teaching
учет 1. measurement, survey, inventory, calculation, tally 2. accounting
учет лесного фонда forest inventory
учет по весу measurement by weight
учет по обьему measurement by volume
учет, последовательный sequential sampling
учет по числу вревен log tally
учет, штучный individual log tally
учреждение establishment, institution
ушан long-eared bat, *Plecotus auritus* L.
ущелье gorge, ravine

Ф

фабрика factory
фабрика, бумажная paper factory
фабрика, мебельная furniture factory
фабрика, фанерная plywood factory, veneer factory
фабрикат manufactured product
фаза phase
фазан, обыкновенный pheasant, *Phasianus colchicus* L.
файрвуд board offcuts, firewood
фальц rebate, groove
фальцгобель rebating plane, grooving plane
фальцевание rebating, grooving
фанера 1. plywood 2. veneer

фанера, венированная veneered plywood
фанера, клееная plywood
фанера, лущеная peeled veneer
фанера, многослойная 1. plywood 2. laminated sheet
фанера, ножевая sliced veneer
фанера, облицовочная face veneer
фанера-переклейка plywood
фанера, пиленая sawn veneer
фанера, рядовая standard plywood
фанера, строганная sliced veneer
фарватер fairway
фасад façade
фаска edge, bevel, face
фауна fauna
фаут defect
фаутность defectiveness
фаутный defective
фашина fascine, faggot
фашинник fascine wood, faggot wood
феллем cork, phellem
феллоген cork cambium, phellogen
фенотип phenotype
феноформ phenological form
ферма 1. farm 2. girder
фестметр solid cubic metre
фиалка violet, *Viola* L.
фибра fibre
фибрилла fibril
фибролит woodchip/cement material
фига fig, *Ficus carica* L.
филенка panel, panelling
филенчатый panelled
филин eagle-owl, *Bubo bubo* L.
филлофаг phyllophage, defoliator
фильтрация filtration, seepage
финик date, *Phoenix* L.
финик, лесной wild date palm, *Phoenix sylvestris* Roxb.
финик, финиконосный date palm, *Phoenix dactylifera* L.
фисташка pistachio, *Pistacia* L.
фисташка, настоящая pistachio, *Pistacia vera* L.
фисташка, туполистная *Pistacia mutica* F. et M.
фисташник pistachio woodland
фитолин 'phytolin' (substance excreted by a plant)
фитонцид phytoncide, antibiotic derived from plants
фитоценоз phytocoenosis
флора flora
флористический floristic
флоэма phloem
фон background, previous cultural measures, the existing situation
фонд fund, stock
фонд, лесной forest resource, total forest area (see Гослесфонд)
фонд, лесокультурный area in need of silvicultural measures
фонд, лесосечный annual volume assigned for cutting
фонды funds, capital, assets
фонды, основные fixed (basic) capital
фонды, производственные productive capital
форвардер forwarder
форма 1. shape, form 2. mould
форма зуба tooth shape
форма крон crown form
форма леса stand constitution, stand storey structure
форма насаждения stand constitution, stand storey structure
формальдегид formaldehyde
формирование formation, shaping
формоизменяемость dimensional instability
формула formula
формула обьема ствола stem volume formula
форсунка sprayer
фосфат phosphate
фосфобактерин phosphobacterin (preparation of *Bacillus megatherium* var. *phosphaticum* used as bacterial fertilizer)
фосфор phosphorus
фотограмметрия photogrammetry
фотография photography
фотосинтез photosynthesis
фотосхемы aerial photo mosaics
фотосъемка photography

франко борт free on board
фрахт freight
фреза 1. moulding cutter 2. rotary cultivator
фреза, концевая dovetail bit
фреза, пазовая slot-mortising bit, grooving cutter
фреза, почвенная rotary cultivator
фреза, фасонная profile cutter
фреза, шпунтовая tonguing cutter
фрезер spindle moulder, moulding machine
фрезерование 1. rotary cutting, spindle moulding 2. rotary cultivating
фрезование, продольное peripheral milling
фрезовать to rotary-cultivate, to spindle-mould
фрезовка 1. rotary cutting, spindle moulding 2. rotary cultivating
фриза parquet strip, parquet board
фуга joint, glue-line
фуганок plane, planer, jointing plane
фугование jointing, planing
фуговать to joint, to plane
фуговка jointing, planing
фумигация fumigation
фунгицид fungicide
фундук 1. filbert, *Corylus maxima* Mill. 2. cultivated variety of hazel
фураж fodder, feed, forage
фурфурол furfural
фут foot
фут, бордсовый board foot
фут, досковый board foot
фут, кубический cubic foot
футляр cabinet, case

X

хак hack (as used in freshening)
хак, химический hack that applies chemical stimulant
халзан golden eagle, *Aquila chrysaetos* L.
халькограф *Pityogenes chalcographus* L.
харвестер harvester
хвоевертка *Evetria* Hb. (*Rhyacionia*) Hb. (see also побеговьюн)
хвоеотделитель conifer foliage stripper
хвоинка needle
хвойник 1. ephedra, *Ephedra* L. 2. young conifer forest
хвойные Coniferae, conifers
хвойный conifer, needle-leafed, coniferous
хворост brushwood
хворостина long rod, switch
хвост tail
хвост верблюжий *Caragana jubata* Poir.
хвостовик drive lug, drive teeth (chainsaw tooth)
хвощ horsetail, *Equisetum* L.
хвощ, болотный *Equisetum palustre* L.
хвощ, лесной *Equisetum sylvaticum* L.
хвощник area of *Equisetum*
хвоя needle-foliage of conifers
хедриг chipping headrig, Chip-N-Saw type of unit
хермес adelgid
хермес, елово-лиственничный *Adelges laricis* Vall.
хермес, желтый *Adelges abietis* L.
хермес, зеленый *Adelges viridis* Ratz.
химера chimaera
химия, лесная 'forest chemistry', wood chemistry (including chemical utilization, resin tapping etc.)
химлесхоз leskhoz specializing in resin tapping
хищник predator
хищный predatory
хлам, лесной slash, logging residue, felling debris
хлопушка bladder campion, *Silene cucubalus* Wibel
хлоропласт chloroplast
хлорофилл chlorophyll
хлыст 1. tree-length log, long log 2. long shoot (for cuttings)
хмелеграб hop-hornbeam, *Ostrya* Scop.
хмелеграб, обыкновенный European hop-hornbeam, *Ostrya carpinifolia* Scop.

хмель hop, *Humulus lupulus* L.
хмыз brushwood
ход passage, course, movement
ход короеда barkbeetle tunnel
ход пилы stroke of a (frame) saw
ход роста course of growth, yield
ход, смоляной resin duct, resin canal
ходовой merchantable
ходуля stilt
хозрасчет business accounting, profit and loss accounting, cost accounting
хозсекция working circle
хозчасть major management unit (e.g. working circle)
хозяин host
хозяин, промежуточный intermediate host
хозяйственный management, economic
хозяйство 1. management unit, e.g. felling series, 'crop-type' (group of similar stands) 2. husbandry, management system
хозяйство, безвершинное pollarding
хозяйство, выборочное selection system
хозяйство, высокоствольное high-forest system
хозяйство, лесное forestry, forest management
хозяйство, непрерывное sustained management
хозяйство, низкоствольное coppice system
хозяйство, охотничье game management
хозяйство, переходное conversion system
хозяйство, периодическое intermittent exploitation
хозяйство, подсечное lopping
хозяйство, подсочное management for resin
хозяйство, постоянное sustained management
хозяйство, сельское farming, agriculture
хозяйство, семенно-лесосечное shelterwood system
хозяйство, сплошно-лесосечное clear-felling system
хозяйство, среднее coppice-with-standards system
холм hill
холмик mound, hillock
холмистый hilly
холод cold
холодостойкость cold hardiness, cold resistance
хомут ring, collar
хомяк hamster, *Cricetus cricetus* L.
хорек, лесной polecat, *Mustela putorius* L.
хорек, черный polecat, *Mustela putorius* L.
хохлатка notodontid
ХПК COD, chemical oxygen demand
хранение 1. preservation, conservation 2. storage
хребет 1. mountain range, ridge 2. spine
хромание пилы irregular running of circular saw
хромосома chromosome
хронометраж timing, time study
хрупкий brittle, fragile
хрущ chafer
хрущ, волосатый *Anoxia pilosa* F.
хрущ, июльский *Polyphylla fullo* L.
хрущ, июньский summer chafer, *Amphimallon solstitialis* L.
хрущ, майский *Melolontha* L.
хрущ, майский (восточный) *Melolontha hippocastani* F.
хрущ, майский (западный) cockchafer, *Melolontha melolontha* L.
хрущ, мраморный *Polyphylla fullo* L.
хрущ, пестрый *Polyphylla fullo* L.
хрущик, садовый garden chafer, *Phyllopertha horticola* L.
хрящ gravel
ХТММ CTMP, chemi-thermomechanical pulp
хурма persimmon, *Diospyros* L.
хурма, кавказская date plum, *Diospyros lotus* L.
хурма, обыкновенная date plum, *Diospyros lotus* L.

Ц

царга frame
царство, растительное vegetable kingdom
ЦБП pulp-and-paper industry
цвести to flower, to bloom
цвет 1. flower 2. colour
цветение flowering, blooming
цветок flower
цветоложе receptacle
цветоножка flower stalk, pedicel, peduncle
цветонос flower stalk, pedicel, peduncle
целина virgin land, virgin soil
целлолигнин lignocellulose
целлюлоза 1. cellulose 2. chemical pulp
целлюлоза, натронная soda pulp
целлюлоза, сульфатная sulphate pulp
целлюлоза, сульфитная sulphite pulp
Целлюлозаимпэкс Pulp Import/Export Agency
целный whole, entire
целостность древесины structural integrity of wood
цель, хозяйственная object of management
цемент, древесный wood cement
цена price, value
цена, договорная contract price
цена, корневая stumpage value
цена, оптовая wholesale price
цена, отпускная selling price, transfer price
цена, попенная stumpage value
цена, продажная selling price
цена, рыночная market price
цена, сдаточная delivery price
цена, таксовая stumpage price
ценность value
ценоз coenosis
центнер 100 kg
центробежный centrifugal
центровка centering (logs on a lathe)
центр тяжести centre of gravity
цепочка просветов pore chain
цепь chain
цех workshop, shop
цех, раскройный re-saw shop (for producing dimension stock)
цикл cycle
цикл, жизненный life cycle
циклевание scraping
цикля scraper blade
цикута cowbane, *Cicuta virosa* L.
цилиндр cylinder, roller, drum
циркуль 1. caliper 2. pair of compasses
циркулярка circular saw
ЦНИИЛГиС Central Research Institute for Forest Genetics and Selection
ЦНИИЛесосплава Central Research Institute for Timber Floating
ЦНИИЛХ Central Research Institute for Forestry
ЦНИИМОД Central Research Institute for Mechanical Conversion of Wood
ЦНИИМЭ Central Research Institute for Mechanization and Electrification in Forest Exploitation
ЦНИЛХИ Central Research Institute for Forest Chemistry
цоколь socle, foundation
ЦЧО Central Chernozem Region
ЦЧР Central Chernozem Region

Ч

чабрец thyme, *Thymus* L. (see also тимьян)
чага *Inonotus obliquus* (Fr.) Pil.
чай tea (see also Иван-чай)
чай, луговой creeping jenny, *Lysimachia nummularia* L.
чан vat, tank
чаран forest-steppe 'landscape' in Yakutia
частица particle
частное quotient
частокол paling, palisade
частота frequency
частота, высокая high frequency
часть part, portion, fraction
часть, составная component, constituent
часть, хозяйственная major management unit, e.g. working circle

чахлый 1. stunted 2. weakly
чашелистник sepal
чашечка calyx
чаща thicket
чеглок hobby, *Falco subbuteo* L.
челен raft section
челнок shuttle
челюстник jaw (grab) loader
челюсть jaw
червец coccid, scale insect
червец, буковый felted beech coccus, *Cryptococcus fagi* Bär.
червец, лаковый lac insect, *Laccifer lacca* Kerr.
червоточина worm-hole
червоточный wormeaten
червь worm
червь, дождевой earthworm
червь, круглый nematode
червь, проволочный wireworm, elaterid larva
червь, шелковичный silkworm
червяк worm
чередование alternation, rotation
черемуха bird-cherry, *Prunus padus* L.
черемша ramsons, *Allium ursinum* L.
черенок cutting, scion
черенок, деревянистый woody cutting
черенок, зеленый green cutting
черенок, корневый root cutting
черенок, листовой leaf cutting
черенок, побеговый shoot cutting
черешня wild cherry, *Prunus avium* L.
черешок petiole
черешчатый petiolate
черкез *Salsola* L., usually *Salsola richteri* Karel.
черневый taiga, coniferous
черника *Vaccinium* L., usually bilberry, whortleberry, *Vaccinium myrtillus* L.
черника, кавказская *Vaccinium arctostaphylos* L.
черничник area of *Vaccinium myrtillus* L. (see бор-черничник)
чернобыльник mugwort, *Artemisia vulgaris* L.
чернозем chernozem
черноклен Tatarian maple, *Acer tataricum* L.
чернокорень, лекарственный hound's-tongue, *Cynoglossum officinale* L.
чернолесье deciduous forest(s), moist deciduous forest in Central Asia
чернолоз bay willow, *Salix pentandra* L.
чернопихтарник Far Eastern fir forest type (*Abies holophylla*)
чернотал bay willow, *Salix pentandra* L.
черта 1. line, feature 2. container recess (resin tapping)
чертополох thistle, *Carduus* L.
четвертина quarter, log quarter
четвертинник quarter, log quarter
чехлик, корневой root cap
чечевичка lenticel
чешуекрылые Lepidoptera
чешуйка scale
чешуя scale
чилига dwarf pea tree, *Caragana frutex* C. Koch
чина, болотная *Lathyrus palustris* L.
чина, лесная *Lathyrus sylvestris* L.
чинар plane, *Platanus* L. (see also платан)
чингиль salt tree, *Halimodendron halodendron* Voss
чинить to repair, to mend
число number
число, видовое form factor
число полнодревесности form factor
число сбега diameter quotient
чистец *Stachys* L.
чистец, болотный *Stachys palustris* L.
чистец, лесной *Stachys sylvatica* L.
чистка cleaning
чистота cleanness, clearness, purity, smoothness
чистотел greater celandine, *Chelidonium majus* L.
чистяк, лютичный lesser celandine, *Ranunculus ficaria* L.
чищоба land cleared by shifting cultivation
член 1. member 2. term
чогон *Salsola subaphylla*

чозения, крупночешуйчатая chosenia, *Chosenia macrolepis* Kom.
чокер choker
чокеровщик chokerman
чубушник mock orange, *Philadelphus* L.
чувствительность susceptibility
чужеземный foreign, exotic
чумаруту pulpwood assortment (for export to Japan)
чурак log, bolt, veneer bolt
чурбан block
чурка small block
чурка, генераторная small block for making producer gas

Ш

шаблон pattern, stencil, gauge
шаблон, разводной saw set, setting gauge
шаблон, угловый angle gauge
шаг 1. pace, step **2.** pitch
шаг зуба tooth pitch
шаг подновки streak interval, distance between freshenings
шаг посадки plant spacing
шагомер pedometer
шадрик wood ash, potash
шайба 1. disk **2.** collar (of circular saw)
шайба, зажимная collar (of circular saw)
шалевка board (1 in (= *c.* 2.5 cm) thick)
шалфей sage, *Salvia* L.
шампиньон common mushroom, *Agaricus campestris* L.
шандора sluice gate, dam beam
шарошка moulding cutter
шаяние burning of sward layers and humus
швелевание low-temperature distillation
шейка neck
шейка, корневая root collar
шейки, рачьи snake-root, *Polygonum bistorta* L.
шейма anchor cable, anchor chain
шелк silk
шелковица mulberry, *Morus* L.
шелковица, белая white mulberry, *Morus alba* L.
шелковица, красная red mulberry, *Morus rubra* L.
шелковица, черная black mulberry, *Morus nigra* L.
шелкопряд, дубовый походный *Thaumetopoea processionea* L.
шелкопряд, златогузка brown-tail moth, *Euproctis chrysorrhoea* L.
шелкопряд, ивовый *Leucoma salicis* L.
шелкопряд, кольчатый lackey moth, *Malacosoma neustria* L.
шелкопряд, лунчатый *Selenephera lunigera* Esp.
шелкопряд -монашенка nun moth, *Lymantria monacha* L.
шелкопряд, непарный gypsy moth, *Lymantria dispar* L.
шелкопряд, сибирский *Dendrolimus sibiricus* Tschv.
шелкопряд, сосновый *Dendrolimus pini* L.
шелкопряд, сосновый походный *Thaumetopoea pinivora* Tr.
шелкопряд, тутовый *Bombyx mori* L.
шелкопрядник forest attacked by *Dendrolimus* (usually *D. sibiricus*)
шеллак shellac
шелуха husk, hull
шелюга желтая willow, *Salix daphnoides* Vill.
шелюга, красная willow, *Salix acutifolia* Willd.
шелюгование planting willow (to fix sands)
шероховатость roughness
шерсть, древесная wood wool
шерхебель rough plane, jack plane
шершень hornet, *Vespa crabro* L.
шест rod, pole
шестерня gear
шибляк 'shiblyak' (deciduous shrub association)
шикша *Empetrum nigrum* L.
шило 1. awl **2.** waney-faced edge of the outer board(s) in sawing through and through
шина 1. tyre **2.** bar

шина, пильная guide bar (of power saw)
шип 1. thorn 2. tenon
шип, зубчатый finger (in finger joint)
шиповник rose, *Rosa* L.
ширина width, breadth
ширина годичного слоя width of annual ring
ширина колеи track width
широколиственный leafy, broadleaved
широкослойный wide-ringed, coarse-ringed
широта latitude
шишечка conelet
шишка cone
шишкосушильня cone-drier
шишкоягода 'berry' of juniper
шкала scale
шкала горимости fire-danger scale
шкала пожарной опасности fire-danger scale
шкант dowel, pin, peg
шкаф cabinet, case
шкив pulley
школа school
школа, высшая high school, 'college'
школа, древесная transplant nursery, transplant section of nursery
школа, лесная transplant nursery, transplant section of nursery
шкура skin, hide
шкурка abrasive paper, sandpaper
шкурка, наждачная emery cloth
шкурка, стеклянная sandpaper
шлаговка fastening a raft together with tackle
шлейф-борона drag harrow
шлем helmet, head-guard
шлемник, обыкновенный skull-cap, *Scutellaria galericulata* L.
шлифование 1. sanding, grinding 2. polishing
шлифовать 1. to sand, to grind 2. to polish
шлифовка 1. sanding, grinding 2. polishing
шлихтование smoothing, dressing
шлюз lock, sluice
шлюпочник assortment used in boat-building
шляпка cap, pileus
шнур cord, string
шнур, мерный measuring cord
шнур, посадный planting cord
шов joint, glue-line
шпала sleeper
шпала, брусковая waney-edged sleeper
шпала, круглая round sleeper
шпала, обрезная square-edged sleeper
шпала, пластинчатая flat (slabbed) sleeper
шпальник sleeper timber
шпангоут rib (of boat)
шпат spat
шпат, полевой feldspar
шпатлевание filling (of wood surfaces)
шпатлевка filler, putty
шпация space between stacked boards in the horizontal plane
шпиговка sowing acorns by pressing into unprepared soil
шпилька 1. dowel, pin, peg 2. small loose knot
шпон veneer, peeled veneer, lamina
шпон-рванина waste veneer scraps
шпонка dowel
шпора spur
шпулька spool, bobbin
шпунт groove, tongue and groove
шпунтование tonguing and grooving
шрам fire scar
штабелер stacker
штабель stack, pile
штакетик fencing wood
штамб stem, clear stem
штамбовый standard (hort.)
штамп stamp, punch
штамповать to stamp, to press, to mould
штанга pole, post
штангенциркуль 1. pole caliper 2. sliding caliper
штивка stowage
штык spade's depth, spit
шуба 1. 'fur coat', muffler effect 2. nurse or accessory species encouraging

straightness and self-pruning of crop trees

шубригель caliper

шурф hole, pit, soil-pit

шырок firewood

шютте, обыкновенное needle-cast caused by *Lophodermium pinastri* Chev.

шютте, снежное snow blight caused by *Phacidium infestans* Karst.

Щ

щавель *Rumex* L.

щавель, малый sheep's sorrel, *Rumex acetosella* agg.

щавель, обыкновенный sorrel, *Rumex acetosa* L.

щан container recess (resin tapping)

щебень rubble, broken stone

щегол goldfinch, *Carduelis carduelis* L.

щелкун elaterid, click-beetle

щелок lye, waste liquor

щелочная почва alkaline soil

щелочность alkalinity

щелочной alkaline

щелочь alkali

щель crack, split

щепа chip(s), particle(s)

щепка chip(s), particle(s)

щетинник bristlegrass, *Setaria* Beauv.

щетка 1. brush 2. group of intergrown dormant buds in wood

щеть, поперечная transverse order (of floating logs)

щипцы pincers, pliers, tongs

щипцы, разводные setting pliers

щит 1. panel, screen, shield 2. load tail-plate (of skidding tractor)

щитовка coccid, scale

щитовник buckler fern, *Dryopteris* Adans.

щитовник, гребенчатый crested buckler fern, *Dryopteris cristata* A. Gr.

щитовник, мужской male fern, *Dryopteris filix mas* L.

щитовник, остистый narrow buckler fern, *Dryopteris spinulosa* Kuntze

щиток bud shield

щуп probe

щур pine grosbeak, *Pinicola enucleator* L.

щучка *Deschampsia caespitosa* (L.) Beauv.

Э

эвкалипт eucalypt, *Eucalyptus* L'Herit.

эдафон edaphon, soil fauna

эдификатор 'edificator' (species playing the main role in forming the environment)

экзот exotic species

эккер cross-staff

эклиметр hypsometer

экология ecology

экосистема ecosystem

эксгаустер exhausting system, dust remover

эксикатор desiccator

эксплуатация utilization, logging, exploitation

экспозиция aspect, exposure

экспорт export(s)

Экспортдрев Roundwood Export Agency

Экспортлес Wood Exporting Agency

эксцентрический eccentric

эктотрофный ectotrophic

эластичность elasticity

элеватор hoist, elevator, lift

электрификация electrification

электропила electric power saw

электропроводность electrical conductivity

элемент element

элемент леса generation, pure even-aged stand

элемент, ситовидный sieve-tube element

элемент, трахеальный tracheary element

эндотрофный endotrophic

энергия energy, power
энергия прорастания germinative energy
энергонасыщенность power/weight ratio
энергосредство power unit
энтомовредитель harmful insect, insect pest
эпидермис epidermis
эпизоотия epizootic
эпифит epiphyte
эрика *Erica* L. (see also вереск)
эрозия erosion
эрозия, овражная serious gully erosion
эрозия, плоскостная sheet erosion
эрозия, поверхностная sheet erosion
эрозия, ручейная rill erosion
эрозия, струйчатая gully erosion
эс S-shaped cleat
эспарто *Stipa tenacissima* L.
эстакада ramp, frame, trestle
этаж storey, layer, stratum (see also ярус)
эталон **1.** standard **2.** sample area
эфедра ephedra, *Ephedra* L.
эфир ether, ester
эффект, парниковый greenhouse effect
эффект, тепловый net thermal effect
эффективность effectiveness, efficiency

Ю

юг south
юмпар jumper, skidding sledge
юность youth
юркий rough

Я

яблоко apple
яблоня, дикая crab apple, *Malus sylvestris* Mill.
яблоня, лесная crab apple, *Malus sylvestris* Mill.
явление **1.** phenomenon **2.** appearance
явнобрачные phanerogams
явор sycamore maple, *Acer pseudoplatanus* L.
ягель reindeer moss, *Cladonia rangiferina* (L.) Wigg.
ягода berry
ягода, снежная snowberry, *Symphoricarpos rivularis* Suks.
ягодник berry-bearing plant
яд poison
ядовитость poisonousness
ядовитый poisonous
ядро **1.** heartwood **2.** nucleus **3.** kernel, core
ядро, клеточное cell nucleus
ядро, красное red heart(wood)
ядро, ложное false heart(wood)
ядрышко nucleolus
язва canker
яйла highland pasture (Crimea)
яйцо egg
якорь anchor
яма hole, pit
яма, парильная steaming pit
яма, посадочная planting hole
ямка small pit, small hole
ямкопатель hole-digger, hole-borer
ярлык label
яровизация vernalization, jarovization
ярус storey, layer, stratum
ярус, верхний upper storey, top storey, overstorey
ярус, древесный tree layer
ярус, искусственный underplanted layer
ярус, кустарниковый shrub layer
ярус, нижний lower storey, understorey
ярусность storeyed structure
ярусный storeyed, layered
ясеневник ash forest
ясень ash, *Fraxinus* L.
ясень, американский white ash, *Fraxinus americana* L.
ясень, железистый tree of heaven, *Ailanthus altissima* Swingle
ясень, зеленый green ash, *Fraxinus pennsylvanica* Marsh.
ясень, маньчжурский Manchurian ash, *Fraxinus mandshurica* Rupr.

ясень, обыкновенный common ash, *Fraxinus excelsior* L.

ясень, пенсильванский green ash, *Fraxinus pennsylvanica* Marsh.

ясень, туркестанский Turkestan ash, *Fraxinus potamophila* Herd.

ясколка *Cerastium* L.

ясменник, пахучий sweet woodruff, *Asperula odorata* L.

яснотка dead-nettle, *Lamium* L.

яснотка, крапчатая spotted dead-nettle, *Lamium maculatum* L.

яснотка, пурпуровая red dead-nettle, *Lamium purpureum* L.

ястреб-перепелятник sparrow-hawk, *Accipiter nisus* L.

ястреб-тетеревятник goshawk, *Accipiter gentilis* L.

ястребника hawkweed, *Hieracium* L.

ятрышник orchis, *Orchis* L.

ящик chest, box, case

ящик, чайный tea-chest

A

aapa (eutrophic wet peatland in northern taiga) аапа
Aaron's rod ухо медвежье, *Verbascum thapsus* L.
abandoned задичалый
abatis засека
abattis засека
ability способность
above-ground надземный
abrasion износ, истирание
abrasion resistance сопротивление износу
abrasive paper шкурка
absenteeism прогул
absorbing root сосательный корень, сосущий корень, всасывающий корень
absorption всасывание, поглощение
absorption, moisture влагопоглощение
absorption, water водопоглощение
abundance обилие
abundant обильный, недефицитный
acacia акация, *Acacia*
acacia, Amur акатник, амурская акация, *Maackia amurensis* Rupr.
academy, forest ЛТА, лесотехническая академия
accessibility доступность
accessible forest доступный лес
accessory species подгон
accident несчастный случай
accident prevention охрана труда, техника безопасности
acclimatization акклиматизация
account счет
accountancy хозяйственный расчет, хозрасчет
accounting учет
accumulation нарастание, наращивание
accuracy точность
accurate точный
acid кислота
acid rain кислотные осадки, кислый дождь
acid soil кислая почва
acidity кислотность
acorn желудь, жолудь
acorn cup плюска
across the grain поперек волокон
act дело
action действие
acuminate заостренный
adaptability приспособляемость
adaptation приспособление
adder's tongue обыкновенный ужовник, *Ophioglossum vulgatum* L.
addition прибавка
adelgid хермес
adhere прирастать
adjacent смежный
adjustment (e.g. of circular saw) выверка, направка
administration управление
administration, forest лесоуправление, лесное управление
admixture примесь
adult взрослый
advance подача
advance felling предварительное пользование, промежуточное пользование, промежуточная рубка, рубка промежуточного пользования
advance growth подрост
advance planting предварительная культура
advanced (progressive) work передовой опыт
adventitious придаточный
adventitious bud придаточная почка
adventitious root придаточный корень, адвентивный корень
adviser, forestry лесной советник
adze тесло
aerial надземный
aerial cableway ВТУ, воздушная трелевочная установка, подвесная установка, канатно-подвесная дорога, воздушная канатная дорога
aerial photo mosaics фотосхемы
aerial photograph аэрофотоснимок
aerial photography аэрофотосъемка, аэрофотография
aerial root воздушный корень
aerial skidding подвесная трелевка
aerial sowing аэросев

aerobic аэробный
aeroplane самолет
aerosol аэрозоль
afforest облесить, залесить, олесить, разводить лес
afforestation разведение леса, залесение, лесоразведение, облесение
afforestation, protective защитное лесоразведение
afforestation order приказ облесения
afforestation without irrigation богарное лесоразведение
after-ripening дозревание
after-treatment доработка
agaric, fly красный мухомор, *Amanita muscaria* L.
age возраст
age, exploitable оборотный возраст, возраст рубка
age, mean средний возраст
age, prevailing господствующий возраст
age class (usually 20 years in European Russia, and 40 years in Siberia) класс возраста, ступень возраста
age-class distribution распределение классов возраста
age determination определение возраста
age limit предельный возраст
age structure возрастный состав
aggregate fruit сборный плод
aggregate ray сложный луч
agricultural сельскохозяйственный
agricultural techniques агротехника
agriculture сельское хозяйство, земледелие
agriculture, slash and burn огневое земледелие, паловое земледелие
agrimony, hemp коноляный посконник, седач, *Eupatorium cannabinum* L.
air воздух
air dispersal полет
air-dry воздушно-сухой
air-dry weight воздушно-сухой вес
air-dry wood воздушно-сухая древесина
air drying атмосферная сушка, воздушная сушка, естественная сушка
air layer воздушный отводок
airborne mensuration officer летнаб-таксатор
aircraft lumber авиапиломатериалы (pl.)
aircraft timber авиалес
alate крылатый
albedo, integrated интегральный албедо
albumin белок
albumin(ous) белковый
alcohol спирт
alcohol, wood древесный спирт
alder ольха, *Alnus* Mill.
alder, bearded бородатая ольха, *Alnus barbata* C.A.M.
alder, common клейкая ольха, черная ольха, *Alnus glutinosa* Gaertn.
alder, dwarf кустарникова ольха, *Alnus fruticosa* Rupr.
alder, Kamchatka камчатская ольха, *Alnus kamtschatica* Kom.
alder, speckled белая ольха, серая ольха, *Alnus incana* Moench
alder buckthorn ломкая крушина, *Rhamnus frangula* L. (*Frangula alnus* Mill.)
alder stand ольховник, ольшаник, ольшитник, ольшинник, ольс, *Alnetum*
alder woodwasp ольховый рогохвост, *Xiphydria camelus* L.
Aleppo pine алепская сосна, *Pinus halepensis* Mill.
alga водоросль
aliform parenchyma крыловидная паренхима
alkali щелочь
alkaline щелочной
alkaline soil щелочная почва
alkalinity щелочность
all-terrain vehicle вездеход
alliance (bot.) союз
allowable cut размер пользования, размер рубки, допустимая рубка
allowable stress допускаемое напряжение
allowance (e.g. for shrinkage) припуск, допуск, прибавка на усушку
alloy сплав
alluvial наносный

alluvial soil наносная почва
alluvium нанос
almond миндаль, обыкновенный миндаль, *Prunus amygdalus* Batsch
almond, dwarf Russian степной миндаль, низкий миндаль, *Prunus tenella* Batsch
almond, spiny колючейший миндаль, *Prunus spinosissima* Franch.
almond willow миндальнолистная ива, *Salix amygdalina* L.
along the grain вдоль волокон
alpine currant альпийская смородина, *Ribes alpinum* L.
alternate вразбежку
alternate clear-strip felling (the unfelled strips 2-3 times wider than the felled) кулисная рубка
alternate pitting очередная поровость
alternate strip felling черезполосная рубка
alternating pressure process цикличный метод
alternation чередование
amelioration (often drainage) мелиорация
amenity planting зеленое строительство
amenity stand зеленое насаждение
American plane западный платан, *Platanus occidentalis* L.
ammonia аммиак
ammonium nitrate аммиачная селитра
ammonium sulphate сернокислый аммоний
amortization амортизация
amount количество
amphibious земноводный
amphibious tractor плавающий трактор
amphibious vehicle вездеход
Amur acacia акатник, амурская акация, *Maackia amurensis* Rupr.
Amur cork tree бархатное дерево, амурский бархат, *Phellodendron amurense* Rupr.
Amur lime амурская липа, *Tilia amurensis* Rupr.
Amur maple приречный клен, *Acer ginnala* Maxim.
analyse разлагать
analysis анализ, разложение
anchor якорь, анкер
anchor cable шейма
anchor chain шейма
anchor root якорный корень
anchor-type scarifier якорный покровосдиратель
andromeda, marsh дубровниколистная андромеда, *Andromeda polifolia* L.
anemone ветреница, ветренница
anemone, wood белая ветреница, *Anemone nemorosa* L.
anemone, yellow wood лютиковая ветреница, *Anemone ranunculoides* L.
anemophilous ветроопыляемый
angelica ангелика, *Angelica* L.
angelica, wild дудник, *Angelica sylvestris* L.
angiospermous покрытосеменный, покрытосемянный
angiosperms покрытосеменные (pl.), покрытосемянные (pl.)
angle угол
angle, (branch) insertion угол врастания
angle, clearance задний угол
angle, cutting передний угол
angle, face угол карры
angle, pressure-bar угол обжима
angle, sharpness угол заострения, угол заточки
angle, sweep-back угол поворота
angle gauge угловый шаблон
angle gauge, Bitterlich's полнотомер Биттерлиха
angle of bevel угол косой заточки
angle of incidence (between direction of tool travel and grain) угол встречи
angle of streak (freshening) угол подновки
animal животное, зверь
anisotropy анизотропия
annual 1. однолетнее растение 2. ежегодник 3. (adj.) однолетний, годичный
annual coupe годичная лесосека
annual increment ежегодный прирост

annual ring годичное кольцо, годичный слой

annual ring width ширина годичного слоя

annual shoot годовой побег

annual turnover годовой оборот

annular кольцеобразный

anobiid точильщик

ant муравей

ant, black carpenter большой черный муравей, *Camponotus herculeanus* L.

ant, common black лесной муравей, *Lasius niger* L.

ant, wood рыжий лесной муравей, *Formica rufa* L.

ant beetle муравьежук, пестряк, *Thanasimus formicarius* L.

ant hill муравейник

antenna усик, ус

anther пыльник

anthropogenic антропоген

antler burnishing сдирание кожицы с рогов

aperture отверстие

aperture, pit отверстие поры

apex верхушка, вершина, вершок

aphid тля, тли (pl.)

apiary пасека, пчельник

apical верхушечный

apical bud центральная почка

apical grafting копулировка, прививка в пенек

apparatus прибор

appearance появление, явление, вид

apple яблоко

apple, crab лесная яблоня, дикая яблоня, *Malus sylvestris* Mill.

appraisal оценка, таксация

appraise оценивать, таксацировать

approach graft прививка сближением

approach grafting аблактировка

apricot абрикос, обыкновенный абрикос, *Prunus armeniaca* L.

apricot, dried урюк

apricot, Manchurian маньчжурский абрикос, *Prunus mandshurica* (Maxim.) Koehne

apricot, Siberian сибирский абрикос, *Prunus sibirica* L.

apron (of dam) рисберм

aquifer водоносный слой

arable land пахота, пашня, пахотные угодья

Araks oak аракский дуб, араксинский дуб, *Quercus araxina* (Trautv.) Grossh.

aralia, Manchurian маньчжурская аралия, *Aralia mandshurica* Rupr. et Max.

arboretum дендрарий

arboriculture арборикультура

arbor-vitae туя, *Thuja* L.

arbor-vitae, Oriental восточная туя, *Thuja orientalis* L.

arc дуга

arch дуга, арка

arch(ed) арочный

archangel, yellow желтый зеленчук, *Gabeobdolon luteum* Huds.

arctic fox песец, *Alopex lagopus* L.

arctiid медведица

area площадь, поверхность, пространство, ареал

area, basal площадь сечения

area, burned пожарище, гарь

area, clear-felled сплошная лесосека

area, cross-sectional площадь поперечного сечения

area, crown projection поверхность кроны

area, cut-over лесосека

area, experimental опытная площадь

area, feeding площадь питания

area, felled вырубка, вырубленная площадь

area, forest лесная площадь

area, forested лесопокрытая площадь

area, non-forest(ed) нелесная площадь, непокрытая лесом площадь

area, sectional площадь сечения

area, unproductive непроизводительная площадь

area under forest облесенность, лесистость, процент лесистости

Arolla pine европейский кедр, *Pinus cembra* L.

arris кант, грань
arrow стрела
arsenic мышьяк
arson поджог
artel артель
artel, forestry лесная артель
artichoke, Jerusalem земляная груша, *Helianthus tuberosus* L.
article изделие
articulation сустав
artificial искусственный
artificial drying искусственная сушка
artificial forest рукотворный лес, плантация
artificial origin искусственное происхождение
artificial regeneration искусственное возобновление, искусственное лесовозобновление
asarabacca европейский копытень, *Asarum europaeum* L.
asbestos board асбокартон
ascending face возходящая карра
ascent подъем
Ascomycetes сумчатые грибы
ash 1. пепел, зола 2. ясень, *Fraxinus* L.
ash, common обыкновенный ясень, *Fraxinus excelsior* L.
ash, green зеленый ясень, пенсильванский ясень, *Fraxinus pennsylvanica* Marsh.
ash, Manchurian маньчжурский ясень, *Fraxinus mandshurica* Rupr.
ash, mountain обыкновенная рябина, *Sorbus aucuparia* L.
ash, Turkestan туркестанский ясень, *Fraxinus potamophila* Herd.
ash, white американский ясень, *Fraxinus americana* L.
ash, wood шадрик
ash content зольность
ash forest ясеневник
ash pit зольник
aspect экспозиция
aspen осина, *Populus tremula* L.
aspen, David's осина Давида, *Populus tremula* var. *davidiana*
aspen forest осинник
'assectator' (natural plant associate of an 'edificator' species) ассектатор
assembling сборка
assembly 1. монтаж 2. набор
assets фонды
assimilable усвояемый, подвижный
assimilate усваивать
assimilation усвоение
assistance to regeneration содействие возобновлению
assistant forester лесной ассистент
associate спутник
associate species сопутствующая порода
association ассоциация
associative сообщественный
assortment сортимент, сорт
Atlas cedar атласский кедр, *Cedrus atlantica* Man.
atmospheric precipitation атмосферные осадки (pl.)
atom, labelled меченый атом
auger бурав
auger, soil почвенный бурав
autoclave автоклав
automatic loader автопогрузчик
automatic volume calculator автокубатурник
automobile автомобиль
autumn frosts осенние заморозки
autumn ploughing зябь
autumn wood осенняя древесина
auxiblast (elongated one-year shoot) ауксибласт
auxiliary вспомогательный
auxiliary underwood выполняющий подлесок
available (of nutrients) усвояемый, подвижный
available nutrients усвояемые минеральные вещества
avalanche лавина, снежный обвал
avens, water речной гравилат, *Geum rivale* L.
avens, wood городской гравилат, *Geum urbanum*
avenue аллея
average 1. среднее 2. (adj.) средний 3. (in insurance) авария

aviation, forestry лесная авиация
awl шило
awning навес
axe топор
axe damage (to wood) заруб
axe head обух топора
axe-man топорник
axil пазуха
axillary пазушный
axillary bud пазушная почка
axis ось, стержень
axle ось
azotobacterin (a preparation of *Azotobacter* spp. used as a bacterial fertilizer) азотобактерин

B

back (of saw tooth) задняя грань
back crossing возвратное скрещивание
back cut валочный пропил
backfire отжиг, встречный огонь, встречный пожар
background фон
backing подкладка
backpack sprayer ранцевый опрыскиватель
bacterium бактерия
badger барсук, *Meles meles* L.
bag мешок
bag boom кошель
bag paper мешочная бумага
bagasse багасса
Baikal/Amur railway БАМ
bait приманка
balance весы (pl.)
balance beam коромысло
bald лысый
bald patch лысина, пролысина
balk брус
balk, waney брус с обзолом
ball (of earth) комок, ком, глыба
ball planting посадка глыбами, посадка с глыбкой, посадка с комом
balloon аэростат
balloon logging аэростатная трелевка
balloon logging installation аэростатная трелевочная установка
balsam бальзам
balsam poplar бальзамический тополь, *Populus balsamifera* L.
bamboo бамбук
band лента
band saw ленточнопильный станок, ленточная пила
bandage бандаж
banded pine weevil **1.** малый сосновый долгоносик, точечная смолевка, *Pissodes notatus* F. **2.** сосновая смолевка, *Pissodes pini* L.
banisters перила
bank вал, берег
bank protection forest берегозащитный лес
bank stabilization укрепление берега
bank vole рыжая лесная полевка, *Clethrionomys glareolus* Schr.
bar шина
bar, guide пильная шина
bar, half-round пластина, половинник
bar, nose прижимая линейка
bar, pressure контрнож, прижимая линейка
bar, short half-round плаха
bar, sighting таксационный прицел, мерная скоба
bar code штриховый код
barb колючка
barbed wire колючая проволока
barberry барбарис, *Berberis* L.
barberry, common обыкновенный барбарис, *Berberis vulgaris* L.
bare **1.** (adj.) голый, лысый **2.** (vb) обнажать
bare fallow черный пар
bare mountain голец
bare patch голье, лысина, пролысина
bare-rooted plant саженец с обнаженными корнями
barge плашкаут, плашкот, баржа
baring the soil оголение почвы
bark **1.** кора **2.** (vb) окорить, обескорить, ошкурить, сдирать кору
bark, fissured трещиноватый кора

bark, rough outer корка
bark, smooth зеркальная кора
bark blaze пролыска
bark-blazer резак
bark chopper корорубка
bark damage, mechanical обшмыг коры, ошмыг коры, обдир коры
bark plate пластина
bark removal (to kill trees) подсерчивание, подсечивание, подсочивание, подчерчивание
bark remover короsниматель
bark scorch ожог коры, солнечный ожог
bark stripping (by elk) погрыз коры, сдирание коры
barkbeetle короед
barkbeetle damage короедина
barkbeetle tunnel ход короеда
barked окоренный
barked patch лысина, пролысина, пролыска
barked wood окоренная древесина
barker корообдирка
barkhan (crescent-shaped shifting sand dune) бархан
barking окорение, окаривание, окоривание, окорка, ошкуривание
barking, batch групповая окорка
barking, clean окорка набело
barking, pneumatic пневмоокорка
barking blade короsниматель
barking iron окорочная лопатка, нож-скобель, скобель, секач
barking machine окорочный станок
barn сарай
barn owl сипуха, *Tyto alba* Scop.
barrel бочка
barrel, dry сухотарная бочка
barrel, wet заливная бочка
barren flower пустоцвет
basal area площадь сечения
basal burr прикорневый кап
basal scar (of an acorn) донце
'basic stocking' (standing vol. per unit of b.a.) элементарный запас
base **1.** основа, основание **2.** база
basic slag томасшлак
basil, wild обыкновенный душевик, пахучка, *Clinopodium vulgare* L.
basin бассейн
basket корзина, кузов
basket willow корзиночник
bast (in general) лыко, луб
bast basket лукошко
bast beetle лубоед
bast cord мочалыга
bast fibre (from lime) мочало
bast matting рогожа
bast stripper лыкодер
bat летучая мышь
bat, long-eared ушан, *Plecotus auritus* L.
batch barking групповая окорка
batch process периодический процесс
bath корыто, ванна
batten лата, драны, дранка, рейка, брусок
battens баттенсы (pl.)
battering ram таран
baulk *see* balk
bay willow пятитычинковая ива, чернолоз, чернотал, *Salix pentandra* L.
beam балка, плащак
bear, brown бурый медведь, *Ursus arctos* L.
bearberry медвежья ягода, толокнянка, *Arctostaphylos uva-ursi* (L.) Spreng.
bearded alder бородатая ольха, *Alnus barbata* C.A.M.
bearing **1.** вкладыш, подшипник **2.** (adj.) опорный
beat лесничество, обход, (of a mounted ranger) объезд
beat up исправлять, подсаживать, пополнять
beater загонщик
beating (a fire) захлестывание
beating up дополнение, пополнение, подсадка
beaver бобр, *Castor fiber* L.
bed **1.** пласт, залежь, гряда **2.** русло, ложе
bed, flower клумба
bed, peat торфяная залежь
bed, plant посадная гряда
bed, raised клумба

bed, river ложе реки
bed, seed семенная гряда
bed, stool маточная плантация
bedstraw подмаренник, *Galium* L.
bee пчела
bee, carpenter древогнезд
bee, leaf-cutter пчела-листорез, *Megachile*
bee-eater пчелояд, *Pernis apivorus* L.
bee-keeper пасечник, пчеловод
bee-keeper, forest бортник
bee-keeping пчеловодство
bee-keeping in forest бортевое пчеловодство, бортничество
beech 1. бук, *Fagus* L. 2. европейский бук, лесной бук, *Fagus sylvatica* L. 3. (adj.) буковый
beech, Crimean крымский бук, *Fagus taurica* Popl.
beech, oriental восточный бук, кавказский бук, *Fagus orientalis* Lipsky
beech forest букняк, бучина, *Fagetum*
beech mast буковый плодокорм
beech nut буковый орешок
beehive улей
beetle жук
beetle, ant муравьежук, пестряк, *Thanasimus formicarius* L.
beetle, bast лубоед
beetle, death watch пестрый точильщик, *Xestobium rufovillosum* De G.
beetle, furniture точильщик
beetle, ground жужелица
beetle, larger pine-shoot большой лесной садовник, *Blastophagus (Myelophilus) piniperda* L.
beetle, lesser pine-shoot малый лесной садовник, *Blastophagus (Myelophilus) minor* Hart.
beetle, longhorn дровосек, усач
beetle, pine-shoot садовник
beetle, powder-post древогрыз
belt полоса, пояс, ремень
belt, safety предохранительный пояс
bench (of land) банкет
bend 1. изгиб, загиб, сгиб, искривление, поворот 2. (of river) лука 3. (vb) гнуть, сгибать
bending гнутье, изгиб, загиб, сгиб
bending, end-grain торцовое гнутье
bending, impact динамический изгиб, ударный изгиб
bending, static статический изгиб
bending strength прочность на изгиб, сопротивление изгибу
bent коленчатый, согнутый
bent grass полевица, *Agrostis* L.
bent-wood furniture гнутая мебель
benzene бензол
benzine бензин
berm берма
berry ягода
berry-bearing plant ягодник
berry catchfly ягодный волдырник, *Cucubalus baccifer* L.
bevel косая заточка, фаска, скос
bevel, angle of угол косой заточки
bevelling скашивание
BHC гексахлоран
biennial 1. двулетнее растение 2. (adj.) двухлетний
bilberry черника, *Vaccinium myrtillus* L.
bill счет
bill of exchange вексель
bill of lading коносамент
billet полено, плашка; кругляк, круглый лес
billhook секач
bin (container for assortments on sorting line) накопитель, лесонакопитель, карман-накопитель
bind связывать
binding обвязка, связывание, связка, вязка, вязание, сплотка
bindweed полевой вьюнок, *Convolvulus arvense* L.
bindweed, black вьюнковый горец, *Polygonum convolvulus* L.
bioassay биоопыт, биотест
biocoenosis биоценоз
biodiversity биоразнообразие, биологическое разнообразие
biogeocoenology биогеоценология
biogeocoenosis биогеоценоз
biological control биологическая защита

'biological drying' of trees подвяливание, биологическая сушка
biological oxygen demand БПК
biomass биомасса
biomass plantation древесносырьевая плантация
biome биом
biosphere биосфера
biotope биотоп
biotype биотип
birch 1. береза, *Betula* L. 2. (adj.) березовый
birch, black даурская береза, черная береза, *Betula dahurica* Pall.
birch, downy пушистая береза, *Betula pubescens* Ehrh.
birch, dwarf 1. карликовая береза, *Betula nana* L. 2. кустарниковая береза *Betula fruticosa* Pall.
birch, iron железная береза, береза Шмидта, *Betula schmidtii* Reg.
birch, Siberian yellow ребристая береза, желтая (далневосточная) береза, *Betula costata* Traut.
birch, silver белая береза, бородавчатая береза, повислая береза, *Betula pendula* Roth. (*Betula verrucosa* Ehrh.)
birch, stone каменная береза *Betula ermanii* Cham.
birch bark береста
birch forest березняк, *Betuletum*
birch mushroom обыкновенный березовик, *Boletus scaber* Bull.
birch sap березовица
bird птица
bird of prey хищная птица
bird-cherry черемуха, *Prunus padus* L.
bird's-eye figure глаз птичий
bird's-nest orchid настоящая гнездовка, *Neottia nidus-avis* Rich.
bisexual двуполый
bison, European зубр, *Bison bonasus* L.
bit сверло
bit, centre центровое сверло
bit, countersink зенкер, зенковка
bit, dovetail концевая фреза
bit, slot-mortising пазовое сверло, пазовая фреза
bit, spoon ложечное сверло
bit, twist спиральное сверло, винтовое сверло
biting flies гнус
Bitterlich's angle gauge полнотомер Биттерлиха
bittern выпь, *Botaurus stellaris* L.
bittersweet сладко-горький паслень, *Solanum dulcamara* L.
black bindweed вьюнковый горец, *Polygonum convolvulus* L.
black birch даурская береза, черная береза *Betula dahurica* Pall.
black broom чернеющий ракитник, *Cytisus nigricans* L.
black carpenter ant большой черный муравей, *Camponotus herculeanus* L.
black grouse тетерев-косач, тетерев, *Lyrurus tetrix* L.
black horehound черный белокудренник, *Ballota nigra* L.
black juniper кара-арча, многоплодный можжевельник, *Juniperus polycarpos* Koch
black locust белая акация, лжеакация, *Robinia pseudoacacia* L.
black mulberry черная шелковица, *Morus nigra* L.
black pine черная сосна, *Pinus nigra* Arn.
black poplar осокорь, черный тополь, *Populus nigra* L.
black poplar stand осокорник
black rat черная крыса, *Rattus rattus* L.
black redstart горихвостка-чернушка, *Phoenicurus ochrurus* Gmel.
black saxaul черный саксаул, *Haloxylon aphyllum* Iljin
black walnut черный орех, *Juglans nigra* L.
black woodpecker желна, черный дятел, *Dryobates martius* L.
blackbird черный дрозд, *Turdus merula* L.
blackcurrant черная смородина, *Ribes nigrum* L.
blackthorn терн, терновик, колючая слива, *Prunus spinosa* L.

bladder campion хлопушка, *Silene cucubalus* Wibel
bladdernut клекачка, *Staphylea* L.
bladdernut, Colchis колхидская клекачка, *Staphylea colchica* Stev.
bladdernut, pinnate перистая клекачка, *Staphylea pinnata* L.
bladder-senna **1.** пузырник, *Colutea* L. **2.** древовидный пузырник, *Colutea arborescens* L.
bladder-senna, eastern восточный пузырник, *Colutea orientalis* Mill.
blade **1.** лезвее, лезвие, нож, лист **2.** (of circular saw) диск **3.** (of band saw) лента
blade, bulldozer бульдозерный отвал
blade, leaf пластинка листа
blade, planer строгальный нож
blade, saw полотнище пилы, полотно
blade, scraper цикля
blade, slicing строгальный нож
blank **1.** заготовка, болванка **2.** пропуск
blanket bog плоское болото
blast взрывать
blasting взрывание
blaze **1.** затеска, пролыска **2.** (vb) залысить, затесывать
bleached (of soil) белесый, беленый
bleached horizon подбел
bleaching отбелка
bleaching powder хлорная известь, белильная известь
bleed through (of glue) пробитие (клея)
bleeding (e.g. of sap) плач
bleeding, gum камедетечение
bleeding, resin смолотечение
bleeding, sap сокоистечение
blight ожог
blind pit слепая пора
blister пузырь
blister rust пузырчатая ржавчина
blizzard вьюга
block **1.** хозяйственная единица **2.** колода, кряж, блок, плаха, обрубок, чурбан
block, fall подцепочный блок
block, parquet штучный паркет
block, paving торец
block, periodic периодная площадь
block, pulley полиспаст
block, saw козлы (pl.)
block, sleeper слипер
block, slide ползун
blockboard столярная плита
blood-sucking flies гнус
bloom **1.** (on fruit) налет **2.** (vb) цвести
blooming цветение
blow удар
blow up взрывать
blowing away дефляция, выдувание
blue Douglas fir сизая лжетсуга, *Pseudotsuga menziesii* var. *glauca*
blue honeysuckle синяя жимолость, *Lonicera coerulea* L.
blue poplar сизолистный тополь, сизый тополь, *Populus pruinosa* Schrenk
blue spruce колючая ель, *Picea pungens* Engel
blue tit лазоревка, *Parus caeruleus* L.
blue willow сизая ива, *Salix glauca* L.
bluestain синь, просинь, синева
bluestained засинелый, посиневший
bluestained wood посиневшая древесина
bluethroat варакушка, *Luscinia svecica* L.
blunting затупление
boar кабан, вепрь, дикая свинья, *Sus scrofa* L.
board **1.** доска, планка, плансон, плита, половица **2.** картон
board, asbestos асбокартон
board, edged обрезная доска
board, extruded экструзионная плита
board, flat-pressed плита плоского прессования
board, insulation изоляционная плита
board, lining-out посадная доска, посадочная доска, сажальный брусок
board, narrow лата, драны, дранка, рейка
board, parquet паркетная доска, фриза
board, planed выстроганная доска
board, roofing кровельная доска
board, semi-hard полутвердая плита
board, skirting плинтус

board, superhard сверхтвердая плита
board, unedged необрезная доска
board, waney обзольная доска
board foot бордсовый фут, досковый фут
board offcuts файрвуд
board thickness высота пропила, пропил (высота)
boat лодка
boatbuilding timber судостройлес
bobak marmot байбак, *Marmota bobac* L.
bobbin катушка, шпулька
BOD БПК
bog болото, верховое болото
bog, hanging висячее болото, склоновое болото
bog, moss моховое болото
bog, peat торфяник, торфяное болото
bog, quaking трясина, зыбунное болото
bog, raised рям, выпуклое болото
bog, transitional переходное болото
bog, upland верховое болото
bog formation заболачивание
bog moss сфагнум, болотный мох, торфяной мох, *Sphagnum* L.
bog oak моренный дуб
bog peat верховой торф
bog soil болотная почва
bog whortleberry голубика, *Vaccinium uliginosum* L.
boggy заболоченный, болотистый
bogland science болотоведение
boiler казан, котел
boiling 1. варка 2. (to assist germination) ошпаривание
bole ствол
Bolle's poplar тополь Болле, *Populus alba* var. *pyramidalis* L. (Bge.)
bolster лага, подушка
bolt полено, плашка, чурак
bone glue костный клей
bone meal костяная мука
bonfire site кострища
book-keeper счетовод
boom бон, запань
boom, bag кошель
boom, fin рея, рей, реевный бон
boom, glance направляюший бон, продольный бон, реевный бон
boom, guiding направляюший бон
boom, holding приемная запань, задерживающий бон, поперечный бон
boom, keyed шпоночный бон
boom, longitudinal продольный бон
boom, needle спицевая запань
boom, single-log однобревенный бон
boom, spoke спицевая запань
boom, storage задерживающий бон
boom, terminal запань
boom, towing кошель
boom, transverse поперечная запань, поперечный бон
boom, trap запань, поперечная запань
boom, wing рея, рей, реевный бон
Bordeaux mixture бордосская жидкость
border кайма, кромка, опушка, бровка
border, forest стена леса
border tree окраинное дерево, краевое дерево
bordered pit окаймленная пора
bore сверлить, буравить
bore through просверливать
boreal forest бореальный лес
borer 1. зонд, сверло, бурав 2. (insect) скрипун, усач
borer, increment приростный бурав
boring высверливание, сверление, зондирование, рассверловка
boring cut (with a power saw) таранный пропил
boring machine сверлильный станок
boron бор
bostrychid капюшонник, ложнокороед
botanical garden ботанический сад
bottleneck узкое место
bottom грунт, дно, подошва
bottomland forest урема
boulder валун
boule буль
bound moisture связанная влага
boundary межа, граница, предел
boundary mark веха
boundary tree пограничное дерево
bow 1. лук 2. продольная покоробленность

bow saw лучковая пила
box 1. ящик, коробка 2. кавказская пальма, самшит, *Buxus sempervirens* L.
box-elder американский клен, ясенелистный клен, *Acer negundo* L.
box set ящичный комплект
box shook тара
box slat коробочная дранка
box veneer коробочная дранка
box wood ящичный кряж, тарный кряж
box wood assortment тарник
brachyblast (short shoot) брахибласт
bracken папоротник-орляк, орляк, *Pteridium aquilinum* Kuhn.
bracket 1. консоль 2. кронштейн
bract прицветник
braid коса
brake shoe тормозная колодка
braking торможение
braking drag волокуша
brambling вьюрок, *Fringilla montifringilla* L.
branch 1. ветвь, ветка (unbranched shoot over 1 year old) 2. отрасль 3. сук, сучок
branch insertion angle угол врастания
branch-pruner, tree-climbing древолазающая ветверезка, древо-лазающая мотоветверезка
branch remover сучкорез
branch stub пенек от сучка, торчок
branch trimmer сучкорезная машина
branched обветвленный, разветвленный
branched knot лапчатый сучок
branches сучья (pl.)
branchiness ветвистость, сучковатость
branching ветвление, обветвление, ответвление, разветвление
branchless безсучковый, бессучковый
branchwood сучняк
branchy обветвленный, суковатый, сучковатый
branchy crown развесистая крона
branchy tamarisk ветвистый гребенщик, *Tamarix ramosissima* Ldb.
brashing овревка сухих ветвей
breadth ширина
break 1. перерыв 2. разрыв
breakage поломка, авария
breakage, crown поломка кроны
breakage, snow снеголом
breakage, wind ветролом, бурелом
breakdown 1. разложение, распад, распадение 2. (of stem or log) раскрой 3. авария
breaking down (a log) раскрой
breaking off (of the stem in felling) скол
breast height (1.3 m) высота груди
breast-height diameter диаметр на высоте груди
breeder селекционер
breeding разведение, селекция, сортоводство
brew жижа
bridge мост
bridge graft прививка мостиком
briquette брикет
bristlegrass щетинник, *Setaria* Beauv.
brittle ломкий, хрупкий
brittleness ломкость
broadcast вразброс
broadcast sowing посев вразброс, сплошной посев
broadcasting разброс, разбрасывание
broadleaved лиственный, широколиственный
broadleaved forest лиственный лес
broadleaved species лиственная порода
broadleaved tree лиственное дерево
broken end облом
broken stone щебень
broken tree полом
broker маклер
bromegrass костер, *Bromus* L.
brood помет
broom 1. метла 2. ракитник, *Cytisus* L.
broom, black чернеющий ракитник, *Cytisus nigricans* L.
broom, Dnieper днепроский ракитник, *Cytisus borysthenicus* Grun.
broom, Russian русский ракитник, *Cytisus ruthenicus* Fisch.
broom, witches' ведьмина метла, метла ведьмина
broomhandle наметельник

brown bear бурый медведь, *Ursus arctos* L.
brown hare заяц-русак, русак, *Lepus europaeus* L.
brown rat серая крыса, пасюк, *Rattus norvegicus* Berk.
brown rot бурая гниль, деструктивная гниль
brown rot of larch бурая гниль (лиственницы), *Fomes officinalis* Fr.
brown rot of oak дубовая губка, бурая гниль (дуба), *Daedalea quercina* Fr.
brown soil бурая почва
brown-tail moth златогузка, *Euproctis chrysorrhoea* L.
brush **1.** коряжник **2.** метла, щетка
brush cutter кусторез, мотокусторез
brushwood заросль, хворост, хмыз
bryophyte мох, мхи (pl.)
bubble пузырь
buckbean трехлистниковая вахта, *Menyanthes trifoliata* L.
bucket ведро
buckler fern щитовник, *Dryopteris* Adans.
buckling продольный изгиб
buckthorn крушина, жестер, жостер, *Rhamnus* L.
buckthorn, alder ломкая крушина, *Rhamnus frangula* L. (*Frangula alnus* Mill.)
buckthorn, common слабительная крушина, *Rhamnus cathartica* L.
buckthorn, evergreen вечнозеленая крушина, *Rhamnus alaternus* L.
buckthorn, sea облепиха, *Hippophae rhamnoides* L.
bud **1.** почка, очко **2.** (vb) окулировать
bud, adventitious придаточная почка
bud, apical центральная почка
bud, axillary пазушная почка
bud, dormant спящая почка
bud, flower цветковая почка, цветочная почка, бутон
bud, lateral боковая почка
bud, leaf листовая почка
bud, terminal верхушечная почка
bud-graft окулировать
bud-grafting окулировка, очкование
bud shield щиток, глазок
budding окулировка, очкование
budding, shield окулировка щитком
buff полировать
buffer stock (of material) страховой задел
buffing полировка
bugle ползучая живучка, *Ajuga reptans* L.
building **1.** строительное дело, строительство **2.** здание, постройка, сооружение
building timber строевой лес, строительный лес, стройлес
bulb луковица
bulk density (e.g. of chips) плотность загрузки
bull бык
bullace тернослив, *Prunus institia* L.
bulldozer бульдозер
bulldozer blade бульдозерный отвал
bullfinch снегирь, *Pyrrhula pyrrhula* L.
bunch (of logs) пачка, пакет лесоматериалов
bunch planting посадка пучками
bunching (logs) пакетование, пакетирование
bundle вязанка, пакет, пучок, связка
bundle, vascular сосудистый пучок
bundle binder (for floated timber) сплоточная машина
bundle of logs пучок бревен, пакет лесоматериалов
bundle-opener разборщик
bundle raft пучковый плот
bundling **1.** обвязка, связывание, связка, вязка, вязание **2.** (logs) пакетование, сплотка
bundling installation сплоточная машина
bung затычка
bunk коник
bunting овсянка, *Emberiza* L.
bunting, corn просянка, *Emberiza calandra* L.
buoy томбуй
buoyancy плавучесть, пловучесть

buoyancy aid подплав
buprestid златка
burn **1.** гарь, пожарище, пал **2.** (vb) гореть, палить, сжигать
burn, controlled управляемый огонь
burn, main выжег, пожог
burn, prescribed управляемый огонь
burned опаленный
burned area пожарище
burning выжигание, горение, сжигание
burning, prescribed выжигание
burning, slash огневая очистка
burnishing (antler) сдирание кожицы с рогов
burnt area гарь
burnt site огнище
burozem (brown earth) бурозем
burr капокорень, кап, наплыв, нарост
burr, basal прикорневый кап
burrow нора
bush **1.** вкладыш, подшипник, гильза, втулка **2.** кустарник, куст, кустарный лес
bush cutter кусторез, мотокусторез
bush formation кущение
bush grass наземный вейник, *Calamagrostis epigeios* Roth
bushy кустовой
business дело, промысел
business accounting хозяйственный расчет, хозрасчет
bustard, great дрофа, *Otis tarda* L.
bustard, little стрепет, *Otis tetrax* L.
butcher's broom иглица, *Ruscus aculeatus* L.
butt **1.** комель, окоренок **2.** (of firearms) приклад **3.** (adj.) комлевой
butt diameter диаметр нижнего отруба, диаметр нижнего отруба, нижний отруб
butt end окоренок, нижний отруб
butt joint притык, соединение впритык
butt log комлевое бревно
butt rot комлевая гниль
butt rot, mottled пестрая напенная гниль, корневая губка, подпар, *Fomes annosus* (Fr.) Cke
butt swelling комлеватость, закомелистость, закомлеватость, корневый наплыв
'butt to top' (i.e. describing loads having butts and tops at both ends) вразнокомелицу
butterfly бабочка
butternut масляный орех, серый орех, *Juglans cinerea* L.
buttress контрфорс
buttress root досковидный корень
buzzard сарыч, *Buteo buteo* L.
by-product побочный продукт

C

cabin (on a log raft) казенка
cabinet шкаф, футляр
cable канат, такелаж, трос
cable railway канатно-рельсовая дорога
cable skidding лебедочная трелевка
cable slide канатный лесоспуск, канатный откать, канатная самотаска
cable-crane кабель-кран, кабельный кран
cableway канатная дорога, подвесная дорога
cableway, aerial ВТУ, воздушная трелевочная установка, подвесная установка, канатно-подвесная дорога, воздушная канатная дорога
cableway, gravity канатный лесоспуск, канатный откать, канатная самотаска
cable-way transporter кабель-кран, кабельный кран
cadastral survey кадастр
cage клетка
calcareous карбонатный, известковый
calcareous soil известковая почва, карбонатная почва
calcium **1.** кальций **2.** (in names of compounds) карбонатный, известковый
calculated cut расчетная лесосека
calculation вычисление, подсчет, расчет, счет, учет
calender каландр
calibration калибрование, тарировка

caliper мерная вилка, циркуль, шубригель
caliper, pole штангенциркуль
caliper, sliding штангенциркуль
calipers кронциркуль
callus каллюс, мозоль, наплыв, нарост
calyx чашечка
cambium камбий
cambium, cork пробковий камбий, феллоген
cambium, storeyed ярусный камбий
camp (for logging) вахта, мастерский участок
camp (in woods) вахта
camphor камфора
campion смолевка, *Silene* L.
campion, bladder хлопушка, *Silene cucubalus* Wibel
'Canadian poplar' канадский тополь, *Populus deltoides* Marsh (but including also various hybrid black poplars)
canal канал
canal, resin смоляной ход
canker рак, язва
canker, enlarging язвенный рак
canker, erumpent swelling опухолевидный рак
canker, open язвенный рак
canker, stem рак ствола
canker, target ступенчатый рак
canopy полог, полог насаждения
canopy closure смыкание крон
canopy density плотность полога, полнота леса, сомкнутость, сомкнутость крон, сомкнутость полога, степень сомкнутости
canopy opening нарушение сомкнутости, перерыв сомкнутости
cant лежень, призма, балка, брус
cant (log with two opposite slabs removed) двухкантный брус
cant (with three straight sides) трехкантный брус
cant hook кондак, вертун, багровый крюк, по"оротный крюк, багор
cant re-setter брусоперекладчик
cantilever консоль
cantilever crane консольный кран
canting, circular бруского-круговой раскрой
canting, segment сегментный раскрой
canting, segment-radial сегментно-радиальной раскрой
canting, segment-tangential сегментно-тангенциальной раскрой
caoutchouc каучук
cap **1.** колпак **2.** шляпка
cap, root корневой чехлик
cap, skidding трелевочный колпак
capacity **1.** вместимость, емкость **2.** мощность **3.** способность
capacity, field полевая влагоемкость, наименьшая влагоемкость, НВ
capacity, germinative всхожесть
capacity, heat теплоемкость
capacity, infiltration водопроницаемость
capacity, load грузоподъемность, нагружаемость
capacity, maximum adsorption максимальная адсорбционная влагоемкость
capacity, maximum water-holding полная влагоемкость
capacity, moisture влагоемкость
capacity, shoot-forming побегопроизводительность
capacity, yield производительная способность
capercaillie глухарь, *Tetrao urogallus* L.
capillarity капиллярность, волосность
capital фонды
capital, fixed (basic) основные фонды
capital, productive производственные фонды
Cappadocian maple яркий клен, *Acer laetum* C.A.M.
capsule коробочка
carabid жужелица
carbohydrate углевод
carbon углерод
carbon dioxide углекислота
carbonic acid углекислота
card карточка
card, punched перфорированная карточка

cardboard картон
cargo груз
carob рожечник, рожковое дерево, *Ceratonia siliqua* L.
carpel плодолистик
carpenter плотник, столяр
carpenter bee (Xylocopidae) древогнезд
carpophore плодонос
carriage каретка, тележка
carriage, frame-saw впередирамная тележка
carriage, lathe суппорт
carriage, log захватывающая тележка
carriage, saw захватывающая тележка
carrion crow черная ворона, *Corvus corone* L.
carrying capacity грузоподъемность
cart вагонетка, воз, телега
cart-load воз
cartography картография
cartridge патрон
carver гравер, резчик
carving ваяние
case **1.** гильза, кожух **2.** ящик, коробка **3.** шкаф, футляр
casein glue казеиновый клей
cask бочка
Caspian honey-locust каспийская гледичия, *Gleditsia caspica* L.
Caspian willow каспийская ива, *Salix caspica* Pall.
castle замок
castor-oil tree, prickly калопанакс, диморфант, *Kalopanax pictus* Nakai (Thunb.)
cat кошка
cat, wild дикая кошка, *Felix silvestris* Schreb.
catch замок, зацепка
catchfly смолевка, *Silene* L.
catchfly, berry ягодный волдырник, *Cucubalus baccifer* L.
catching лов, ловля
catchment лог, водосбор
catchment-protection zone водоохранная зона
category (for survey work, forests are divided into 4 accuracy categories, based on % crop utilization) разряд
category I (all wood sold) разряд I
category II (all wood from later thinnings and principal felling sold) разряд II
category III (75% of wood from principal felling utilized) разряд III
category IV (25-75% of wood from principal felling utilized) разряд IV
caterpillar гусеница
catkin сережка
cat's foot кошачьи лапки, *Antennaria dioica* (L.) Gaertn.
Caucasian elm граболистная дзельква, *Zelkova carpinifolia* Dipp.
Caucasian fir пихта Нордманна, кавказская пихта, *Abies nordmanniana* Spach.
Caucasian hornbeam кавказский граб, *Carpinus caucasica* Grossh.
Caucasian lime кавказская липа, *Tilia caucasica* Rupr.
Caucasian nettle-tree кавказский каркас, *Celtis caucasica* Willd.
Caucasian oak крупнопыльниковый дуб, восточный дуб, *Quercus macranthera* F. et M.
Caucasian rhododendron кавказский рододендрон, *Rhododendron caucasicum* Pall.
Caucasian wing-nut кавказская лапина, *Pterocarya caucasica* C.A.M.
caul поддон
cavity дупло, полость
cavity, pit полость поры
cavity, resiniferous смоляное вместилище, смоловместилище
cecidomyid галлица
cedar кедр (often *Pinus sibirica* (but see below))
cedar, Atlas атласский кедр, *Cedrus atlantica* Man.
cedar, incense речной кедр, *Libocedrus decurrens* Torr.
cedar of Lebanon ливанский кедр, *Cedrus libani* Loud.
celandine, greater чистотел, *Chelidonium majus* L.

celandine, lesser лютичный чистяк, *Ranunculus ficaria* L.
cell 1. камера 2. клетка, клеточка
cell, procumbent ray лучевая лежачая клетка
cell, septate parenchyma паренхимная перегородчатая клетка
cell, sheath кроющая клетка
cell, sieve ситовидная клетка
cell, stone каменистая клетка
cell, tile черепитчатая клетка
cell, upright ray лучевая стоячая клетка
cell nucleus клеточное ядро
cell plate клеточная пластинка
cell sap клеточный сок
cell wall оболочка клетки, стенка клетки
cellular tissue клеточная ткань
cellulose клетчатка, целлюлоза
cement, wood древесный цемент
centaury золототысячник, *Centaurium* Hill
centering (logs on a lathe) центровка
centre bit центровое сверло
centre of gravity центр тяжести
centrifugal центробежный
cerambycid дровосек, усач
certificate паспорт
certification сертификация
chafer хрущ
chafer, garden садовый хрущик, *Phyllopertha horticola* L.
chafer, rose золотистая бронзовка, *Cetonia aurata* L.
chafer, summer нехрущ, июньский хрущ, *Amphimallon solstitialis* L.
chaffinch зяблик, *Fringilla coelebs* L.
chain цепь, зубчатый венец
chain, pore цепочка просветов
chain-flail weeder цепной осветлитель
chain mortiser цепнодолбежный станок
chain saw цепная пила
chaining (as a means of felling) тумервал
chair стул
chalk мел
chalky soil меловая почва
chamber камера
chamber, drying сушильная камера
chamber, mating брачная камера
chamber, pit камера поры
chamfer фаска, скос
chamfering скашивание
change смена
channel 1. желоб 2. канал 3. коридор 4. русло, ложе
channel, pit канал поры
channel, root корневина
channel, timber-floating сплавной канал
chanterelle настоящая лисичка, *Cantharellus cibaris* Fr.
character свойство
character, silvicultural лесоводственное свойство
characteristic признак
charcoal древесный уголь
charcoal, lump уголь-сырец
charcoal burner углежег
charcoal burning углежжение
charge загружать
charring wood (for preservation) обугливание дерева
chase 1. гонка 2. гонять
check 1. контроль 2. трещина, трещина усушки
check, growth задержка в росте
check, internal свищ, внутренняя трещина
check, seasoning сухобойна
check, shrinkage трещина усушки
check, surface поверхностная трещина
checking растрескивание
chemical control химическая борьба
chemical oxygen demand ХПК
chemical pulp целлюлоза
chemigroundwood древесная химическая масса
chemistry, forest лесная химия, лесохимия
chemistry, wood лесная химия, лесохимия
chemi-thermomechanical pulp ХТММ
chernozem чернозем
cherry вишня
cherry, Cornelian кизил, *Cornus mas* L.
cherry, ground вишарник, кустарниковая вишня, степная вишня, *Prunus fruticosa* Pall.

cherry, mahaleb кучина, антипка, *Prunus mahaleb* L.
cherry, sour обыкновенная вишня, *Prunus cerasus* L.
cherry, wild черешня, птичья вишня, *Prunus avium* L.
cherry-laurel лавровишня, *Prunus laurocerasus* L.
chest ящик, коробка
chestnut каштан
chestnut, horse конский каштан, *Aesculus hippocastanum* L.
chestnut, sweet сьедобный каштан, благородный каштан, посевной каштан, *Castanea sativa* Mill.
chestnut-leafed oak каштанолистный дуб, *Quercus castaneifolia* C.A.M.
chestnut soil каштановая почва
chickweed мокрица, *Stellaria media* (L.) Vill.
chickweed wintergreen европейский седмичник, *Trientalis europaea* L.
chief **1.** начальник **2.** (adj.) главный, заведующий
chimaera химера
'Chinese beard' (inverted V-shaped swelling on stems, associated with branch stub occlusions) бровка
Chinese elm мелколистный вяз, *Ulmus parvifolia* Jacq.
Chinese poplar китайский тополь, *Populus simonii* Carr.
chip **1.** стружка, дробленка **2.** (vb) рубить
chip formation стружкообразование
chip mat стружечный ковер
chipboard ДСП, древесностружечные плиты
chip-breaker стружколоматель
chipmunk, Siberian бурундук, *Eutamias sibiricus* Laxm.
Chip-N-Saw unit хедриг, фрезернопильный агрегат
chipped grain задранное волокно
chipper рубительная машина, дробилка, дробильная машина, рубильный станок
chipper tooth Г-образный зуб, режкообразный зуб, фуговочный зуб
chipper-canter headrig фрезернобрусующий станок
chipping дробление, раздробление
chipping headrig хедриг, фрезернопильный агрегат
chip(s) щепа, щепка
chips, industrial технологическая щепа
chisel долото, зубило, стамеска
chiselling долбление
chlorophyll хлорофилл
chlorophyll-carotene paste хлорофиллокаротиновая паста
chloroplast хлоропласт
choice выбор
choker чокер
chokerman чокеровщик
choker-release man отцепщик
chop изрубить, рубить
chopper **1.** косарь, тяпка **2.** рубильный станок
chopper, bark корорубка
chosenia крупночешуйчатая чозения, *Chosenia macrolepis* Kom.
Christmas tree рождественская елка
Christ-thorn держи-дерево, *Paliurus spina-christi* Mill.
chromosome хромосома
chrysomelid листоед
chrysopid златоглазок
chuck патрон
chute древоспуск, лесоскат, лесоспуск, спускной желоб, лоток, скат
chute, dry сухий лоток
chute, log бревноспуск
cinnamon-brown soil коричневая почва
circle круг
circle, working хозяйственная секция, хозсекция
circular saw дисковая пила, круглая пила, круглопильный станок, циркулярка
circulation круговорот, обращение
circumference круг, обхват, окружность
city forest городской лес
cladding облицовка, обшивка
clamp вайма, зажим, скоба, тиски (pl.)
class класс, разряд, ранг, степень, ступень

class, age класс возраста, ступень возраста
class, average site-quality изобонитет
class, diameter класс толщины, ступень толщины
class, height класс высоты, ступень высоты
class, height quality разряд высоты
class, Kraft (dominance) класс Крафта
class, management класс хозяйства
class, merchantability класс товарности
class, natural diameter естественная ступень толщины
class, quality класс добротности
class, site бонитет, класс бонитета
classification классификация
claw коготь
clay глина, ил
clay deposition оглинивание
clay soil глинозем
clayey глинистый, илистый
clayey soil илистая почва, глинистая почва
clay-loam 1. суглинок 2. (adj.) суглинистый
clay-loam soil суглинистая почва
clean (vb) очищать, прочистить
clean barking окорка набело
cleaning очистка, очищение, уход за составом, прочистка
cleaning/weeding (usually to improve composition, in stands of age 11-20 years) рубка прочистки, прочистка
cleanly начисто
cleanness чистота
clear felling сплошная лесосека, сплошная рубка
clear specimen чистый образец
clear stem штамб
clearance, forest корчевание леса, лесоочистка, сводка леса
clearance, ground клиренс
clearance, stump корчевание пней
clearance angle задний угол
cleared land подсечка, подсека, сеча
cleared strip просека
clear-felled area сплошная лесосека
clear-felling system сплошно-лесосечное хозяйство
clearing 1. очистка, очищение 2. перелесье, поляна, прогалина
clearness чистота
clearwing стеклянница
clearwing, hornet тополевая большая стеклянница, *Sesia apiformis* L.
cleat, S-shaped эс
cleavability раскалываемость, расколимость
cleavable колкий
cleave колоть, раскалывать, раскалывать
cleaver колун, тяпка
cleaving колка, расколка, раскалывание
cleaving hammer расколочный молот
cleaving resistance сопротивление раскалыванию
cleft fuelwood колотые дрова
cleft graft прививка в расщеп
cleft timber колотый лес
cleft wood плашник, плашняк
click-beetle щелкун
cliff скала
climate климат
climatic type климатип
climax forest климаксный лес
climbing irons лазы (pl.), кошки (pl.), когти (pl.)
clinometer уклономер, кренометр
clip подстригать
clipping стижка
clod комок, ком, глыба
clonal seed orchard клоновая плантация
clone 1. клон 2. сортообразец
clone bank архиво-маточная плантация, клоновый архив
close canopy (vb) расти в сомкнутости, смыкаться, сомкнуться
closed сомкнутый
closed forest сомкнутый лес
closed stand сомкнутое насаждение
close-ribbed (herring-bone) face ребристая карра
closure, degree of плотность полога, полнота леса, сомкнутость, сомкну-

тость крон, сомкнутость полога, степень сомкнутости
closure of canopy смыкание крон
clothes одежда
clothes, working спецодежда
clothing одежда
cloudberry морошка, *Rubus chamaemorus* L.
cloudiness облачность
clover трилистник, клевер, *Trifolium* L.
club дубина
clubmoss плаун, *Lycopodium* L.
clubmoss, common булавовидный плаун, *Lycopodium clavatum* L.
clubmoss, fir плаун -баранец, *Lycopodium selago* L.
clump гнездо, куртина, группа
clump of broadleaves (planted in conifers) ремиза
cluster, pore группа просветов
coal уголь
coal tar каменноугольная смола
coal tit московка, *Parus ater* L.
coalesce срастаться
coalescing сращивание, срастание
coarse крупный, суровый
coarse soil крупнозем
coarse-ringed крупнослойный, широкослойный
coastal navigation каботаж
coat кожура, оболочка
cob (maize) початок
coccid 1. червец 2. щитовка
coccinellid коровка
coccus, felted beech буковый войлочник, буковый червец, *Cryptococcus fagi* Bär.
cockchafer майский (западный) хрущ, *Melolontha melolontha* L.
cock's foot сборная ежа, *Dactylus glomerata* L.
COD ХПК
code, forest лесной кодекс
co-dominant согосподствующий
coenosis ценоз
cohesion, soil связность почвы
Colchis bladdernut колхидская клекачка, *Staphylea colchica* Stev.
cold холод
cold hardiness холодостойкость
cold resistance холодостойкость
cold-setting glue клей холодного отверждения
Coleoptera жуки (pl.), жесткокрылые (pl.)
collapsible разборный
collar 1. кольцо, хомут 2. (of circular saw) шайба, зажимная шайба 3. шейка
collar (root) корневая шейка
collection сбор, собирание
collection, resin собирание живицы
collection, seed сбор семян, семенозаготовка
collective коллективный
collective farm колхоз
collector ditch водосборная канава
Collembola вилохвостки (pl.)
collenchyma колленхима
colonization (by vegetation) зарастание, заселение, население, расселение
colonize зарастать
colonizing перерастание
colophony канифоль, гарпиус
colour краска, окраска, цвет
colouring крашение, окраска
coltsfoot мать-и-мачеха, *Tussilago farfara* L.
columbine обыкновенный водосбор, *Aquilegia vulgaris* L.
column столб
combine комбайн, комбинат
combine, forest лесокомбинат
combine harvester комбайн
combustibility загораемость, горючесть, сгораемость
combustion горение
combustion, spontaneous самовоспламенение
Comecon СЭВ, Совет Экономической Взаимопомощи
commercial forest хозяйственный лес
comminuted wood измельченная древесина
comminution измельчение
commodity exchange биржа

common alder клейкая ольха, черная ольха, *Alnus glutinosa* Gaertn.

common ash обыкновенный ясень, *Fraxinus excelsior* L.

common barberry обыкновенный барбарис, *Berberis vulgaris* L.

common black ant лесной муравей, *Lasius niger* L.

common buckthorn слабительная крушина, *Rhamnus cathartica* L.

common clubmoss булавовидный плаун, *Lycopodium clavatum* L.

common dandelion лекарственный одуванчик, *Taraxacum officinale* agg.

common elder черная бузина, *Sambucus nigra* L.

'common elm' обыкновенный вяз, usually *Ulmus laevis* Pall.

common fumitory лекарственная дымянка, *Fumaria officinalis* L.

common furniture beetle мебельный точильщик, *Anobium punctatum* De G.

common hazel обыкновенная лещина, *Corylus avellana* L.

common jujube унаби, *Zizyphus jujuba* Mill.

common juniper обыкновенный можжевельник, *Juniperus communis* L.

common mallow лесной просвирник, *Malva sylvestris* L.

common mushroom шампиньон, *Agaricus campestris* L.

common myrtle обыкновенный мирт, *Myrtus communis* L.

common osier лоза корзинная, корзиночная ива, прутовидная ива, *Salix viminalis* L.

common pear обыкновенная груша, *Pyrus communis* L.

common polypody сладкокорень, *Polypodium vulgare* L.

common privet обыкновенная бирючина, *Ligustrum vulgare* L.

common shrew лесная землеройка, *Sorex araneus*

common speedwell лекарственная вероника, *Veronica officinalis* L.

common stag-beetle жук-олень, *Lucanus cervus* L.

common tormentil узик, лесная лапчатка, прямостоящая лапчатка, *Potentilla erecta* (L.) Räusch.

common vole обыкновенная полевка, *Microtus arvalis* Pall.

common wintergreen малая грушанка, *Pyrola minor* L.

common yew ягодный тисс, европейский тисс, *Taxus baccata* L.

communal forest общинный лес

communication сообщение

community сообщество

community forest общинный лес

compaction уплотнение

compactness плотность

compartment квартал

compartment boundary line квартальная линия

compartment kiln камера периодического действия, сушилка периодического действия

compartment map планшет

compass компас, буссоль

compass point страна света

compasses (pair of) циркуль

competition конкуренция

competitor конкурент

complete cultivation сплошная обработка

complete enumeration подеревный перечет, сплошной перечет

complex **1.** комплекс **2.** (adj.) комплексный, сложный

component деталь, составная часть

composite сложный

composition состав

composition (of stand) состав насаждения

compost компост

compound соединение

compound fruit сложный плод

compound NPK fertilizer нитроаммофоска

compreg laminate древеснослоистый пластик, ДСП (древеснослойстые пластики)

compress сжимать

compressed прессованный

compressed wood прессованная древесина, ДП (древесина прессованная), спрессованная древесина
compressing прессование
compressing, contour контурное прессование
compressing, uni-axial одноосное прессование
compression обжим, сжатие
compression parallel to grain сжатие вдоль волокон
compression perpendicular to grain сжатие поперек волокон, смятие
compression strength прочность на сжатие, сопротивление сжатию
compression wood крень, креневая древесина, кремень, кремнина, креноватость
concave saw шаровая пила
concession (timber) закрепление лесосырьевой базы
concrete, reinforced железобетон
concrete, wood арболит, древесный бетон
condenser kiln конденсационная сушилка
condition состояние
condition, shipping-dry транспортная влажность
conditioning акклиматизация
conditions условия (pl.)
conditions, environmental внешние условия
conditions, growth условия произрастания
conditions, site условия произрастания
conducting root проводящий корень
conducting tissue проводящая ткань
conduction, moisture влагопроводность
conductivity проводимость (usually electrical or thermal, but sometimes used for the flow of fluids through wood)
conductivity, electrical электропроводность
conductivity, thermal теплопроводимость, теплопроводность
cone 1. конус 2. шишка
cone(s), immature озимь
cone-drier шишкосушильня
conelet шишечка
confluent parenchyma сомкнутокрыловидная паренхима
conifer хвойное дерево, хвойная порода
conifer foliage stripper хвоеотделитель
conifer forest хвойный лес
conifer scrub крень, кренюжник
Coniferae иглолистные (pl.), хвойные (pl.)
coniferous хвойный, иглолистный
coniferous species хвойная порода
coniferous tree хвойное дерево
conifers иглолистные (pl.), хвойные (pl.)
connecting rod дышло, тяга
connection соединение
conservation консервирование, хранение, сохранение
conservation, nature охрана природа
constancy постоянство
constant постоянный
constituent составная часть
constitution сложение
constitution, stand форма насаждения, форма леса
construction конструкция, постройка, сооружение, строение, строительство
constructional строевой
constructional timber строевой лес, строительный лес, стройлес
consumption потребление
contact, crown соприкосновение крон
contact drier контактная сушилка
contact pesticide контактный пестицид
container вместилище, накопитель, карман-накопитель, лесонакопитель
container recess (resin tapping) черта, щан
containerized plant саженец с закрытыми корнями
containing streak (to prevent spread of chemical stimulants) предохранительная подновка
content(s) содержание
contiguous смежный
continuous непрерывный, сплошной
continuous forest дауервальд

continuous process непрерывный процесс
contour контур
contour compressing контурное прессование
contour line горизонталь
contouring оконтуровка
contract price договорная цена
contractor подрядчик
control 1. контроль 2. управление 3. борьба
control measure мера борбы
control method контрольный метод, метод контроля
controlled burn управляемый огонь
convection kiln конвекционная сушилка
conventional условный
conversion обработка, обрабатывание, переработка, превращение, преобразование, разработка, реконструкция
conversion, primary первичная обработка
conversion, wood деревопереработка, разработка леса
conversion point разделочная площадка
conversion system переходное хозяйство
convert (vb) заготовить, обработать
converting factor переводной коэффициент
conveying перевозка
conveyor конвейер, самотаска
cooking варка
cooking, hydrotropic гидротропная варка
cooking acid (pulping) варочная кислота
cooper бондарь, бочар
cooperage бондарная, бондарня
cooperage stave бочарная клепка
cooperage wood бочарный лес
co-operative кооператив
cooper's workshop бондарная, бондарня
copal копал
coppice низкоствольник
coppice forest низкоствольный лес
coppice growth поросль
coppice regeneration порослевое возобновление
coppice shoot порослевина, побег от пня
coppice shoots пневая поросль, поросль от пня
coppice stand низкоствольник
coppice system низкоствольное хозяйство
coppice tree низкоствольник
coppice with standards средний лес
coppice-with-standards system среднее хозяйство
coppicing посадка на пень
copse перелесок
copying lathe копировальный станок
coral-root ладьян, *Corallorhiza trifida* Chatel
coral-wort луковичная зубянка, *Dentaria bulbifera* L.
cord 1. поленница 2. шнур
cord, measuring мерный шнур
cord, planting посадный шнур
cordate сердцевидный
cordon кордон
corduroy road бревенчатая дорога, гать, поперечная стлань
core 1. (of veneer bolt) карандаш 2. стержень 3. ядро
core, increment приростный керн
cork феллем, пробка
cork cambium пробковий камбий, феллоген
cork layer пробковый слой
cork oak пробковый дуб, *Quercus suber* L.
cork tree (Amur) бархатное дерево, амурский бархат, *Phellodendron amurense* Rupr.
corn bunting просянка, *Emberiza calandra* L.
Cornelian cherry кизил, *Cornus mas* L.
cornice карниз
corolla венчик
correction поправка
corridor коридор
corrosion коррозия
corrugated (herring-bone) face рифленая карра
corrugation гофр

Corsican pine калабрийская сосна, *Pinus brutia* Ten.
cortex кора
cossid древоточец
cost(s) стоимость
cost accounting хозяйственный расчет, хозрасчет
cost price себестоимость
cotoneaster кизильник, *Cotoneaster* Ehrh.
cotton-grass пушица, *Eriophorum* L.
cotyledon семядоля, семенодоля
couch-grass ползучий пырей, *Agropyron repens* Beauv.
'coulisse' (the broader unfelled strip between strip fellings) кулиса
coulter сошник
coulter, skim дерносним, дерноснима-тель
counter счетчик
counter fire отжиг, встречный огонь, встречный пожар
countersink bit зенкер, зенковка
counter-wedge контрклин
coupe лесосека
coupe, annual годичная лесосека
coupe demarcation закладка лесосеки, отвод лесосеки
cover **1.** колпак, накладка, покров, покрышка **2.** (vb) покрывать засыпать
covered покрытый
covering **1.** (e.g. seeds) заделка, заделывание **2.** крыша, кровля **3.** облицовка **4.** покрывание, покрышка
cow parsley лесной купырь, *Anthriscus sylvestris* (L.) Bernh.
cow parsnip борщевик, *Heracleum* L.
cowbane цикута, ядовитый вех, *Cicuta virosa* L.
cowberry брусника, *Vaccinium vitis-idaea* L.
cowl колпак
co-worker сотрудник
cow-wheat марьянник, *Melampyrum* L.
crab apple лесная яблоня, дикая яблоня, *Malus sylvestris* Mill.
crack **1.** раскол, щель **2.** трещина
crack willow ломкая ива, *Salix fragilis* L.
cracking растрескивание
cradle **1.** коник **2.** люлька
craftsman кустарь
cramp скоба, струбцинка
cranberry болотная клюква, *Oxycoccus palustris* Pers.
crane кран
crane, cantilever консольный кран
crane, gantry козловый кран
crane, hydraulic гидрокран
crane, jib поворотный кран, стреловой кран
crane, overhead travelling мостовой кран
crane, tower башенный кран
crank кривошип
crawler tractor гусеничный трактор
creaming **1.** (for particularly valuable assortments) приисковая рубка, выборочная рубка на прииск, рубка на прииск **2.** (to a diameter limit) рубка с вершка, подневольная рубка, выборочная рубка с вершка, выборочная подневольная рубка
creep ползучесть
creeping jenny луговой чай, *Lysimachia nummularia* L.
creeping soft-grass мягкий бухарник, *Holcus mollis* L.
creeping willow ползучая ива, *Salix reptans* Rupr.
crenate городчатый
creosote креозот
crest (in tree felling) недоруб, недопил
crested buckler fern гребенчатый щитовник, *Dryopteris cristata* A. Gr.
crested tit хохлатая синица, *Parus cristatus* L.
crew бригада
cribwork dam ряжевая плотина
Crimean beech крымский бук, *Fagus taurica* Popl.
Crimean linden крымская липа, *Tilia euchlora* K. Koch
Crimean pine крымская сосна, *Pinus nigra* var. *caramanica*
crimp гофр

crook 1. (used in boatbuilding) кница, кокорная кница, кокора, кривой лес 2. кривизна
crooked извилистый, корявый, коленчатый, перекошенный
crooked growth корявый рост
crooked-grown извилистый, корявый
crop 1. культура 2. урожай
crop density плотность насаждения, густота насаждения, полнота леса
crop failure неурожай
crop rotation (agricultural) плодосмен, севооборот
cross 1. помесь 2. (vb) скрещивать
cross-bar перекладина, поперечина
crossbill клест-еловик, *Loxia curvirostra* L.
cross-cut (into logs) раскряжевать
cross-cut saw поперечная пила
cross-cut to length торцевать
cross-cutting перерезание, перерезывание, поперечное пиление, поперечная распиловка, разделка, раскряжевка
cross-cutting to length торцовая распиловка, торцовка, торцевание
cross-ditch перекоп
cross-field перекрест, поле перекреста
cross-grain поперек волокон
cross-leaved heath болотный вереск, *Erica tetralix* L.
cross-piece перекладина, поперечина, траверса, траверс, перемычка
cross-pollination перекрестное опыление
cross-section сечение, поперечный разрез, торцовый разрез
cross-sectional area площадь поперечного сечения
cross-staff эккер
crossing скрещивание
crossing, back возвратное скрещивание
crossing, distant отдаленное скрещивание
crossing, interspecific межвидное скрещивание
crossing, intraspecific внутривидное скрещивание
crossing, reciprocal обратное скрещивание
crossing, remote отдаленное скрещивание
crossing sleeper переводный брус
crow ворона
crow, carrion черная ворона, *Corvus corone* L.
crow, hooded серая ворона, *Corvus cornix* L.
crowbar лом, аншпуг
crowberry черная водяника, шикша, *Empetrum nigrum* L.
crowded сгущенный
crowding out вытеснение
crown крона, макушка
crown breakage поломка кроны
crown contact соприкосновение крон
crown cover поверхность кроны
crown fire верховой пожар, вершинный пожар, повальный пожар
crown form форма крон
crown length протяженность кроны
crown projection проекция кроны
crown projection area поверхность кроны
crown thinning верховой метод ухода, верховое прореживание, проходная верховая рубка
crown vetch разноцветный вязель, *Coronilla varia* L.
Cruciferae крестоцветные (pl.)
cruise таксацировать
cruiser таксатор
cruising таксация
crumbling выкрошивание
crusher дробильная машина
crushing strength прочность на сжатие, сопротивление сжатию
crust корка
cryptogam споровое растение, тайнобрачное растение
cryptogams тайнобрачные (pl.)
cubic content кубатура
cubic foot кубический фут
cubic metre кубометр, кубический метр
cubic metre, solid плотный кубометр, плотный метр, фестметр

cubic metre, stacked складочный кубометр, объемный метр, складочный метр
cubical rot призматическая гниль
cuckoo кукушка, *Cuculus canoris* L.
cudgel дубина
cudweed, wood лесной бессмертник, *Gnaphalium sylvaticum* L.
cull выбраковать
culling (by shooting) отстрел
culm стебель
cultivar культивар
cultivate выращивать, обработать, разводить
cultivated обработанный
cultivation 1. культура 2. обработка, обрабатывание
cultivation, complete сплошная обработка
cultivation, shifting огневое земледелие, паловое земледелие, лесное земледелие, подсечное земледелие
cultivation (soil) обработка почвы, уход за почвы
cultivation history (of a site) агрофон
cultivator культиватор, груббер
cultivator (rotary) фреза, почвенная фреза
culture культура
cuneiform клиновидный
cup (in resin tapping) приемник
cupping поперечная покоробленность
cupule плюска
curculionid долгоносик, слоник
curl (leaf) курчавость листьев
curly кудрявый, свилеватый
curly grain свиль, свилеватость
currant смородина, *Ribes* L.
currant, alpine альпийская смородина, *Ribes alpinum* L.
currant, black черная смородина, *Ribes nigrum* L.
currant, golden золотиситая смородина, *Ribes aureum* Pursh.
currant, red красная смородина, *Ribes rubrum* L.
current поток, течение, ток
current (annual) increment текущий прирост
current-maker потокообразователь, ускоритель
current-reducer гаситель скорости течения
curve кривая, поворот
curved извилистый, корявый, согнутый
cushion подушка
cushion (leaf) листовая подушечка
cut 1. лесопользование, лесоэксплуатация, рубка, пользование лесом, лесосека 2. резать, рубить, разрезывать, вырезать, нарубать
cut (allowable) размер пользования, размер рубки, допустимая рубка
cut (calculated) расчетная лесосека
cut (permissible) размер пользования, размер рубки, допустимая рубка
cut (planned) сметная вырубка, сметная рубка, расчетная лесосека
cut area заруб
cut back отсекать
cut back to stump посадить на пень, сажать на пень
cut down валить, вырубать, сваливать, подсекать
cut in врезать
cut off обрезать, обрубать, отрубать, отсекать
cut out валить, вырубать, сваливать
cut-over area лесосека, заруб
cuticle надкожица, кутикула
cutter резец, секач
cutter-block ножевой вал
cutter head ножевая головка, фрезерная головка
cutter profile профиль резца
cutter-raker подрезающий зуб
cutter tooth резающий зуб
cutting 1. вырубка, валка, вырубание, рубка 2. выемка 3. подрезка 4. черенок
cutting angle (of rotary cutter) передний угол
cutting back срезка на обратный рост
cutting edge режущая кромка
cutting force усилие резания

cutting off обрезка, обрезание, срезание, срезывание
cutting point острие
cutting resistance сопротивление резанию
cutting section (Wagner's Blendersaumschlag system) лесорубочное звено
cutting speed скорость резания
cutting through прорубка
cycle круговорот, цикл
cylinder валок, цилиндр
cylinder saw кольцевая пила, цилиндрическая пила
cypress **1.** кипарис, *Cupressus* Tourn. **2.** кипарисовик, *Chamaecyparis* Spach.
cypress, Mediterranean обыкновенный кипарис, *Cupressus sempervirens* L.
cypress, swamp болотный кипарис, *Taxodium distichum* Rich.
Czekanow's larch чекановская лиственница, *Larix czekanowskii* Szaf.

D

Dahurian juniper даурский можжевельник, *Juniperus dahurica* Pall.
Dahurian larch даурская лиственница, *Larix gmelinii* Litv.
Dahurian rhododendron даурский рододендрон, *Rhododendron dauricum* L.
dam гать, запруда, перемычка, плотина
dam beam шандора
dam drive сплавной попуск
damage вред, повреждение, убыток
damming запруживание
damping off полегание
damson алыча, *Prunus divaricata* Led.
dandelion, common лекарственный одуванчик, *Taraxacum officinale* agg.
danewort зеленик, *Sambucus ebulus* L.
dark heart темнина
date palm финиконосный финик, *Phoenix dactylifera* L.
date palm, wild лесной финик, *Phoenix sylvestris* Roxb.
date plum обыкновенная хурма, кавказская хурма, *Diospyros lotus* L.
Dauerwald дауервальд
David's aspen осина Давида, *Populus tremula* var. *davidiana*
day-labour поденщина
day-labourer поденщик
day shift денная смена
day wage поденная плата
day work поденная работа
d.b.h. диаметр на высоте груди
dead knot несросшийся сучок
dead standing сухостойный
dead standing forest сухостойный лес
dead standing tree(s) сухостойное дерево, сухоподстой, сухостой, подстой, подсушина
dead wood мертвый лес
deadly nightshade белладонна, *Atropa belladonna* L.
deadman мертвяк
dead-nettle яснотка, *Lamium* L.
dead-nettle, red пурпуровая яснотка, *Lamium purpureum* L.
dead-nettle, spotted крапчатая яснотка, *Lamium maculatum* L.
dead-nettle, white глухая крапиво, белая яснотка, *Lamium album* L.
deal frame рама второго ряда
deals дильсы (pl.)
death-watch beetle пестрый точильщик, *Xestobium rufovillosum* De G.
debarker окорочная машина, окорочный станок
de-brancher сучкорезная машина
debris отпад
debris (felling) лесной хлам, лесосечные остатки, порубочные остатки, лесорубочные остатки (pl.)
debris (floating) карча
decapitate обезвершинивать
decare декар
decay **1.** гниение, загнивание, гниль, истлевание, истление **2.** (vb) сгнивать
decay, incipient **1.** темнина **2.** дряблость
decayed гнилой
deciduous опадающий
deciduous forest летнезеленый лес

deciphering расшифровка
decking настил, стлань, слань
decking plank палубная доска
decline отмирание
declining перестойный
decompose разлагать
decomposition истлевание, истление, разложение, распад, распадение
decorative paper overlay текстурная бумага
decrease убыль
decrepitude ветхость
deer олень
deer, fallow лань, *Dama dama* L.
deer, Far Eastern red изюбрь, *Cervus elaphus xanthopygos* M.Edw.
deer, musk кабарга, *Moschus moschiferus* L.
deer, red благородный олень, *Cervus elaphus* L.
deer, roe козуля, косуля, *Capreolus capreolus* L.
deer, Siberian red марал, *Cervus elaphus sibiricus* Sev.
deer, sika пятнистый олень, *Cervus nippon* Temm.
defect недостаток, порок, фаут
defective фаутный
defective timber фаутный лес
defective tree фаутное дерево
defective wood фаутная древесина, порочная древесина
defectiveness фаутность
defibration дефибрирование
defibrator дефибратор
deficiency **1.** (nutrient) голодание **2.** недостаток
deficient дефицитный
deflection прогиб
defoliator филлофаг
deforest обезлесить
deforestation корчевание леса, лесоочистка, обезлесение
deforested обезлесенный
deformation деформация
deformed уродливый
degeneration вырождение, перерождение
degradation деградация
dehiscent fruit растрескивающийся плод
dehydrated обезвоженный
de-limber сучкорезная машина
delivery доставка, сдача
delivery price сдаточная цена
demand потребность
demanding требовательный
demarcation межевание, размеживание
demonstration forest учебная дача
dendrochronology дендрохронология
dendroclimatology дендроклиматология
dendrology дендрология
dendrometer дендрометр
dendrometry дендрометрия
denitrification денитрификация
dense густой
densified прессованный
density **1.** объемный вес **2.** густота, плотность
density, bulk плотность загрузки
density (canopy) плотность полога, полнота леса, сомкнутость, сомкнутость крон, сомкнутость полога, степень сомкнутости
density (crop) плотность насаждения, густота насаждения, полнота леса
density, nominal условная плотность
density, population плотность населения
density, soil плотность почвы
density, stand **1.** плотность насаждения, густота насаждения, полнота леса **2.** степень полноты
density, wood плотность древесины
dentate пильчатый, зубчатый
deodar гималайский кедр, *Cedrus deodara* (Roxb.) Loud.
department ведомство
dependence зависимость
depletion of forest лесоистощение
deposit месторождение, осадок
depot склад
depot, industrial log нижний склад
depot for primary conversion лесной склад
depreciation (allowance) амортизация

depth глубина
depth gauge (chainsaw teeth) ограничительный зуб, ограничитель, выступ-ограничитель
depth of cut (height of kerf) высота пропила, пропил (высота)
derrick деррик
descending face нисходящая карра
description описание
description, mensurational таксационное описание
desert пустыня
desertification опустынивание
desiccation высушивание, высыхание
desiccator эксикатор
destroy разрушать
destruction истребление, опустошение, разрушение, уничтожение
detour объезд
detritus нанос
devastation опустошение
develop освоить, развивать
development демутация, освоение, развитие
device прибор
dew роса
dew point точка росы
dewberry ожина, сизая ежевика, *Rubus caesius* L.
de-wing (seeds) обескрыливать
de-winging (of seeds) обескрыливание
diameter 1. диаметр, толщина 2. (used in loose sense) отруб
diameter, breast-height диаметр на высоте груди
diameter, butt диаметр нижнего отруба
diameter, mean средний диаметр
diameter, mid срединный диаметр
diameter, top диаметр верхнего отруба, верхний отруб
diameter class класс толщины, ступень толщины
diameter growth рост в толщину
diameter increment прирост по диаметру, прирост по тольщине
diameter over bark диаметр с корой, диаметр в коре
diameter quotient число сбега
diameter ruler мерная скоба
diameter under bark диаметр без коры
dibble сажальный кол
dicotyledonous двусеменодольный
dicotyledons двудольные (pl.)
die back отмирать
die hammer штамповальный молот, номеровальный молот
die off отмирать, усыхать
die-back усыхание, отмирание
difference разность
differential, selection селекционная разница
diffuse light рассеянный свет
diffuse parenchyma диффузная паренхима
diffused light рассеянный свет
diffuse-porous рассеяннопоровый, рассеяннососудистый
diffuse-porous wood рассеяннопоровая древесина
diffusivity температуропроводность
dig копать
dig up выкапывать
digester варочный котел
digger body винтовый отвал
digger plough body винтовый плуг
digging, deep перештыковка
digging up выкапывание, выкопка
digitate пальчатый
digitizer кодировщик
dike перемычка, плотина
dilapidation ветхость
dimension мера, размер
dimension saw прирезный станок
dimension stock заготовка, болванка
dimensional instability формоизменяемость
dioecious двудомный
diopter диоптр
Diptera двукрылые (pl.)
direct extraction (without skidding) безтрелевочная вывозка
direction направление
directional felling направленная валка
disbudding пасынкование
discharging выгрузка
discoloration обесцвечивание

disease болезнь
disease, pathological болезнь, патологическая
disinfectant протравитель, протрава
disinfection дезинфекция, протравливание
disjunctive parenchyma разъединенная паренхима
disjunctive tracheid разъединенная трахеида
disk диск, круг, шайба
disk (stem) кружок от ствола
dismantle разбирать
dispatch отправка
dispersal распространение, полет
dispersal (seed) разлет семян, распространение семян
dispersal by wind распространение ветром
displacement **1.** вытеснение **2.** сдбиг
disposition размещение, распоряжение, расположение
dissolve растворять
distance дальность, дистанция, расстояние
distance meter дальномер
distance piece (in frame saws) прокладка, разлучка
distant дистанционный
distant crossing отдаленное скрещивание
distil гонять
distillation гонка, выгонка, перегонка
distorted перекошенный
distortion искривление
distortion, shoot искривление побегов
distribution распределение
distribution, age-class распределение классов возраста
distribution, species распределение пород
distributor распределитель
district местность, область, округ, участок
disturbance нарушение
disturbance, growth расстройство роста
ditch канава, ров, траншея
ditch digger канавокопатель
divergence расхождение
diversity разнообразие
divide водораздельная линия, водораздел
division **1.** (of an area into subcompartments) выдел участка **2.** деление, отделение, разделение
Dnieper broom днепроский ракитник, *Cytisus borysthenicus* Grun.
docking saw торцовочная пила, торцовочный станок
dodder повилика, *Cuscuta* L.
dog rose роза собачья, *Rosa canina* L.
dog's mercury многолетняя перелеска, *Mercurialis perennis* L.
dogs (of sawmill carriage) клещи (pl.)
dogwood красный дерен, глог, свидина, *Cornus sanguinea* L.
dogwood, Tatarian татарский дерен, *Cornus tatarica* Mill.
dominance господство, господствование
dominant господствующий, доминирующий
dominant species господствующая порода
dominant tree господствующее дерево
dominate господствовать
dominating господствующий, доминирующий
door дверь
door jamb дверной косяк
door post дверная стойка
dormancy покой, покой семян
dormant спящий
dormant bud спящая почка
dormouse орешниковая соня, *Muscardinus avellanarius* L.
dormouse, fat соня-полчок, *Glis glis* L.
dormouse, forest лесная соня, *Dyromys nitedula* Pall.
dormouse, garden садовая соня, *Eliomys quercinus* L.
double (of flowers) махровый
double pith двойная сердцевина
double-cut streak сдвоенная подновка
double-entry table таблица с двумя входами

double-mouldboard plough двухотвальный плуг
Douglas fir тиссолистная лжетсуга, дугласия, дугласовая пихта, *Pseudotsuga menziesii* Mirb.
Douglas fir, blue сизая лжетсуга, *Pseudotsuga menziesii* var. *glauca*
Douglas fir, green зеленая лжетсуга, *Pseudotsuga menziesii* var. *viridis*
Douglas fir, grey серая лжетсуга, *Pseudotsuga menziesii* var. *caesia*
dove голубь, *Columba* L.
dove, stock клинтух, *Columba oenas* L.
dove, turtle горлица, *Streptopelia turtur* L.
dovetail лапа-сковородень, сковородень
dovetail bit концевая фреза
dovetail joint соединение на ласточкин хвост
dowel шпилька, шкант, нагель, шпонка
dowelled joint шкантовое соединение
dowel-maker шкантовый станок
down пух
downy birch пушистая береза, *Betula pubescens* Ehrh.
downy oak пушистый дуб, *Quercus pubescens* Willd.
downpour ливень
downstream нижний бьеф
drag anchor кошка
drag brake волокуша
drag harrow шлейф-борона, волокуша
dragging волочение, подтаскивание
dragging in втаскивание
dragging out вытаска
drain 1. водосток, осушитель, кювет 2. (vb) осушать
drainage дренирование, дренаж, осушение, осушка
drainage ditch сточная канава, осушительная канава, водоотводная канава
drainage system дренажная сеть
draining дренирование, дренаж, осушение, осушка
draught осадка
drawing board чертежная доска
drawing instruments готовальня
draw-knife струг, окорочная лопатка, нож-скобель, скобель
dressed обработанный
dressing шлихтование
dried высушенный
drier сушильня, сушилка
drill 1. сверло, сверлильный станок, бурав 2. бороздка 3. сеялка 4. (vb) сверлить, буравить
drill (seed) 1. сеялка 2. посевная бороздка
drill marker бороздник
drill sowing бороздковый посев
drill through просверливать
drilling высверливание, сверление
drive 1. гонять 2. передача, привод
drive lug (chain-saw tooth) хвостовик
drive sprocket (chain saw) звездочка
drive tooth (chain-saw tooth) хвостовик
driving (of wood) сгон, гонка
driving force двигательная сила
drought засуха
drought resistance засухоустойчивость
drought tolerance засухоустойчивость
droughty засушливый
'drowning' (of seedlings) вымокание
drum цилиндр, барабан
drum-type mover барабанный ускоритель
drunken saw пьяная пила
drupe косточковый плод, костянка
dry сохнуть
dry barrel сухотарная бочка
dry chute сухий лоток
dry distillation сухая перегонка, термолиз
dry matter сухое вещество
dry out усыхать
dry period засуха
dry pruning обрезка засохших ветвей, обрезка сухих ветвей
dry stave сухотарная клепка
dry up ссыхаться
dry valley суходол
dry wind суховей
dry wood высушенный лес
drying высушивание, высыхание, сушение, сушка

drying chamber сушильная камера
drying out усыхание
drying schedule режим сушки
drying up высушивание, высыхание, засыхание, осушение, осушка, пересыхание
duct труба, трубка
duct (resin) смоляной ход
dulling затупление
dumper опрокидыватель
dune дюна
dune fixing укрепление дюн
dung помет
durability прочность, стойкость
durability of wood сохраняемость древесины
durmast oak сидячецветный дуб, скальный дуб, зимний дуб, *Quercus sessiliflora* Salisb. (*Quercus petraea* (Matt.) Liebl.)
dust пыль
dust remover эксгаустер
duster опыливатель
dusting опудривание, опыливание, распыление
Dutch elm disease голландская болезнь
dwarf карликовый
dwarf alder кустарникова ольха, *Alnus fruticosa* Rupr.
dwarf birch березка, ерник, карликовая береза, березовый стланец, *Betula nana* L.
dwarf pea tree кустарная карагана, чилига, *Caragana frutex* Koch
dwarf Russian almond степной миндаль, низкий миндаль, *Prunus tenella* Batsch
dwarf tree карликовое дерево
dwarf willow травянистая ива, *Salix herbacea* L.
dye краситель, красящее вещество, краска
dyeing крашение
dyer's greenweed красильный дрок, *Genista tinctoria* L.
dyestuff краситель, красящее вещество
dying высушивание, высыхание, гибель
dying off вымирание, отмирание, усыхание
dying tree отмирающее дерево

E

eagle, golden халзан, беркут, *Aquila chrysaetos* L.
eagle, short-toed змееяд, *Circaetus gallicus* Gmel.
eagle-owl филин, *Bubo bubo* L.
early скороспелый
early frosts осенние заморозки
early wood весенняя древесина, ранняя древесина
early-maturing скороспелый
earnings заработок
earth **1.** грунт **2.** земля **3.** почва
earth road грунтовая дорога
earth up окучивать
earthing up окучивание
earthworm дождевой червь
East Siberian fir белокорая пихта, *Abies nephrolepsis* Maxim.
eastern balsam poplar бальзамический тополь, *Populus tacamahaca* Mill.
eastern bladder-senna восточный пузырник, *Colutea orientalis* Mill.
eastern hawthorn восточный боярышник, *Crataegus orientalis* Pall.
eastern honeysuckle кавказская жимолость, *Lonicera orientalis* Lam.
eastern hornbeam грабинник, восточный граб, *Carpinus orientalis* Mill.
eating поедание
eating up поедание
ebony черное дерево, эбеновое дерево, *Diospyros ebenum* König
eccentric эксцентрический
ecology экология
economic хозяйственный
economic maturity (yielding highest income) хозяйственная спелость, экономическая спелость
ecosystem экосистема
ectotrophic эктотрофный
edaphon эдафон

edge 1. кант, грань, ребро, фаска, скос 2. кромка, лезвее, лезвие, острие 3. опушка, бровка
edge, cutting режущая кромка
edge, forest стена леса
edged обрезной
edged board обрезная доска
edge-gluing ребросклейка
edge-gluer ребросклеивающий станок
edge-jointer кромкофуговальный станок
edger обрезная пила, обрезной станок
edging обрезка, обрезание, обрезка досок, ребровая распиловка
education воспитание
effectiveness эффективность
efficiency коэффициент полезного действия, эффективность
efflorescence зацветание
effluent сточная вода, сток
egg яйцо
ejecting сбрасывание, сброска
elasticity упругость, эластичность
elasticity, modulus of модуль упругости
elaterid щелкун
elaterid larva проволочник, проволочный червь
Eldar pine эльдарская сосна, *Pinus eldarica* Medwed.
elder бузина, *Sambucus* L.
elder, common черная бузина, *Sambucus nigra* L.
elder, ground обыкновенная сныть, *Aegopodium podagraria* L.
elder, red красная бузина, *Sambucus racemosa* L.
electric power saw электропила
electrical conductivity электропроводность
electrification электрификация
element элемент
element, sieve-tube ситовидный элемент
element, trace микроэлемент
element, tracheary трахеальный элемент
elevator элеватор, подъемник, лесокатка
elfin woodland криволесье
elimination отбор
elk лось, *Alces alces* L.
elm илмь, вяз, берест, *Ulmus* L.
elm, Caucasian граболистная дзельква, *Zelkova carpinifolia* Dipp.
elm, Chinese мелколистный вяз, *Ulmus parvifolia* Jacq.
elm, common обыкновенный вяз, usually *Ulmus laevis* Pall.
elm, European white гладкий вяз, *Ulmus laevis* Pall.
elm, Siberian 1. приземистый вяз, карагач, ильмовник, *Ulmus pumila* L. 2. перистоветвистый вяз, мелколистный вяз, *Ulmus pumila* L. var. *arborea* Litv. 3. карагач, перистовет-вистый вяз, *Ulmus pinnato-ramosa* Dieck.
elm, smooth берест, лиственный вяз, листоватый вяз, *Ulmus carpinifolia* Gled. (*U. foliacea* Gilib.)
elm, wych горный илмь, шершавый вяз, *Ulmus glabra* Huds. (*U. scabra* Mill.)
elm forest ильмовник, ильмняк
elm stand вязник, вязовник, *Ulmetum*
eluvial horizon горизонт вымывания
embankment вал, насыпь
embryo зародыш
emergent маяк
emery наждак
emery cloth наждачная шкурка
emission выделение, отделение
emissions, industrial промышленные выбросы
empty дуплистый, пустой
enclose городить, обгородить
enclosed knot закрытый сучок, заросший сучок, вросший сучок
enclosing огораживание
enclosure городьба, загораживание, изгородь
end кончик, торец, отруб
end grain торцовая поверхность, торцовый разрез, торец
end surface торцовая поверхность
end-grain bending торцовое гнутье
endotrophic эндотрофный
ends (pl.) дилены
end-trimmer концеравнитель
end-trimming saw торцовочный станок

endurance выносливость
energy энергия
energy, germinative энергия прорастания
engine двигатель, паровоз
engineer инженер
engineering, genetic генетическая инженерия
engineering, hydraulic гидротехника
engraver 1. гравер 2. резчик
enlarging canker язвенный рак
enterprise предприятие
enterprise, integrated forest комплексное лесное предприятие, КЛП
entire сплошной, целный
enumeration перечет, таксация, перечислительная таксация
enumeration, complete подеревный перечет, сплошной перечет
enumeration, partial частичный перечет, выборочный перечет
enumeration, strip ленточный перечет
enumeration officer таксатор
enumeration survey перечислительная таксация
envelope оболочка
environment среда, внешняя среда
environmental conditions внешние условия
ephedra хвойник, эфедра, *Ephedra* L.
epicormic shoot волчок, водяной побег, стволовой побег
epicotyl надсеменодольное колено
epidermis кожица, эпидермис
epiphyte эпифит
epizootic эпизоотия
equilibrium равновесие
equilibrium moisture content равновесная влажность
equipment оборудование, снасть, техника
Ericaceae вересковые (pl.)
Erisyphaceae мучнисторосяные грибы
erosion эрозия
erosion, gully размывание, размыв, струйчатая эрозия
erosion, rill ручейная эрозия
erosion, sheet смыв, смывание, плоскостная эрозия, поверхностная эрозия
erosion, wind дефляция, выдувание
error ошибка, погрешность
erumpent swelling canker опухолевидный рак
esker сельга
essential oil эфирное масло
establishment 1. учреждение, заведение 2. закладка, заложение
estate усадьба, лесная дача
ester эфир
estimate оценивать
estimation оценка, смета
estimation, ocular глазомер, глазомерная оценка
estuary лиман
ether эфир
eucalypt эвкалипт, *Eucalyptus* L'Herit.
European bison зубр, *Bison bonasus* L.
European hop-hornbeam обыкновенный хмелеграб, *Ostrya carpinifolia* Scop.
European larch европейская лиственница, *Larix decidua* Mill.
European mink норка, *Mustela lutreola* L.
European spindle tree европейский бересклет, *Euonymus europaeus* L.
European white elm гладкий вяз, *Ulmus laevis* Pall.
eutrophic peatland низинное болото, низовое болото
evaluate оценивать
evaluation оценка
evaporation испарение
even-aged одновозрастный
even-aged stand одновозрастное насаждение
ever-bearing ремонтантный
evergreen зимнозеленый, вечнозеленый, неопадающий
evergreen buckthorn вечнозеленая крушина, *Rhamnus alaternus* L.
evergreen tree вечнозеленое дерево
exacting требовательный
excavation выемка

excess избыток
exchange 1. обмен, смена 2. биржа
exchange acidity обменная кислотность
exchange of experience обмен опытом
exchangeable обменный
exciter возбудитель
excrescence вырост
excretion выделение
exhausting system эксгаустер
exhaustion истощение
exhaustion of forest лесоистощение
exhaustion of soil утомление почвы
exotic иноземный, чужеземный
exotic species экзот
expenditure затрата
expense затрата
experience опыт
experiment испытание, опыт
experimental area опытная площадь
experimental field опытное поле
experimental forest district опытное лесничество
experimental plot опытная площадь
experimentation опытное дело
expert специалист
exploitable age оборотный возраст, возраст рубка
exploitable volume (the volume of wood suitable by size and quality for harvesting) эксплуатационный запас
exploitation разработка, эксплуатация
exploitation, forest лесопользование, лесоэксплуатация, лесоразработка
exploited forest освоенный лес, элсплуатируемый лес
explosion взрыв
export 1. вывоз 2. вывозить
export(s) экспорт
exposure экспозиция
extensibility растяжимость
extension отросток
extension root ростовый корень
extensive ('concentrated') clear felling концентрированная рубка, сплошно-концентрированная рубка
external внешний, наружный
external dimensions габарит
external environment внешняя среда
extinction вымирание
extinguish (fire) тушить (пожар)
extinguisher, fire огнетушитель
extinguishing тушение
extract 1. вывозить 2. (seed) вылущивать 3. (n.) вытяжка
extraction 1. вытаска, добывание 2. извлечение 3. (of seeds) лущение
extraction, direct безтрелевочная вывозка
extraction costs стоимость вывозки
extraction road лесовозная дорога
extruded board экструзионная плита
eye глаз, глазок, очко, коуш
eye ring коуш
eyelet глазок
eyepiece окуляр

F

façade фасад
face 1. грань 2. (of saw tooth) передняя грань 3. (resin tapping) карра 4. фаска, скос 5. (of a hammer) бой
face angle угол карры
face area, actual зеркало карры
face delimitation (in resin tapping) оконтуровка
face/diameter ratio (in resin tapping) нагрузка
face/girth ratio (in resin tapping) нагрузка
face site карра
face siting разметка карр
face veneer облицовочная фанера
facing накладка, облицовка
factory завод, фабрика
faggot фашина
faggot wood фашинник
failure излом, разрушение
fairlead (e.g. of winch drum) направляющая
fairway фарватер
falcon, peregrine сапсан, сокол, *Falco peregrinus* Tunst.
falcon, red-footed кобчик, *Falco vespertinus* L.

fall опадение, листопад, опад
fall behind отставать
fall block подцепочный блок
fallen fruit падалица
falling (off) опадение
fallow залежь, пар, перелог
fallow, bare черный пар
fallow deer лань, *Dama dama* L.
fallow field поле под паром
false fruit ложный плод
false heart(wood) ложное ядро
false larva (of sawflies) ложногусеница
family семья, семейство
family tree родовое дерево
farm усадьба, ферма
farm forest крестьянский лес
farmer крестьянин
farming сельское хозяйство, земледелие
fascine фашина
fascine road фашинная дорога
fascine wood фашинник
fat жир
fat dormouse соня-полчок, *Glis glis* L.
fathom (English, 6 ft (= *c.* 1.8 m); Russian, 7 ft (= *c.* 2.1 m) сажень
fathom, piled кубическая сажень
fatigue усталость
fauna фауна, (of soil) эдафон
feather-grass ковыль, *Stipa* L.
feature признак, черта
feed **1.** фураж, корм, кормовое вещество **2.** подача
feed rack кормушка
feed trough кормушка
feed up жировать
feeding area (i.e. growing space) площадь питания
feldspar полевой шпат
fell валить, вырубать, сваливать, рубить, срубать
felled вырубленный, поваленный, срубленный
felled area вырубка, вырубленная площадь
felled timber срубленный лес
feller вальщик, древоруб, лесоруб, рубщик
feller-buncher валочно-пакетирующая машина
feller-skidder ВТМ, валочно-трелевочная машина
felling **1.** вырубка, валка, вырубание, повал, пользование, рубка, срубка **2.** (in shelterwood system) прием
felling, advance предварительное пользование, промежуточное пользование, промежуточная рубка, рубка промежуточного пользования
felling, alternate clear-strip кулисная рубка
felling, alternate strip черезполосная рубка
felling, clear сплошная лесосека, сплошная рубка
felling, combined individual and group selection комплексно-выборочная рубка
felling, combined with thinning комплексная рубка
felling, conversion рубка переформирования
felling, directional направленная валка
felling, extensive ('concentrated') clear концентрированная рубка, сплошно-концентрированная рубка
felling, final окончательная рубка, очистительная рубка, очистная рубка
felling, group гнездовая рубка, котловинная рубка
felling, group selection группово-выборочная рубка
felling, illegal порубка
felling, increment простор, рубка простора
felling, intermediate предварительное пользование, промежуточное пользование, промежуточная рубка, рубка промежуточного пользования, рубка ухода
felling, landscape ландшафтная рубка
felling, localized clear сплошно-лесосечная рубка
felling, narrow strip каемчатая рубка
felling, preparatory (seeding) подготовительная рубка

felling, principal главное пользование, главная рубка, рубка главного пользования

felling, regeneration возобновительная рубка

felling, salvage санитарная рубка

felling, sanitation санитарная рубка

felling, secondary вторичная рубка, осветительная рубка

felling, seeding обсеменительная лесосека, обсеменительная рубка, семенная рубка

felling, selection выборочная рубка, выборочная добровольная рубка, добровольно-выборочная рубка

felling, shelterwood постепенная рубка

felling, strip каемчатая лесосека, полосная рубка

felling, successive последовательная рубка

felling, tending рубка ухода

felling, uniform shelterwood семенно-лесосечная рубка

felling, wavy-fronted бухтообразная рубка

felling admitting light светлая рубка

felling and primary conversion заготовка, лесозаготовка

felling area лесосека, площадь вырубки, площадь лесосеки, площадь пользования

felling area, planned сметная лесосека

felling cut валочный пропил

felling debris лесной хлам, лесосечные остатки, порубочные остатки, лесорубочные остатки (pl.)

felling licence лесорубочный билет

felling machine валочная машина

felling permit лесорубочный билет

felling plan ведомость рубкам, план рубок

felling season период рубки леса

felling sequence 1. очередь рубки 2. примыкание лесосек

felling series лесосечный ряд, лесорубочная сечь

felling site вырубка, вырубленная площадь

felling system ведение рубки, способ рубки

felling technique техника валки

fellings, interval between оборот хозяйства

fellings, order of последовательность лесосек

felloe колесный косяк, обод

felt войлок

felted beech coccus буковый войлочник, буковый червец, *Cryptococcus fagi* Bär.

female 1. самка 2. женский

fen 1. топь, низинное болото, низовое болото 2. (adj.) топяный

fence 1. городьба, забор, изгородь 2. (vb) городить, обгородить

fence post столб изгороди

fencing загораживание

fencing in огораживание

fencing wood штакетик

fender козырек

fenny топяный

fermentation брожение

fern папоротник

fern, buckler щитовник, *Dryopteris* Adans.

fern, crested buckler гребенчатый щитовник, *Dryopteris cristata* A. Gr.

fern, lady женский кочедыжник, женский папоротник, *Athyrium filix-femina* Roth.

fern, male мужской папоротник, мужской щитовник, *Dryopteris filix mas* L.

fern, narrow buckler остистый щитовник, *Dryopteris spinulosa* Kuntze

fern, shield многорядник, *Polystichum* Roth.

ferns папоротниковые (pl.)

ferruginous soil железистая почва

fertile плодовитый, плодородный

fertility плодородие

fertilization оплодотворение

fertilizer удобрение

fertilizing удобрение

fescue овсяница, *Festuca* L.

fescue, sheep's типчак, овечья овсяница, *Festuca ovina* L.
fibre волокно, мочка, фибра
fibre, bast мочало
fibre, libriform волокно либриформа
fibre, phloem лубяное волокно
fibre, wood древесное волокно
fibre saturation point предел насыщения волокна, точка насыщения волокна
fibre tracheid волокнистая трахеида
fibre waste скоп
fibreboards ДВП (древесноволокнистые плиты)
fibril волоконце, фибрилла
fibrous мочковатый, волокнистый
field 1. поле 2. полевой
field capacity полевая влагоемкость, наименьшая влагоемкость, НВ
field germination грунтовая всхожесть
field maple полевой клен, паклен, *Acer campestre* L.
field of vision поле зрения
field shelterbelt полезащитная полоса
field-mouse, striped полевая мышь, *Apodemus agrarius* Pall.
fig инжир, фига, смоковница, *Ficus carica* L.
figure рисунок, узор, текстура
figured узорчатый
figured wood узорчатая древесина
figwort шишковатый норичник, *Scrophularia nodosa* L.
filbert крупная лещина, фундук, *Corylus maxima* Mill.
file 1. пила, напильник, напилок, подпильник, подпилок 2. (vb) опиливать, пилить
filing опиловка, пиление
filings опилки (pl.)
fill насыпь
fill gaps пополнять
fill in засыпать
filler шпатлевка
fillet багет
filling 1. насыпь 2. (of wood surfaces) шпатлевание
filling gaps дополнение, пополнение
film пленка
film glue пленочный клей
filter paper фильтровальная бумага
filtration фильтрация
fin boom рея, рей, реевный бон
final crop tree 1. дерево будущего 2. избранник
final felling окончательная рубка, очистительная рубка, очистная рубка
final stage окончательная стадия
final yield главное пользование
financial maturity финансовая спелость
financial return рента
fine earth мелкозем
fine-fibred мелковолокнистый
fine-grained мелковолокнистый
fines мелочь
finger (in a finger-joint) зубчатый шип
finger joint соединение на зубчатый шип, клиновидное соединение, зубчатое соединение
finishing доработка, отделка
fir 1. пихта, *Abies* Mill. 2. пихтовый
fir, Caucasian пихта Нордманна, кавказская пихта, *Abies nordmanniana* Spach.
fir, Douglas тиссолистная лжетсуга, дугласия, дугласовая пихта, *Pseudotsuga menziesii* Mirb.
fir, East Siberian белокорая пихта, *Abies nephrolepsis* Maxim.
fir, Kamchatka грациозная пихта, камчатская пихта, тонкая пихта, *Abies gracilis* Kom.
fir, Manchurian цельнолистная пихта, *Abies holophylla* Maxim.
fir, Sakhalin сахалинская пихта, *Abies sachalinensis* Mast.
fir, Siberian сибирская пихта, *Abies sibirica* Ledeb.
fir, silver европейская пихта, белая пихта, гребеньчатая пихта, *Abies alba* Mill.
fir balsam пихтовый бальзам
fir clubmoss плаун-баранец, *Lycopodium selago* L.
fir forest пихтарник, пихтач, *Abietum*
fir oil пихтовое масло

fir/spruce forest пихтоельник
fire огонь, пожар
fire, back отжиг, встречный огонь, встречный пожар
fire, counter отжиг, встречный огонь, встречный пожар
fire, crown верховой пожар, вершинный пожар, повальный пожар
fire, forest лесной пожар
fire, ground подземный пожар, почвенный пожар
fire, peat торфяной пожар
fire, running беглый пожар
fire, spot пятнистый пожар
fire, surface наземный пожар, низовой пожар
fire break противопожарная полоса, заградительная полоса, противопожарный разрыв
fire-break maker полосопрокладыватель
fire-damaged forest горельник, горелый лес
fire danger горимость, пожарная опасность
fire-danger index показатель пожарной опасности
fire-danger index, Nesterov's комплексный гидротермический показатель
fire-danger scale шкала горимости, шкала пожарной опасности
fire extinguisher огнетушитель
fire line противопожарная полоса, заградительная полоса, противопожарный разрыв
fire protection пожарная охрана
fire-resistant огнестойкий, огнеупорный
fire retardant антипирен
fire risk пожарная опасность, опасность пожара
fire scar подсушина, пожарная подсушина, шрам
fire service пожарное дело
fire tower пожарная вышка, пожарная каланча
firecrest красноголовый королек, *Regulus ignicapillus* Temm.
firewood дровяная древесина, дрова (pl.), файрвуд, шырок
firewood tree дровяное дерево
firewood-cleaving machine дровокольный станок
firing топка
fish glue рыбий клей
fission расщепление
fissured (of bark) трещиноватый
fissured bark трещиноватый кора
fitter слесарь
five-year plan пятилетка
fixed (basic) capital основные фонды
fixed price такса
fixed stumpage price попенная плата
fixing закрепление, укрепление
fixing dunes укрепление дюн
flaker щепорубительная машина
flame пыл
flat car платформа
flat sawing тангенциальная распиловка
flat (slabbed) sleeper пластинчатая шпала
flat-pressed board плита плоского прессования
flaw свищ
fleck (of light) блик
fleck (pith) сердцевинное повторение, сердцевинное пятно
fleet (e.g. of vehicles) парк
flesh (of fruit) мякоть
flexibility гибкость
flight полет
flight map планшет
flint кремень
flitch ванчес, брус
float **1.** поплавок, томбуй **2.** (vb) сплавлять
floatable сплавной
floated timber сплавной лес
floating сплавное дело, сплав, лесосплав
floating dam лесосплавная плотина
floating debris (e.g. trees with roots) карча
floating depot рейд, сплавной рейд
floating hook багор

floating of loose wood лесосплав россыпью, моль, молевой сплав, сплав россыпью
floating of wood сплав леса
floating operator плотовщик
floating river (for driving or rafting) сплавная река
floating season навигация, навигационный период
flock гурт, стадо
flood затопление, наводнение, паводок, половодье
flooding затопление, наводнение
floodland плавня
floodplain пойма
floodplain forest заливной лес, пойменный лес
floodplain soil пойменная почва
floor пол; (forest) лесная подстилка
floor board половица, доска пола, половая доска, настилочная доска
flooring настил, стлань, слань
flora флора
floristic флористический
flotation oil флотационное масло
flotsam аварийная древесина
flour мука
flour, vitamin витаминная мука, хвойно-витаминная мука
flour, wood древесная мука
flower цвет, цветок
flower bed клумба
flower bud цветковая почка, цветочная почка, бутон
flower-bud formation бутонизация
flower stalk цветонос, цветоножка
flowering цветение
flowering plant явнобрачное растение, цветочное растение
flowing поточный
fluff пух
fluid жидкость
fluidity текучесть
flume водяной лесоспуск, лоток, мокрый лоток, водный лоток
flushing листораспускание, распускание
flute ройка, рукава
fluting ройка
fly муха
fly agaric красный мухомор, *Amanita muscaria* L.
fly honeysuckle обыкновенная жимолость, лесная жимолость, *Lonicera xylosteum* L.
flycatcher мухоловка
flycatcher, pied мухоловка-пеструшка, *Muscicapa hypoleuca* Pall.
flycatcher, spotted серая мухоловка, *Muscicapa striata* Pall.
flycatcher, white-collared мухоловка-белошейка, *Muscicapa albicollis* Temm.
focus (of infection or infestation) очаг
fodder фураж, корм, кормовое вещество
fodder plant кормовое растение
fodder yeasts кормовые дрожжи (pl.)
fog туман
folding rule складной метр
foliage зелень, листва
foliation облиствение
foothill(s) предгорье
forage фураж, корм, кормовое вещество
fore-and-aft road лежневая дорога
forecast(ing), weather предсказание погоды
foreign trade внешняя торговля
foreman мастер
forest 1. лес 2. лесной
forest, accessible доступный лес
forest, artificial рукотворный лес, плантация
forest, bank protection берегозащитный лес
forest, bottom-land урема, тугай
forest, broadleaved лиственный лес
forest, centrally managed лес государственного значения
forest, city городской лес
forest, climax климаксный лес
forest, closed сомкнутый лес
forest, commercial хозяйственный лес
forest, communal общинный лес
forest, community общинный лес
forest, conifer хвойный лес
forest, continuous дауервальд
forest, coppice низкоствольный лес

forest, dead standing сухостойный лес
forest, deciduous летнезеленый лес
forest(s), deciduous чернолесье
forest, demonstration учебная дача
forest, depletion of лесоистощение
forest, exhaustion of лесоистощение
forest, exploited освоенный лес, эксплуатируемый лес
forest, farm крестьянский лес
forest, fire-damaged горельник, горелый лес
forest, floodplain заливной лес, пойменный лес
forest, fuelwood дровяной лес
forest, gallery галлерейный лес
forest, high семенной лес, высокоствольник, высокоствольный лес, семенное насаждение
forest, inaccessible недоступный лес
forest, kolkhoz колхозный лес
forest, managed хозяйственный лес
forest, mature спелый лес
forest, mixed смешанный лес
forest, montane горный лес
forest, multi-storeyed сложный лес
forest, normal нормальный лес
forest, open редкий лес, редколесье, светлый лес
forest, opening in межлесье
forest, overmature перестройиый лес
forest, planting of насаждение леса
forest, primary девственный лес, первобытный лес, стихийный лес, пралес
forest, private частновладельческий лес, частный лес
forest, protection защитный лес, охранный лес
forest, rain дождевой лес
forest, ravine байрачный лес
forest, reserved заказник
forest, riparian тургай, урема
forest, scrub кустарный лес
forest, secondary вторичный лес
forest, selection выборочный лес
forest, single-storeyed простой лес
forest, sovkhoz совхозный лес
forest, sparse редкий лес, редколесье
forest, state государственный лес
forest, suburban пригородный лес
forest, sustained-yield непрерывно-производительный лес
forest, training учебная дача
forest, unexploited неосвоенный лес, резервный лес
forest, uniform high лесосечный лес
forest, urban городской лес
forest, virgin девственный лес, первобытный лес, стихийный лес, пралес
forest, water-conservation водоохранный лес
forest, water-regulating водорегулирующий лес
forest academy ЛТА, лесотехническая академия
forest administration лесоуправление, лесное управление
forest approaching maturity приспевающий лес
forest area лесная площадь
forest area (total) лесной фонд
forest bee-keeper бортник
forest bee-keeping бортевое пчеловодство, бортничество
forest border стена леса
forest botany лесоботаника
forest chemical industry (resin tapping, chemical utilization of wood, etc.) лесохимическая промышленность
forest chemistry лесная химия, лесохимия
forest clearance корчевание леса, лесоочистка, сводка леса
forest code лесной кодекс
forest combine (large integrated processing plant) лесокомбинат
forest concession лесосырьевая база
forest cover percent облесенность, лесистость, процент лесистости
forest destruction лесоистребление
forest district **1.** лесная дача (subdivision of lesnichestvo) **2.** лесничество (subdivision of leskhoz) **3.** лесоучасток, лесной округ
forest dormouse лесная соня, *Dyromys nitedula* Pall.

forest economic region лесоэкономический район, ЛЭР
forest edge стена леса
forest estate лесная дача
forest experiment station ЛОС, лесная опытная станция
forest exploitation лесопользование, лесоэксплуатация, лесоразработка
forest fire лесной пожар
forest fire statistics горимость
forest floor лесная подстилка
forest-forming species лесообразователь
forest grazing лесная пастьба
forest improvement мелиорация леса
forest in use освоенный лес, элсплуатируемый лес
forest industry лесная промышленность
forest-industry complex ЛПК, лесопромышленный комплекс
forest inspector лесной ревизор
forest inventory лесная таксация, учет лесного фонда
forest lands лесные угодья
forest law лесной закон, лесное право
forest log depot верхний склад
forest management **1.** лесное хозяйство, лесное дело **2.** (in the strict sense) лесоустройство
forest manager лесоустроитель
forest massif лесной массиф
forest nursery лесной питомник
forest orchard (of wild fruit and nut trees) лесосад, лесной сад
forest originating from coppice (or suckers) порослевый лес
forest outlier колок, островный лес, колочный лес, остров
forest outliers колки (pl.)
forest owner лесовладелец
forest park лесопарк
forest plant лесное растение
forest plot лесоучасток
forest plough лесной плуг
forest policy лесная политика
forest produce (minor) побочный продукт
forest product(s) лесной материал, лесоматериал(ы), лесопродукция
forest property лесная собственность
forest protection лесозащита, лесоохрана, лесоохранение, охрана леса
forest regulation лесоустройство
forest reserve заповедный лес
forest resource лесной фонд
forest right лесное право
forest road лесная дорога
forest science лесоведение
forest seed orchard ЛСП, лесосеменная плантация
forest service лесная служба
forest soil лесная почва
forest stand лесонасаждение, лесное насаждение
forest statistics лесная статистика
forest steppe лесостепь
forest strip лесная полоса
forest strip (protective) лесозащитная полоса
forest tending уход за лесом
forest tree лесное дерево
forest tundra лесотундра
forest type тип леса
forest typology лесная типология
forest under management хозяйственный лес
forest warden лесничий, лесник
forested лесистый, облесенный, покрытый лесом
forested area лесопокрытая площадь
forester лесничий, лесник, лесохозяйственник, лесовод
forester, assistant лесной ассистент
forester, junior (subordinate) лесной ассессор
forester, practising лесной практикант
forestry **1.** лесное хозяйство, лесное дело **2.** (adj.) лесохозяйственный
forestry adviser лесной советник
forestry artel лесная артель
forestry aviation лесная авиация
forestry power source ЛЭС, лесное энергетическое средство
forestry technical school лесной техникум
forget-me-not незабудка, *Myosotis* L.
fork вилка, развилина

forked вилообразный, вильчатый
forked tree двойчатка, развилка
forking развилистый
fork-lift truck автопогрузчик с вилочным захватом
form 1. форма 2. вид
form factor видовое число, число полнодревесности
form height видовая высота
form quotient коэффициент формы
formaldehyde формальдегид
formation 1. образование, формирование 2. свита
formation of forest лесообразование
formula (stem volume) формула обьема ствола
forwarder сортиментовоз, движитель, подборщик-сортиментовоз, погрузочно-транспортная машина, форвардер
forwarding wood подача леса
fossil ископаемый
foundation основа, основание, цоколь
foundation pit котлован
fountain ключ
four cutter четырехсторонний станок
fox лисица, *Vulpes vulpes* L.
fox, arctic песец, *Alopex lagopus* L.
fox-coloured sawfly сосновый рыжий пилильщик, сосновый пилильщик, *Diprion sertifer* Geoffr.
fox-tail лисохвост, *Alopecurus* L.
fraction часть
fracture излом
fragile ломкий, хрупкий
fragrant poplar душистый тополь, *Populus suaveolens* Fisch.
frame(work) каркас, остов
frame re-saw рама второго ряда, распускная рама
frame saw лесорама, рамная пила, рама, лесопильная рама, пильная рама
frame saw (for logs) рама первого ряда
frame saw (for deals) рама второго ряда
frame saw, horizontal горизонтальная рама
frame saw, vertical вертикальная рама
frame-saw log carriage впередирамная тележка
frame-saw operator рамщик
frass древоточина, буровая мука
fraying 1. обшмыг коры, ошмыг коры, обдир коры 2. сдирание кожицы с рогов
free свободный
free moisture свободная влага
free on board франко борт
freezing замораживание, замерзание, оледенение
freezing point точка замерзания
freight фрахт
freight turnover (e.g. of a road) грузооборот
frequency 1. встречаемость 2. частота
frequency, high высокая частота
frequency curve кривая частоты
fresh свежий
fresh (i.e. fairly moist) soil свежая почва
freshening (in resin tapping) вздымка, рана-подсочка, карроподновка, подновка
freshening interval пауза
freshening tool струг
freshly felled свежесрубленный
freshly felled tree свежесрубленное дерево
fret saw лобзик, ажурная пила, прорезная пила, лобзиковый станок
friable рыхлый
friction трение
friction resistance сопротивление трению
frighten off отпугивать
fringe бахрома
frog лягушка, *Rana* L.
frond вайя
front (of saw tooth) передняя грань
frost мороз
frost, glazed гололедица
frost, hoar изморозь
frost, morning утренник
frost crack морозобой(на), зяблина, морозная трещина, морозобойная трещина

frost hardiness морозоупорность, морозостойкость
frost-hardy морозостойкий
frost heave мерзлотное пучение, выжимание
frost resistance морозоупорность, морозостойкость
frost-resistant морозостойкий
frost rib морозобойный нарост
frosts заморозки (pl.)
frosts, autumn осенние заморозки
frosts, early осенние заморозки
frosts, late весенние заморозки
frosts, spring весенние заморозки
frozen мерзлый
frozen ground мерзлота
fruit плод
fruit, aggregate сборный плод
fruit, compound сложный плод
fruit, dehiscent растрескивающийся плод
fruit, fallen падалица
fruit, false ложный плод
fruit, stone косточковый плод, костянка
fruit stalk плодоножка
fruit tree плодовое дерево
fruitful плодовитый, плодородный, урожайный
fruitfulness урожайность
fruiting 1. плодоношение 2. (adj.) плодоносный
fruiting body плодовое тело
fuel горючее, горючее вещество, горючий материал, топливо
fuelwood дровяная древесина, дрова (pl.), дровяной лес
fuelwood forest дровяной лес
full seed полноценная семя
full-cell process метод полного поглощения
fullness of seeds полнозернистость семян
fully stocked полный
fully stocked stand полное насаждение
fumigation фумигация, окуривание
fuming морение
fumitory, common лекарственная дымянка, *Fumaria officinalis* L.
function зависимость
fund(s) фонд(ы)
fungi, lower низшие грибы
fungi, mould плесневые грибы
fungi, pileate шляпочные грибы
fungi, rust ржавчинные грибы
fungi, wood-destroying дереворазрушающие грибы
fungi, wood-staining деревоокрашивающие грибы
Fungi imperfecti несовершенные грибы
fungicide протравитель, протрава, фунгицид
fungus гриб, трут, трутовик
fur мех, пушнина
fur coat (muffler effect) шуба
furfural фурфурол
furnace печь, топка
furniture мебель
furniture beetle точильщик
furniture beetle, common мебельный точильщик, *Anobium punctatum* De G.
furniture factory мебельная фабрика
furrow борозда, бороздка
furrow planting бороздовая посадка
furrow slice пласт
furrow sowing бороздовый посев
fusiform веретенообразный

G

gale, sweet обыкновенный восковник, *Myrica gale* L.
galipot галипот
gall чернильный орех, галл, чернильный орешок
gallery forest галлерейный лес
gall-gnat галлица
game дичь
game management охотничье хозяйство
gang бригада
gang, mechanized комплексная бригада
gang of workers артель рабочих
ganger бригадир, мастер
gantry crane козловый кран

gap голье, котловина, пустое место, прогалина, просвет, пропуск, окно
gappy stand насаждение с просветами
garden сад
garden chafer садовый хрущик, *Phyllopertha horticola* L.
garden dormouse садовая соня, *Eliomys quercinus* L.
gardener садовник
gas газ; (producer) генераторный газ
gas generator газогенератор
gas resistance газоустойчивость
gate ворота (pl.)
gateway ворота (pl.)
gathering сбор
gauge **1.** датчик **2.** калибр, шаблон
gauge, angle угловый шаблон
gauge, depth ограничительный зуб, ограничитель
gauge, non-adjustable предельный калибр
gauge, rain дождемер
gauge, resistance strain датчик сопротивления
gauge, seed семеномер
gauge, setting разводной шаблон
gauge, thickness толщиномер
gene ген
gene bank генофонд
gene resource(s) генофонд
genealogy родословие
generation **1.** поколение **2.** элемент леса
generator tar генераторная смола
generic родовой
genetic diversity генетическое разнообразие
genetic engineering генетическая инженерия
genetic gain генетическая прибыль
genome геном
genotype генотип
gentian горечавка, *Gentiana* L.
genus род
geocoenosis геоценоз
geographic information system географическая информационная система, ГИС
geometrid пяденица
Georgian honeysuckle грузинская жимолость, *Lonicera iberica* M.B.
Georgian oak грузинский дуб, *Quercus iberica* Stev.
gerbil песчянка
germander дубровник, *Teucrium* L.
germinate всходить, прорастать
germinated seedling росток
germinating seedling проросток
germination прорастание
germination (field) грунтовая всхожесть
germination percent процент всхожести
germinative capacity всхожесть
germinative energy энергия прорастания
giant sequoia мамонтовое дерево, гигантская секвоя, *Sequoia gigantea* Decne.
gibberellic acid ГК
gimlet буравчик
gipsy-wort европейский зюзник, *Lycopus europaeus* L
girder ферма
girdle **1.** кольцевать **2.** кольцо
girdling кольцевание, окольцовывание **2.** (to kill trees in shifting cultivation) подсечка
girdling trees (before felling) подвяливание на корню
girth обхват, окружность
GIS географическая информационная система, ГИС
glaciation оледенение
glacier ледник
glade перелесье, поляна
glance boom направляюший бон, продольный бон, реевный бон
gland железа, железка
glasshouse парник, теплица
glazed frost гололедица
gleyed soil оглееная почва
gleying оглеение
global positioning system глобальная система позиционирования, ГПС
glue **1.** клей **2.** (vb) склеивать
glue, bone костный клей
glue, casein казеиновый клей

glue, cold-setting клей холодного отверждения
glue, film пленочный клей
glue, fish рыбий клей
glue, hide мездровый клей
glue, hot-setting клей горячего отверждения
glue, phenol-formaldehyde фенольно-формальдегидный клей
glue, thermoplastic термореактивный клей, термопластичный клей
glue, thermo-setting термореактивный клей, термопластичный клей
glue, urea карбамидный клей
glue, urea-formaldehyde мочевино-формальдегидный клей
glue joint клеевое соединение
glueline клеевая нить, шов, фуга
glueline defect непроклей
glue-spreader клеенаноситель
gluing склейка
glutton россомаха, *Gulo gulo* L.
goat (female) коза
goat (male) козел
goat-leaf honeysuckle каприфоль, жимолость-каприфоль, *Lonicera caprifolium* L.
goat moth пахучий древоточец, ивовый древоточец, *Cossus cossus* L.
goat willow козья ива, ракита, бредина, *Salix caprea* L.
goat's beard луговой козлобородник, *Tragopogon pratensis* L.
goldcrest желтоголовый королек, *Regulus regulus* L.
golden-chain золотой дождь, *Laburnum anagyroides* Med.
golden currant золотиситая смородина, *Ribes aureum* Pursh.
golden eagle халзан, беркут, *Aquila chrysaetos* L.
golden-rod золотарник, золотая розга, *Solidago virgaurea* L.
goldfinch щегол, *Carduelis carduelis* L.
goniometer угломер
goods товар
goods, rejected бракованный товар
goods, unsorted бессортный товар
gooseberry крыжовник, *Ribes grossularia* L.
goosegrass цепкий подмаренник, *Galium aparine* L.
gorge ущелье
goshawk ястреб-тетеревятник, *Accipiter gentilis* L.
gouge-type tooth Г-образный зуб, режкообразный зуб, фуговочный зуб
goutweed обыкновенная сныть, *Aegopodium podagraria* L.
GPS глобальная система позиционирования, ГПС
grab грейфер, захват
graceful tamarisk изящный гребенщик, *Tamarix gracilis* Willd.
gradation градация, ступень
grade **1.** сортимент, сорт **2.** (vb) сортировать, браковать
grader подборщик, бракер
gradient уклон, наклон
gradient, ruling руководящий подъем
grading **1.** планировка **2.** сортировка, браковка
grading, strength прочностная сортировка
grading, stress силовая сортировка
gradual постепенный, ступенчатый
graft **1.** прививание, прививка, прививок **2.** (vb) прививать
graft, approach прививка сближением
graft, bridge прививка мостиком
graft, cleft прививка в расщеп
graft, pith-to-cambium прививка сердцевиной на камбий
graft, rind прививка за кору
graft, saddle прививка седлом
graft, side прививка сбоку
graft, wedge прививка в вырез
grafter прививальщик
grafting прививание, прививка
grafting, apical копулировка, прививка в пенек
grafting, approach аблактировка
grafting, splice копулировка, прививка в пенек
grafting, wedge инкрустация
grafting, whip простая копулировка

grafting, whip-and-tongue копулировка с язычком, улучшенная копулировка
grafting knife садовый нож
grafting wax древесный воск, садовая смола
grain 1. волокно, узор, текстура, рисунок 2. зерно
grain, across the поперек волокон
grain, along the вдоль волокон
grain, chipped задранное волокно
grain, curly свиль, свилеватость
grain, end торцовая поверхность, торцовый разрез, торец
grain, irregular свиль, свилеватость
grain, parallel to the вдоль волокон
grain, perpendicular to the поперек волокон
grain, slope of наклон волокон
grain, spiral косослой, косослойность
grain, torn вырванное волокно
grain, wavy свиль, свилеватость
grain size зернистость
Gramineae злаки, злаковые
granularity зернистость
graph график
grapple гидроманипулятор, манипулятор, захват
grass злак, мурава, злаковое растение, трава
grass cover травяной покров
grass sward залежь, мурава
grasses злаки, злаковые
grassy травянистый
gravel хрящ
graveyard (test area) полигон
gravity сила тяжести, тяжесть
gravity, centre of центр тяжести
gravity, specific удельный вес
gravity cableway канатный лесоспуск, канатный откать, канатная самотаска
grazing пастьба
grazing (forest) лесная пастьба
grease жир, мазь
grease band ловчее кольцо
grease-proof paper жиростойкая бумага
great bustard дрофа, *Otis tarda* L.
great tit большая синица, *Parus major* L.
greater celandine чистотел, *Chelidonium majus* L.
greater spotted woodpecker большой пестрый дятел, *Dryobates major* Brehm
grebe поганка, *Podiceps* L.
Grecian juniper высокий можжевельник, *Juniperus excelsa* Bieb.
green свежесрубленный, сырой
green ash зеленый ясень, пенсильванский ясень, *Fraxinus pennsylvanica* Marsh.
green belt зеленый пояс
green cutting зеленый черенок
green Douglas fir зеленая лжетсуга, *Pseudotsuga menziesii* var. *viridis*
green manure зеленое удобрение
green manure crop сидерационная культура, сидерат
green manuring сидерация
green pruning обрезка зеленых ветвей
green-stuff зелень
green wood сырая древесина
green woodpecker зеленый дятел, *Picus viridis* L.
greenbark maple зеленокорый клен, *Acer tegmentosum* Maxim.
greenery зелень
greenhouse теплица, парник
greenhouse effect парниковый эффект
greenweed, dyer's красильный дрок, *Genista tinctoria* L.
grey desert soil серозем
grey Douglas fir серая лжетсуга, *Pseudotsuga menziesii* var. *caesia*
grey poplar сереющий тополь, серый тополь, *Populus canescens* L.
grey soil серая почва
grey willow серая ива, пепельная ива, *Salix cinerea* L.
grid решетка
grill решетка
grind шлифовать
grinder дефибрер
grinding 1. дефибрирование 2. заточка, шлифование, шлифовка
groove долбеж, долбежка, дорожка, желоб, паз, фальц, шпунт

groover пазовальный станок
grooving фальцевание
grooving cutter пазовое сверло, пазовая фреза
grooving plane дорожник, фальцгобель
grosbeak, pine щур, *Pinicola enucleator* L.
gross income валовой доход
ground beetle жужелица
ground cherry вишарник, кустарниковая вишня, степная вишня, *Prunus fruticosa* Pall.
ground clearance (of vehicle) клиренс
ground cover живой покров, напочвенный покров
ground elder обыкновенная сныть, *Aegopodium podagraria* L.
ground fire подземный пожар, почвенный пожар
ground ivy плющевидная будра, *Glechoma hederacea* L.
ground peat торфяная пыль
ground skidding наземная трелевка
groundsel обыкновенный крестовник, *Senecio vulgaris* L.
groundwater грунтовая вода
groundwater level уровень грунтовой воды
groundwood, white белая древесная масса
groundwood pulp древесная масса
group **1.** куртина, группа **2.** (in group system) котловина
group felling гнездовая рубка, котловинная рубка
group I forests (shelterbelts, amenity and greenbelt forests, etc., where tending and regeneration are the prime objects of management) группа I (первая)
group II forests (forests of commercial and water-conservation importance in less well-forested regions, where cut should not exceed m.a.i.) группа II (вторая)
group III forests (all other forests, i.e. production forests, where logging is the prime concern) группа III (третья)
group regeneration куртиное возобновление
group selection felling группово-выборочная рубка
grouse, black тетерев-косач, тетерев, *Lyrurus tetrix* L.
grouse, willow белая куропатка, *Lagopus lagopus* L.
grove роща
grove (sacred) мольбище
grow прирастать, произрастать, расти
grow together срастаться
grow up подрастать
grower растениевод
growing произрастание
growing in врастание
growing over перерастание
growing point конус нарастания, точка роста
growing season вегетационный период
growing space жизненное пространство
growing stock запас
growing stock, normal нормальный запас
growing-stock increment прирост по запасу
growing-stock target целевой запас
growing together сращивание, срастание
growing wild дикорастущий
growth нарастание, наращивание, прирост, произрастание, рост
growth check задержка в росте
growth conditions условия произрастания
growth course ход роста
growth curve кривая хода роста
growth disturbance расстройство роста
growth ring годичное кольцо, годичный слой
growth substance ростовое вещество
grub up выкорчевывать
grubber культиватор, груббер, корчевальная машина
grubbing up корчевка, корчевание, выкорчевывание, раскорчевка
guaiacol гваякол
guard **1.** охрана, страж, сторож **2.** (vb) охранять
guelder rose калина, *Viburnum opulus* L.

guide (e.g. of band-saw) направляющая
guide bar (of power saw) пильная шина
guiding boom направляюший бон
gullet (of saw tooth) впадина, междузубие, пазуха
gully овраг, промоина
gully erosion размывание, размыв, струйчатая эрозия
gum камедь
gum bleeding камедетечение
gum turpentine живичный скипидар
gummosis гуммоз
gunpowder порох
gust of wind порыв ветра
gutta-percha (tree) гутта(перча), гуттаперчевое дерево, *Eucommia ulmoides* Oliv.
gutter **1.** (e.g. in resin tapping) желобок **2.** канава
guy растяжка
gymnospermous голосеменный, голосемянный
gymnosperms голосеменные (pl.), голосемянные (pl.)
gypsy moth непарный шелкопряд, *Lymantria dispar* L.

H

habit габитус
habitat местообитание, местопроизрастание, родина
hack (as used in freshening) хак
hackberry каркас, *Celtis* L.
hack-saw ножевка, ножовка, ножевая пила
hail град
hail damage градобитие, градобой
hailstone градина
hair moss обыкновенный кукушкин-лен, обыкновенный политрихум, *Polytrichum commune* Hedw.
half пол-, полу-
half-barked полуокоренный
half-finished product заготовка, болванка
half-round bar пластина, половинник
half-sib полусиб
half-stuff полумасса
hammer молоток, молот
hammer, cleaving расколочный молот
hammer, die штамповальный молот, номеровальный молот
hammer, marking маркировочный молот, разметочный молот
hammer, numbering нумератор
hammer hardening (saw blades) наклеп, проковка, рихтовка
hamster хомяк, *Cricetus cricetus* L.
hand saw ручная пила
hand work ручная работа
handicraftsman кустарь
handle рукоятка, ручка
hang (of vertical frame saw) перевешивание пилы, наклон пилы, уклон пилы, свес
hanging bog висячее болото, склоновое болото
hard alloy твердый сплав
hard-alloy tip пластинка твердого сплава
hard wood твердая древесина
hardboard твердая плита
hardened resin серка
hardening закаливание, закалка
hardiness выносливость, стойкость
hardness твердость
hardwood лиственная древесина
hard-wooded broadleaved твердолиственный
hard-wooded broadleaved species твердолиственная порода
hard-wooded species твердая порода
hardy выносливый
hare заяц, *Lepus* L.
hare, brown заяц-русак, русак, *Lepus europaeus* L.
hare, mountain заяц-беляк, беляк, *Lepus timidus* L.
hare's-tail влагалищная пушица, *Eriophorum vaginatum* L.
harm вред
harmful вредительный, вредный
harmful insect энтомовредитель, вредное насекомое

harness сбруя, упряжь
harrow борона
harrowing боронение
Hartwiss oak дуб Гартвиса, *Quercus hartwissiana* Stev.
harvest 1. уборка, урожай 2. (vb) заготовить
harvest (seed) урожай семян
harvest (timber) урожай древесины
harvest mouse мышь-малютка, *Micromys minutus* Pall.
harvester лесной комбайн, харвестер
harvesting добывание, заготовка, лесозаготовка, сбор
hastate willow копьевидная ива, *Salix hastata* L.
hatch люк
hatchet топорик
hatchway люк
haul line тяговой канат
hauling возка
hautbois strawberry клубника, *Fragaria moschata* Duch.
hawfinch дубонос, *Coccothraustes coccothraustes* L.
hawk moth бражник
hawk-owl ястребиная сова, *Surnia ulula* L.
hawkweed ястребника, *Hieracium* L.
hawser причал
hawthorn 1. боярышник, *Crataegus* L. 2. однопестичный боярышник, *Crataegus monogyna* Jacq. 3. пятипестичный боярышник, *Crataegus pentagyna* W.K.
hawthorn, eastern восточный боярышник, *Crataegus orientalis* Pall.
hawthorn, Midland обыкновенный боярышник, колючий боярышник, *Crataegus oxyacantha* L.
hay сено
hay-making сенокос
haze мгла, марь
hazel 1. лещина, орешник, *Corylus* L. 2. разнолистная лещина, *Corylus heterophylla* Fisch.
hazel, common обыкновенная лещина, *Corylus avellana* L.
hazel, Japanese лещина Зибольда, *Corylus sieboldiana* Bl.
hazel, Manchurian маньчжурская лещина, *Corylus mandshurica* Max.
hazel, Turkish медвежий орех, древовидная лещина, медвежий орешник, *Corylus colurna* L.
hazel wood лещинник, орешник
hazel-hen рябчик, *Tetrastes bonasia* L.
head frame saw рама первого ряда
head of axe обух топора
head saw головная пила, головной древнопильный станок
head spar головая мачта
head-guard шлем
headings (for casks) донник
headrig, chipper-canter фрезерно-брусующий станок
headrig, chipping хедриг, фрезернопильный агрегат
heal over зарастать
healing over зарастание
heap (esp. for burning) костер, куча
heap up окучивать
heaping up окучивание
heart, dark темнина
heart rot сердцевинная гниль, центральная гниль
heart shake (extending the length of the material) метик, метиковая трещина, сердцевинная трещина
heart shake (in a single vertical plane) согласный метик
heart shake, spiral несогласный метик
heartwood ядро, ядровая древесина
heart(wood), false ложное ядро
heart(wood), red красное ядро
heat жар, жара, тепло, теплота
heat, specific теплоемкость
heat capacity теплоемкость
heat-loving теплолюбивый
heat resistance теплостойкость
heath, cross-leaved болотный вереск, *Erica tetralix* L.
heath, tree древовидный вереск, *Erica arborea* L.
heather вереск, *Calluna vulgaris* L. (Hull.)

heave (frost) мерзлотное пучение, выжимание
heaven, tree of айлант, *Ailanthus altissima*
hectare гектар
hedge 1. изгородь 2. (vb) городить, огородить
hedgehog обыкновенный еж, *Erinaceus europaeus* L.
hedge-sparrow лесная завирушка, *Prunella modularis* L.
heel (of cutting) пятка
heel in прикапывать
heeling in прикопка
height высота
height, breast высота груди
height, form видовая высота
height, mean средняя высота
height, mid срединная высота
height, stand высота насаждения
height class класс высоты, ступень высоты
height curve кривая высот
height/diameter ratio стройность
height growth рост в высоту
height increment прирост по высоте
height quality class (defined by the height/diameter curve) разряд высоты
height series разряд высоты
height table таблица высот
height to top высота вершины
helmet шлем
helve топорище
hemicellulose гемицеллюлоза
Hemiptera полужесткокрылые (pl.)
hemlock болиголов, *Conium maculatum* L.
hemp конопля, *Cannabis sativa* L.
hemp agrimony коноляный посконник, седач, *Eupatorium cannabinum* L.
hemp-nettle пикульник, *Galeopsis* L.
henbane черная белена, *Hyoscyamus niger* L.
Hepaticae печеночные мхи
herb травянистое растение, трава
herb Paris вороний глаз, *Paris quadrifolia* L.
herb Robert Роберта герань, *Geranium robertianum* L.
herbaceous травянистый
herbaceous cover травяной покров
herbaceous plant травянистое растение
herbarium гербарий
herbicide гербицид, протравитель, протрава
herbivorous травоядный
herd гурт, стадо
hereditary наследственный
heredity наследственность
heritability наследуемость
herringbone (design) елочка
heterogeneous разнородный
hew (roughly) затесывать, обтесывать, отесывать, тесать
hewing обтеска, тесание, теска
hewing off стесывание
hewn обтесанный
hewn blank заготовка, болванка
hewn timber обтесанный лес, тесанный лес
hibernating зимующий
hibernation зимование, зимняя спячка
hide шкура, кожа
hide glue мездровый клей
high forest (i.e. of seedling origin) семенной лес, высокоствольник, высокоствольный лес, семенное насаждение
high-forest system высокоствольное хозяйство
high frequency высокая частота
high-frequency field поле высокой частоты
high-frequency kiln высокочастотная сушилка
high-grade высокосортный
high-rigger верхолаз
high school высшая школа
high-seat лабаз, помост
high water паводок, половодье
higher plant высшее растение
hill холм
hill(ock) бугор, кочка
hilly холмистый
hoar frost изморозь, иней
hobby чеглок, *Falco subbuteo* L.

hoe мотыга
hog рубительная машина, дробилка
hogging дробление, раздробление
hoist элеватор, подъемник, самотаска
hoisting winch подъемная лебедка
hold over передержать
holding boom приемная запань, задерживающий бон, поперечный бон
hole-borer ямкопатель
hole-borer (powered) мотобур
hole-digger ямкопатель
hole planting посадка в ямки
hole sowing посев в ямки
hollow mortising chisel полое долото
hollow tree дуплистое дерево
hollow-ground saw строгальная пила
holly 1. падуб, *Ilex* L.**2.** остролистный падуб, *Ilex aquifolium* L.
holly, wrinkled морщинистый падуб, *Ilex rugosa* Fr. Schmidt
holm oak каменный дуб, *Quercus ilex* L.
holy-grass душистая зубровка, *Hierochloe odorata* (L.) Beauv.
homeland родина
honey fungus опенок, *Armillaria mellea* Quel.
honey-bee медоносная пчела, *Apis mellifera* L.
honeycomb rot ситовина
honeydew медвяная роса, медовая роса
honey-locust **1.** гледичия, *Gleditsia* L. **2.** обыкновенная гледичия, *Gleditsia triacanthos* L.
honey-locust, Caspian каспийская гледичия, *Gleditsia caspica* L.
honeysuckle жимолость *Lonicera* L.
honeysuckle, blue синяя жимолость, *Lonicera coerulea* L.
honeysuckle, eastern кавказская жимолость, *Lonicera orientalis* Lam.
honeysuckle, fly обыкновенная жимолость, лесная жимолость, *Lonicera xylosteum* L.
honeysuckle, Georgian грузинская жимолость, *Lonicera iberica* M.B.
honeysuckle, goat-leaf каприфоль, жимолость-каприфоль, *Lonicera caprifolium* L.
honeysuckle, Tatarian татарская жимолость, *Lonicera tartarica* L.
hooded crow серая ворона, *Corvus cornix* L.
hook **1.** зацепка, крюк **2.** (of saw) передний угол
hook (canthook) кондак, вертун, багровый крюк, поворотный крюк, багор
hoop обруч
hoopoe удод, *Upupa epops* L.
hop хмель, *Humulus lupulus* L.
hop-hornbeam хмелеграб, *Ostrya* Scop.
hop-hornbeam, European обыкновенный хмелеграб, *Ostrya carpinifolia* Scop.
hopper накопитель, карман-накопитель, лесонакопитель
horehound, black черный белокудренник, *Ballota nigra* L.
horizon горизонт
horizon, eluvial горизонт вымывания
horizon, illuvial горизонт вмывания
horizon, ploughed пахотный горизонт, пахотный слой
horizon, soil горизонт почвы
horizontal frame saw горизонтальная рама
hornbeam **1.** граб, *Carpinus* L. **2.** обыкновенный граб, *Carpinus betulus* L.
hornbeam, Caucasian кавказский граб, *Carpinus caucasica* Grossh.
hornbeam, eastern грабинник, восточный граб, *Carpinus orientalis* Mill.
hornet шершень, *Vespa crabro* L.
hornet clearwing тополевая большая стеклянница, *Sesia apiformis* L.
horse chestnut конский каштан, *Aesculus hippocastanum* L.
horse power лошадиная сила
horse-mint лесная мята, длиннолистная мята, *Mentha longifolia* (L.) Huds.
horsetail хвощ, *Equisetum* L.
host хозяин
host, intermediate промежуточный хозяин
host plant растение-хозяин
hot-and-cold bath treatment метод горяче-холодной ванны

hotbed парник
hothouse теплица
hot-pressed sheet пластик, древесный пластик
hot-setting glue клей горячего отверждения
hound's-tongue лекарственный чернокорень, *Cynoglossum officinale* L.
house longhorn домовый черный усач, *Hylotrupes bajulus* L.
house mouse домовая мышь, *Mus musculus* L.
house sparrow домовый воробей, *Passer domesticus* L.
humic acid гуминовая кислота
humidity влажность
humidity, relative относительная влажность
hummock кочка
hummocky tundra кочковатая тундра
humus перегной, гумус
humus, mild сладкий гумус, мягкий гумус, нейтральный гумус
humus, raw кислый гумус, грубый гумус, сырой гумус, мор
humus layer гумусовый слой
hunter охотник
hunting охотничье дело, охота, лов, ловля
hunting knife охотничий кортик
hurricane ураган
husbandry хозяйство
husk шелуха
hybrid гибрид, помесь, сортообразец, ублюдок, бастард
hybridization гибридизация, скрещивание
hydraulic crane гидрокран
hydraulic engineering гидротехника
hydraulic felling wedge гидроклин
hydroelectric station гидростанция
hydrology гидрология
hydrolysis гидролиз
hydrotropic гидротропный
hydrotropic cooking (with Na xylene-sulphonate solution) гидротропная варка
hygrometer влагомер
hygroscopicity гидрофильность, гигроскопичность
Hymenomycetes грибы гименомицеты
Hymenoptera перепончатокрылые (pl.)
hypocotyl подсеменодольное колено
hypsometer высотомер, эклиметр

I

ice-bound season (of rivers) рекостав
ice crust гололедица
ice-free season (of rivers) рекоплав
ice road ледяная дорога
icing обледенение
idle time простой
ignition зажигание, загорание
illegal felling порубка
illumination освещение
illuvial horizon горизонт вмывания
immature недозрелый, незрелый, неспелый
immaturity недозрелость
impact удар
impact bending динамический изгиб, ударный изгиб
impact strength прочность на удар, сопротивление удару
impact work ударная работа
impermeable 1. (of shelterbelts) непродуваемый 2. непроницаемый
impermeable shelterbelt непродуваемая полоса
impermeable to water водонепроницаемый
implement орудие
import ввоз
importation ввоз
impoverishment истощение, оскудение
impregnate пропитывать
impregnated wood пропитанная древесина
impregnating tank пропиточный котел
impregnation пропитывание, пропитка
impregnation of standing tree(s) пропитка на корню
improvement 1. мелиорация, облагораживание 2. селекция

improvement cutting улучшительная рубка
inaccessible forest недоступный лес
incense cedar речной кедр, *Libocedrus decurrens* Torr.
incidence встречаемость
incidence, angle of угол встречи
incipient decay темнина, дряблость
incipient deterioration (of felled beech) задыхание
incising накалывание
incision надрез, накол, нарезка
inclination уклон, наклон, наклонение
included phloem включенный луб
included sapwood внутренняя заболонь, двойная заболонь
income доход
income, gross валовой доход
income, net чистый доход
increase нарастание, наращивание, прибавка, приращение
increment приращение, прирост
increment, annual ежегодный прирост
increment, current (annual) текущий прирост
increment, diameter прирост по диаметру, прирост по толщине
increment, growing stock прирост по запасу
increment, height прирост по высоте
increment, light световой прирост
increment, mean (annual) средний прирост
increment, normal нормальный прирост
increment, periodic (annual) периодический прирост
increment, stand прирост насаждения
increment, value прирост по ценности
increment, volume массовый прирост, прирост по массе, прирост по объеме
increment borer приростный бурав
increment core приростный керн
increment determination определение прироста
increment felling простор, рубка простора
increment percent процент прироста
indented зазубренный
index показатель
index, fire-danger показатель пожарной опасности
index, site показатель местообитания
indicating percent указательный процент
indicator plant растение-индикатор
indicator stand указательное насаждение
indigenous туземный
individual log tally штучный учет
individual tree отдельное дерево
inducer возбудитель
industrial chips технологическая щепа
industrial combine комбинат
industrial log depot (usually at railway or river) нижний склад
industry промышленность, производство
infected зараженный
infection заражение
infection, degree of зараженность
infestation поражение, нападение
infiltration capacity водопроницаемость
inflammability воспламеняемость
inflammable огнеопасный
inflorescence соцветие
infrastructure инфраструктура
infructescence соплодие
ingrowing врастание
inheritance наследство
inherited наследственный
inhibition торможение
initial инийнальная клетка, начальная клетка
injurious вредительный, вредный
injurious organism вредитель
injury вред, повреждение, ранение
inner внутренний
innovator новатор
inoculate прививать
inoculation прививание, прививка
insect насекомое
insect, leaf-eating листогрызущее насекомое
insect, needle-eating хвоегрызущее насекомое

insect, useful полезное насекомое
insect outbreak вспышка
insect pest энтомовредитель, вредное насекомое
insecticide инсектицид
insectivorous насекомоядный
inserted tooth вставной зуб
insolation инсоляция
inspect браковать
inspection контроль, ревизия, браковка
inspector надзиратель, бракер, лесной ревизор
instability, dimensional формоизменяемость
installation 1. монтаж 2. сооружение, установка
institution учреждение, заведение
instruction учение
insulation изолирование
insulation board изоляционная плита
insurance страхование
integrated комплексный
integrated albedo интегральный албедо
integrated forest enterprise комплексное лесное предприятие, КЛП
integrated mill комбинат
interaction взаимодействие
intercellular межклеточный
intercellular layer межклетная пластинка
intercellular space межклетник, межклеточное пространство
interception of precipitation задержание осадков, перехватка осадков
interface, vertical межкарровая перемычка, ремень
intergeneric межродовой
intergrown knot живой сучок, сросшийся сучок
interlayer прослойка
intermediate felling (any felling, e.g. cleaning or thinning, between stand establishment and the principal felling(s)) предварительное пользование, промежуточное пользование, промежуточная рубка, рубка промежуточного пользования, рубка ухода
intermediate host промежуточный хозяин
intermediate plan (5- or 7-year) перспективный план
intermittent exploitation периодическое хозяйство
intermittent feed толчковая подача
intermittent process периодический процесс
internal внутренний
internal check свищ, внутренняя трещина
internal phloem внутренный луб
internal shake внутренняя трещина
internal stress внутреннее напряжение
internode междоузлие
interpretation (of photos) дешифрирование, расшифровка
interruption перерыв, разрыв
interspecific межвидовой
interspecific crossing межвидовое скрещивание
interval between fellings (in selective logging) оборот хозяйства
intraspecific внутривидовой
intraspecific crossing внутривидовое скрещивание
introduction внедрение
inundation затопление, наводнение
inventory запись, инвентарь, опись, таксация, учет
inventory, forest лесная таксация, учет лесного фонда
iron железо
iron pan ортштейн
iron wood, Persian боккаут, *Parrotia persica* C.A.M.
irregular неправильный, неравномерный
irregular grain свиль, свилеватость
irrigate орошать, поливать
irrigation ирригация, обводнение, оводнение, орошение, поливка, полив
irrigation ditch оросительная канава, арык
island остров
isolation 1. изоляция, изолирование 2. свободное стояние

isthmus перешеек
IUFRO МСЛИО, МСЛНИО
ivy плющ, *Hedera helix* L.

J

jack домкрат
jack plane шерхебель
jack snipe гаршнеп, *Lymnocryptes minimus* Brünn.
jackdaw галка, *Corvus monedula* L.
jamb косяк
jamb, door дверной косяк
jamb, window оконный косяк
Japanese hazel лещина Зибольда, *Corylus sieboldiana* Bl.
Japanese larch тонкочешуйчатая лиственница, японская лиственница, *Larix leptolepis* Murr.
Japanese walnut орех Зибольда *Juglans sieboldiana* Maxim.
Japanese yew восточный тисс, *Taxus cuspidata* Sieb. et Zucc.
jarovization яровизация
jaw челюсть
jaw (grab) loader челюстник
jay сойка, *Garrulus glandarius* L.
jerboa (Dipodidae) тушканчик
Jerusalem artichoke земляная груша, *Helianthus tuberosus* L.
jib стрела
jib crane поворотный кран, стреловой кран
joiner столяр
joinery saw столярная пила
joinery wood столярный лес
joint соединение, стык, сустав, шов, фуга
joint, butt притык, соединение впритык
joint, dovetail соединение на ласточкин хвост
joint, dowelled шкантовое соединение
joint, finger соединение на зубчатый шип, клиновидное соединение, зубчатое соединение
joint, glue клеевое соединение
joint, scarf соединение вполдерева, соединение на ус, усовое соединение
joint, tenon шиповое соединение
joint, tongue-and-groove соединение гребнем в паз, шпунтовое соединение, соединение шпунтом в фальц
joint venture совместное предприятие
jointer пригоночный станок, фуговальный станок
jointing прифуговка, сплотка, фуговка, фугование
jointing plane фуганок
joist лага
Judas tree иудино дерево, *Cercis siliquastrum* L.
juice пасок, сок
juicy сочный
jujube, common унаби, *Zizyphus jujuba* Mill.
jumper юмпар
junction узел
jungle дебри, джунгли
junior (subordinate) forester лесной ассессор
juniper можжевельник, *Juniperus* L.
juniper, black кара-арча, многоплодный можжевельник, *Juniperus polycarpos* Koch
juniper, common обыкновенный можжевельник, *Juniperus communis* L.
juniper, Dahurian даурский можжевельник, *Juniperus dahurica* Pall.
juniper, Grecian высокий можжевельник, *Juniperus excelsa* Bieb.
juniper, prickly красный можжевельник, *Juniperus oxycedrus* L.
juniper, semiglobose саур-арча, полушаровидный можжевельник, *Juniperus semiglobosa* Rgl.
juniper, Seravshan кара-арча, зеравшанский можжевельник, *Juniperus seravschanica* Kom.
juniper, Siberian сибирский можжевельник, *Juniperus sibirica* Burgsd.
juniper, stinking вонючий можжевельник, тяжелопахутий можжевельник, *Juniperus foetidissima* Willd.

juniper, Turcoman туркменский можжевельник, *Juniperus turcomanica* B.Fed.

juniper, Turkestan туркестанский можжевельник, тасва-арча, урюк-арча, балх-арча, *Juniperus turkestanica* Kom.

K

Kamchatka alder камчатская ольха, *Alnus kamtschatica* Kom.

Kamchatka fir грациозная пихта, камчатская пихта, тонкая пихта, *Abies gracilis* Kom.

Kamchatka rhododendron камчатский рододендрон, *Rhododendron camtschaticum* Pall.

Karelian birch карельская береза, *Betula verrucosa* f. *carelica*

Kashgar tamarisk жестковолистный гребенщик, опушенный гребенщик, *Tamarix hispida* Willd.

katsura tree багрянник, *Cercidiphyllum japonicum* Sieb. et Zucc.

kerf пропил

kerf bottom подошва пропила

kerf loss потеря на пропил

kernel ядро

kestrel обыкновенная пустельга, *Falco tinnunculus* L.

key 1. ключ 2. (in tree felling) недоруб, недопил

keyed boom шпоночный бон

kicking (e.g. from a conveyor) сбрасывание, сброска

kiln древосушка, лесосушилка, камера, сушильная камера, сушильня, сушилка

kiln, compartment камера периодического действия, сушилка периодического действия

kiln, condenser конденсационная сушилка

kiln, convection конвекционная сушилка

kiln, high-frequency высокочастотная сушилка

kiln, progressive камера непрерывного действия, сушилка непрерывного действия

kiln, recirculating external-fan ежекционная сушилка

kiln, seed семеносушильня, семяносушильня

kiln, tunnel туннельная сушилка

kiln, ventilating вентиляционная сушилка

kiln, VHF СВЧ-камера

kiln drying камерная сушка

knapsack sprayer ранцевый опрыскиватель

knee кница, кокорная кница, кокора, кривой лес

knife нож, резец

knot сук, сучок, сучья (pl.)

knot, branched лапчатый сучок

knot, dead несросшийся сучок

knot, enclosed закрытый сучок, заросший сучок, вросший сучок

knot, intergrown живой сучок, сросшийся сучок

knot, large splay пасынок

knot, live живой сучок, сросшийся сучок

knot, loose выпадающий сучок

knot, punk табачный сучок, крапивный сучок

knot, resinous splay роговой сучок

knot, round круглый сучок

knot, small loose шпилька

knot, sound здоровый сучок, плотный сучок

knot, splay продолговатый сучок, сшивный сучок

knot, tight плотный сучок

knot, unsound гнилой сучок, загнивший сучок

knot, visible открытый сучок

knot, wing лапчатый сучок

knot-free несучковатый, безсучковый, бессучковый

knot hole дыра, свищ

knottiness сучковатость

knotty суковатый, сучковатый
knotty timber сучковатый лес
Kolesov's planting iron меч Колесова
kolkhoz колхоз
kolkhoz forest колхозный лес
Korean pine маньчшурская кедровая сосна, корейский кедр, маньчжурский кедр, корейская сосна, *Pinus koraiensis* Sieb. et Zucc.
Kraft tree dominance class класс Крафта
Kraft tree dominance classification классификация Крафта
krasnozem краснозем
Kurile larch курильская лиственница, *Larix gmelinii* var. *japonica* Pilg.

L

label ярлык
labelled меченый
labelled atom меченый атом
labour работа, труд
labour-intensive трудоемкий
labour protection охрана труда
lac insect лаковый червец, *Laccifer lacca* Kerr.
lackey moth кольчатый шелкопряд, *Malacosoma neustria* L.
lacquer лак
lacquering лакировка
ladder лестница
lady fern женский кочедыжник, женский папоротник, *Athyrium filix-femina* Roth.
lady's bedstraw настоящий подмаренник, *Galium verum* L.
lag behind отставать
lair логовище
lake raft озерный плот
lamella, middle срединная пластинка
lamina 1. пластинка **2.** шпон
laminate, compreg древеснослоистый пластик, ДСП (древеснослоистые пластики)
laminated многослойный
laminated densified sheet(s) древеснослоистый пластик, ДСП (древеснослоистые пластики)
laminated sheet многослойная фанера
Lammas growth вторичный прирост (в высоту)
Lammas shoot Иван-побег, летний побег, побег Иванов, вторичный прирост (в высоту)
lanceolate ланцетовидный
land(s) угодья (pl.)
land management землеустройство
land use землепользование
landing площадка, склад, складосвал, становище, стойбище
landing, lower нижний склад
landing, riverside катище
landing, upper верхний склад
landing, waterside пристань
landscape ландшафт
landscape felling ландшафтная рубка
landslide обвал, оползень
landslip обвал, оползень
larch 1. лиственница, *Larix* Mill. **2.** лиственничный
larch, Czekanow's чекановская лиственница, *Larix czekanowskii* Szaf.
larch, Dahurian даурская лиственница, *Larix gmelinii* Litv.
larch, European европейская лиственница, *Larix decidua* Mill.
larch, Japanese тонкочешуйчатая лиственница, японская лиственница, *Larix leptolepis* Murr.
larch, Kurile курильская лиственница, *Larix gmelinii* var. *japonica* Pilg.
larch, Siberian сибирская лиственница, *Larix sibirica* Led.
larch, Sukachev's лиственница Сукачева, *Larix sukaczewii* Dyl.
larch forest листвяг, *Laricetum*
larch leaf-miner лиственничная чехликовая моль, *Coleophora laricella* Hb.
larch shoot-borer лиственничная побеговая моль, *Argyresthia laevigatella* H.S.
large timber (usually over 7 cm diam.) толстомерный лес, крупномер, круп-

ная древесина, крупный лес, толстомер

large-leafed linden крупнолистная липа, *Tilia platyphyllos* Scop.

larger wintergreen круглолистная грушанка, *Pyrola rotundifolia* L.

larva гусеница, личинка

lash охлестывать

Lasiocampidae колонопряды (pl.)

last (for shoes) колодка

late frosts весенние заморозки

late wood летняя древесина, осенняя древесина, поздняя древесина

lateral (cambial) area боковая поверхность

lateral bud боковая почка

lateral shoot боковой побег

latex tube млечник, млечная трубка

lath лата, дрань, дранка, рейка, штукатурная дранка

lathe carriage суппорт

lathe operator токарь

lathing обрешетка

latitude широта

lattice решетка

launch катер

launcher (wood) скатчик

launching (loose wood into water) сбрасывание, сброска, срывка, скатка леса

laurel благородный лавр, *Laurus nobilis* L.

laurel-leafed poplar лавролистный тополь, *Populus laurifolia* Ldb.

law право

law (forest) лесной закон, лесное право

layer 1. (in vegetation) этаж, ярус, слой 2. (vegetative propagation) отводок

layered ярусный

layering размножение отводками

laying out закладка, заложение

lay-up (veneer) укладка листов шпона

leached soil палевая почва

leaching вымывание, выщелачивание

leader 1. передовик, руководитель 2. верхушечный побег, вершинный побег, конечный побег

leaf лист

leaf arrangement листорасположение, размещение листьев

leaf blade пластинка листа

leaf bud листовая почка

leaf curl курчавость листьев

leaf cushion листовая подушечка

leaf cutting листовой черенок

leaf fall листопад

leaf mosaic мозаика листьев

leaf-cutter bee пчела-листорез, *Megachile*

leaf-eating insect листогрызущее насекомое

leaflet листочек

leaf-miner минер, минирующее насекомое

leaf-miner, larch лиственничная чехликовая моль, *Coleophora laricella* Hb.

leaf-roller трубковерт

leafy лиственный, широколиственный

lease аренда

Leguminosae бобовые (pl.)

leguminous бобовый

leguminous plant бобовое растение

lemming, wood лесной лемминг, *Myopus schisticolor* Lill.

length 1. длина 2. отрезок

Leningrad (Petrograd) standard ленинградский стандарт

lenticel чечевичка

leopard moth въедливая древесница, *Zeuzera pyrina* L.

Lepidoptera чешуекрылые (pl.)

leskhoz (an administrative forestry enterprise) лесное хозяйство, лесхоз

leskhoz, experimental опытный лесхоз

leskhoz, mechanized мехлесхоз

leskhozzag (enterprise combining the features of a leskhoz and a lespromkhoz) лесхоззаг

lespromkhoz (an administrative forestry enterprise, chiefly concerned with logging) леспромхоз, лесное промышленное хозяйство

lessee арендатор

lesser butterfly orchid двулистная любка, *Platanthera bifolia* (Cust.) Rich.

lesser celandine лютичный чистяк, *Ranunculus ficaria* L.
lesser periwinkle малый барвинок, *Vinca minor* L.
lesser pine-shoot beetle малый лесной садовник, *Blastophagus (Myelophilus) minor* Hart.
lesser spotted woodpecker малый пестрый дятел, *Dryobates minor* L.
lessivé палевая почва
lessor арендодатель
level off выравнивать
levelling нивелировка, планировка
levelling instrument нивелир
lever рычаг, аншпуг
leveret зайчик
ley залежь, перелог
liane лиана
liberate освобождать
libriform fibre волокно либриформа
lichen лишай, лишайник
lichen tundra ягельная тундра
life cycle жизненный цикл
lift элеватор, подъемник
lift truck автопогрузчик
lifting (plants) выкапывание, выкопка
light свет
light, diffuse(d) рассеянный свет
light (flotation) oil всплывное масло
light increment световой прирост
light soil легкая почва
light-demanding светолюбивый
light-demanding species светолюбивая порода
light-loving светолюбивый
light-resistant светоустойчивый, световыносливый
light-tolerant светоустойчивый, световыносливый
lighter лихтер, плашкаут, плашкот, баржа
lightning молния
lightning shake громобойна
ligneous деревянистый, деревянный
lignification деревянение, деревенение, одеревенение
lignified одревесневший
lignify деревенеть
lignin лигнин
lignocellulose целлолигнин, лигноцеллюлоза
lignum vitae бакаут, бакаутовое дерево, *Guaiacum officinale* L.
lilac сирень, *Syringa vulgaris* L.
lily-of-the-valley ландыш, *Convallaria majalis* L.
lime **1.** известь **2.** липа, *Tilia* L. (*see also* linden)
lime, Amur амурская липа, *Tilia amurensis* Rupr.
lime, Caucasian кавказская липа, *Tilia caucasica* Rupr.
lime, red красная липа, *Tilia rubra* DC.
lime, milk of известковое молоко
lime, slaked гашенная известь
lime wood липняк, *Tilietum*
limestone известняк
liming известкование
limit межа, граница, предел
limp tamarisk рыхлый гребенщик, *Tamarix laxa* Willd.
limy карбонатный, известковый
linden липа, *Tilia* L. (*see also* lime)
linden, Crimean крымская липа, *Tilia euchlora* K. Koch
linden, large-leafed крупнолистная липа, *Tilia platyphyllos* Scop.
linden, Manchurian маньчжурская липа, *Tilia mandshurica* Rupr.
linden, Siberian сибирская липа, *Tilia sibirica* Bayer
linden, silver войлочная липа, серебристая липа, *Tilia tomentosa* Moench
linden, small-leafed мелколистная липа, *Tilia cordata* Mill.
line **1.** канат, трос **2.** линия, ряд, черта
line, fire противопожарная полоса, заградительная полоса, противопожарный разрыв
line, haul тяговой канат
line, load грузовая марка
line, narrow-gauge узкоколейка
line, operating тяговой канат
line, overhead подвесная дорога
line, Plimsoll грузовая марка

line, railway колея
line, return возвратный канат
line, shadow теневая линия
line, sight визир, визирная линия
line, sighting визир, визирная линия
line, slip линия скольжения
line, survey визир, визирная линия
line, valley тальвег
line intersect линейное пересечение
line of cut линия реза
line out перешколивать, пересаживать
line planting рядовая посадка
line sowing рядовой посев
line thinning линейная рубка
linear линейный
lining подкладка
lining out перешколивание, пересадка, пересаживание
lining-out board посадная доска, посадочная доска, сажальный брусок
link звено
lip 1. козырек 2. (in resin tapping) крампон 3. свес
liquid жидкость, жижа
liquid manure навозная жижа
list ведомость, опись, перечень, список
litmus paper лакмусовая бумага
litter опад, отпад, мертвый покров
litter fall опад
litter raking сгребание подстилки
little bustard стрепет, *Otis tetrax* L.
live knot живой сучок, сросшийся сучок
liverworts печеночные мхи
livestock скот
lixiviation выщелачивание
load 1. груз, нагрузка, загрузка 2. (vb) грузить, нагружать, погружать
load capacity грузоподъемность, нагружаемость
load line грузовая марка
load tail-plate (of skidding tractor) щит
loader грузчик, автопогрузчик
loader, jaw (grab) челюстник
loading загрузка, нагрузка, погрузка
loading platform погрузочная площадка
loading ramp погрузочная площадка
loam суглинистая почва, суглинок
loam (sandy) супесчаный, супесь
loamy суглинистый
lobe лопасть
lobed лопастной
local местный
local volume table местная массовая таблица
locality местность, местонахождение
location survey трассирование
lock 1. замок 2. шлюз
locomotive паровоз
locust, black белая акация, лжеакация, *Robinia pseudoacacia* L.
lodged tree зависшее дерево
lodging полегание
loess лесс
loess soil лессовая почва
log бревно, колода, кряж, блок, плаха, отрубок, плашка, полено, чурак
log, butt комлевое бревно
log, long 1. долготье 2. хлыст 3. бревно долготье
log, peeler фанерный кряж
log, short коротье
log, sleeper шпальный кряж
log, sunken коряга, кокорник, кокорье, кокорвяга
log, top вершинник
log, tree-length хлыст
log, veneer фанерный кряж
log bunch пакет лесоматериалов
log bundle пакет лесоматериалов
log carriage захватывающая тележка
log chute бревноспуск
log conveyor лесотаска, бревнотаска
log deck (in sawmill) казенка
log depot (in forest) верхний склад
log dump лесосклад
log ejector бревносбрасыватель
log elevator лесокатка
log frame рама первого ряда
log jam 1. (not blocking the stream) залом 2. затор, пыж
log kicker отсекатель
log (or bundle) mover ускоритель
log pond лесохранилище, бассейн, котлован
log quarter четвертина, четвертинник
log rollway бревнокатка

log sorter отсекатель
log storage place лесосклад
log tally учет по числу бревен
log unloader бревносвал
log volume table таблица объемов бревен
log washer обмыватель бревен
logger лесозаготовитель
logging заготовка, лесозаготовка, эксплуатация
logging and conversion лесоразработка
logging camp вахта, мастерский участок
logging debris лесосечные отходы
logging foreman мастер лесозаготовок
logging industry лесозаготовительная промышленность
logging residues лесной хлам, лесосечные остатки, порубочные остатки, лесорубочные остатки
logging terminal нижний склад
log-jam breaker пыжедержатель, пыжелом
log-kicker сбрасыватель
Lombardy poplar пирамидальный тополь, *Populus nigra* var. *italica* L. (Duroi), (*P. pyramidalis* Roz.)
London plane кленолистный платан, *Platanus acerifolia* Willd.
long-eared bat ушан, *Plecotus auritus* L.
long-eared owl ушастая сова, *Asio otus* L.
long-fibred длинноволокнистый
long log долготье, бревно долготье, хлыст
long pitwood руддолготье
long rod хворостина
long shoot (for cuttings) хлыст
long-stemmed длинноствольный
long-tailed field mouse лесная мышь, *Apodemus silvaticus* L.
long-tailed tit длиннохвостая синица, *Aegithalos caudatus* L.
long timber долготье, длиномерный лес
longevity долговечность
longhorn beetle дровосек, усач
longhorn, house домовый черный усач, *Hylotrupes bajulus* L.
longitude долгота
longitudinal boom продольный бон
longitudinal section продольный разрез
look-out tower сторожевая башня, пожарная вышка, пожарная каланча
loop петля
looper пяденица
looper, pine сосновая пяденица, *Bupalus piniarius* L.
loose-floated timber молевой лес, сплавляемый россыпью лес, молевая древесина
loose floating лесосплав россыпью, моль, молевой сплав, сплав россыпью
loose knot выпадающий сучок
loose soil рыхлая почва
loose stacking рыхлая кладка
loosen взрыхлять, разрыхлять
loosener, soil взрыхлитель
loosening (of soil) взрыхление,. рыхление почвы
loosestrife, purple иволистный дербенник, *Lythrum salicaria* L.
loosestrife, yellow обыкновенный вербейник, *Lysimachia vulgaris* L.
lop and top вершины и сучья (pl.)
lopped tree подсечное дерево
lopping подсечное хозяйство
lorry грузовик, автолесовоз, автомобиль, лесовоз
lorry road автомобильная дорога
lorry transport автотранспорт
loss потеря, убыток
lot партия
low-growing приземистый
low-lying site плаккат
low place низина
low thinning низовой метод ухода, низовое прореживание, проходная низовая рубка
lower fungi низшие грибы
lower landing нижний склад
lower reach нижиий бьеф
lower storey нижний ярус
lowering (of wood downhill) спуск
lowland низина, низменность

Lowry empty-cell process метод полуограниченного поглощения (по Лауру)
low-temperature distillation швелевание
lubricant смазка
lucerne люцерна
lumber древесина, лес, пиломатериалы (pl.), пиленый лес, пиленый материал, пилопродукция
lumber package пакет пиломатериалов
lumen полость клетки
lump комок, ком, глыба
lump charcoal уголь-сырец
lung-wort медуница, *Pulmonaria* L.
lupin лупин, люпин, *Lupinus* L.
lupin, perennial многолетний люпин, *Lupinus polyphyllus* Lindl.
lustre блеск
lyctid древогрыз
lye щелок
lymantriid волнянка
lyme grass песчаный волоснец, *Elymus arenarius* L.
lymexylonid сверлильщик, корабельщик, сверлило
lynx рысь, *Lynx lynx* L.

M

Macedonian pine румелийская сосна, *Pinus peuce* Grisob.
machete секач
machine машина, станок
machine tool станок
magnolia магнолия, *Magnolia* L.
magpie сорока, *Pica pica* L.
mahaleb cherry кучина, антипка, *Prunus mahaleb* L.
maidenhair tree гинкго, *Ginkgo biloba* L.
main burn (in shifting cultivation) выжег, пожог
main drain коллектор
main line (road, rail or water) магистраль
main root главный корень, скелетный корень
main stand главное насаждение
maintenance ремонт
maize кукуруза, *Zea mays* L.
maize cob початок
male **1.** мужской **2.** самец
male fern мужской папоротник, мужской щитовник, *Dryopteris filix mas* L.
mallet колотушка, деревянный молот
mallow, common лесной просвирник, *Malva sylvestris* L.
mallow, marsh лекарственный алтей, *Althaea officinalis* L.
mammal млекопитающее
management ведение, управление
management class класс хозяйства
management system хозяйство
management unit (e.g. felling series, 'crop-type') хозяйственная единица, хозяйство
manager заведующий
manager (forest) лесоустроитель
Manchurian apricot маньчжурский абрикос, *Prunus mandshurica* (Maxim.) Koehne
Manchurian aralia маньчжурская аралия, *Aralia mandshurica* Rupr. et Max.
Manchurian ash маньчжурский ясень, *Fraxinus mandshurica* Rupr.
Manchurian fir цельнолистная пихта, *Abies holophylla* Maxim.
Manchurian hazel маньчжурская лещина, *Corylus mandshurica* Max.
Manchurian linden маньчжурская липа, *Tilia mandshurica* Rupr.
Manchurian maple маньчжурский клен, *Acer mandshuricum* Maxim.
Manchurian walnut маньчжурский орех, *Juglans mandshurica* Maxim.
mantle (soil) почвенный покров
manufacture **1.** выработка, выпуск **2.** производство **3.** (vb) производить
manufactured product фабрикат
manure **1.** навоз, помет, удобрение **2.** (vb) удобрять
manure, green зеленое удобрение
manure, liquid навозная жижа
manuring унаваживание, навожение, удобрение

map карта, план
map, compartment планшет
map, flight планшет
map, stock план лесонасаждений
maple клен, *Acer* L.
maple, Amur приречный клен, *Acer ginnala* Maxim.
maple, Cappadocian яркий клен, *Acer laetum* C.A.M.
maple, field полевой клен, паклен, *Acer campestre* L.
maple, greenbark зеленокорый клен, *Acer tegmentosum* Maxim.
maple, Manchurian маньчжурский клен, *Acer mandshuricum* Maxim.
maple, Norway остролистный клен, *Acer platanoides* L.
maple, small-leafed мелколистный клен, *Acer mono* Maxim.
maple, sycamore явор, клен явор, сикомор, *Acer pseudoplatanus* L.
maple, Tatarian татарский клен, неклен, черноклен, *Acer tataricum* L.
maple, velvet бархатный клен, *Acer velutinum* Boiss.
maple forest кленовник
maple sap кленовица
marble мрамор
marbled мраморный
margin окраина, опушка
marginal tree окраинное дерево, краевое дерево
marigold, marsh болотная калужница, *Caltha palustris* L.
maritime pine приморская сосна, *Pinus pinaster* Sol.
marjoram обыкновенная душица, *Origanum vulgare* L.
mark **1.** марка, знак, клеймо, метка **2.** (vb) заклеймить, залысить, метить, намечать, отмечать, отметить
marked меченый
marked tree заклейменное дерево
marker разметчик
market рынок, сбыт
market price рыночная цена
marketable товарный
marking клеймение, маркировка, намечание, отметка, разметка
marking hammer маркировочный молот, разметочный молот
marking iron клеймо
marking out отведение
marl рухляк, мергель, рыхляк
marmot сурок, *Marmota* L.
marmot, bobak байбак, *Marmota bobac* L.
marmot, Siberian тарбаган, *Marmota sibirica* Radde
marsh болото
marsh andromeda дубровниколистная андромеда, *Andromeda polifolia* L.
marsh mallow лекарственный алтей, *Althaea officinalis* L.
marsh marigold болотная калужница, *Caltha palustris* L.
marshy заболоченный, болотистый
marten куница, *Martes* L.
marten, pine лесная куница, *Martes martes* L.
marten, stone каменная куница, белодушка, *Martes foina* Erx.
mass масса
mass reproduction массовое размножение
massif (forest) лесной массиф
mast **1.** мачта **2.** плодокорм
mast timber мачтовый лес
mast year семенной год, урожайный год
mastic **1.** мастика **2.** мастиковое дерево, *Pistacia lentiscus* L.
masut (fuel oil) мазут
mat (of dry grass etc.) войлок
match **1.** спичка **2.** (adj.) спичечный
match splint спичечная соломка
match stick спичечная соломка
match stock спичечный кряж
material вещество, материал
maternal материнский
mat-grass мычка, торчастый белоус, *Nardus stricta* L.
mating chamber брачная камера
mattock мотыга
mature созрелый, спелый, спелый к рубке

mature forest спелый лес
mature stand спелое насаждение
mature wood спелая древесина
maturing назревание, созревание
maturity возмужалость, зрелость, спелость
maturity, economic хозяйственная спелость, экономическая спелость
maturity, financial финансовая спелость
maturity, natural естественная спелость
maturity, physical физическая спелость
maturity, qualitative качественная спелость
maturity, quantitative количественная спелость
maturity, technical техническая спелость
maturity for regeneration (giving best natural regeneration) возобновительная спелость
Maximovich poplar тополь Максимовича, *Populus maximowiczii* Henry
maximum adsorption capacity максимальная адсорбционная влагоемкость
maximum water-holding capacity полная влагоемкость
May lily двулистный майник, *Maianthemum bifolium* (L.) F.W. Schmidt
MDF МДФ
meadow луг
meadow grass луговой мятлик, *Poa pratensis* L.
meadow soil луговая почва
meadow-grass мятлик, *Poa* L.
meadow-sweet вязолистный лабазник, вязолистная таболга, *Filipendula ulmaria* (L.) Maxim.
meal мука
mean 1. среднее 2. (adj.) средний
mean age средний возраст
mean (annual) increment средний прирост
mean diameter средний диаметр
mean height средняя высота
mean tree среднее дерево
measure 1. мера, мероприятие 2. (vb) вымерять, мерить
measure off отмерять
measurement измерение, учет
measurement by volume учет по обьему
measurement by weight учет по весу
measuring измерение
measuring cord мерный шнур
measuring lath мерный брусок
measuring rod мерная доска, мерная рейка
measuring stick 1. мерная рейка 2. (for diameter, height, plot demarcation, etc.) трость таксатора
measuring tape мерная лента, рулетка, мерная тесьма
mechanical damage to bark обшмыг коры, ошмыг коры, обдир коры
mechanical pulp древесная масса
mechanized gang комплексная бригада
mechanized leskhoz мехлесхоз
medicinal лекарственный
medicinal plant лечебное растение, лекарственное растение
mediterranean средиземноморский
Mediterranean cypress обыкновенный кипарис, *Cupressus sempervirens* L.
medium среда
medlar мушмула, *Mespilus germanica* L.
melick, wood одноцветная перловка, *Melica uniflora* Retz.
mellowness of soil спелость почвы
melt таять
melt water талая вода
mensuration таксация, лесная таксация
mensuration officer (airborne) летнаб-таксатор
mensurational description таксационное описание
merchandise товар
merchantability ликвидность
merchantability class (determined by the % of timber in the standing volume) класс товарности
merchantability of stand (classified by % timber in total standing volume) товарность древостоя
merchantable ликвидный, рыночный, товарный, ходовой
merchantable material ликвид

merchantable wood ликвидная древесина
meristem меристема, образовательная ткань
mesophile мезофильный
mesophilous мезофильный
mesophily мезофильность
mesophyll мезофилл
mesophyllous мезофилловый
mesophyte мезофит
mesotrophic мезотрофный
metabolism обмен веществ
metamorphosis метаморфоз
meter счетчик
méthode du contrôle контрольный метод, метод контроля
metre метр
metre, cubic кубометр, кубический метр
metre, running погонный метр
metre, square квадратный метр
metric ton (1000 kg) тонна
mezereon волчеягодник, волчье лыко, *Daphne mezereum* L.
microclimate микроклимат
microflora микрофлора
micro-relief микрорельеф
mid срединный
mid diameter срединный диаметр
mid height срединная высота
middle **1.** середина **2.** (adj.) срединный
middle-aged средневозрастный
middle lamella срединная пластинка
Midland hawthorn обыкновенный боярышник, колючий боярышник, *Crataegus oxyacantha* L.
migration расселение
mild humus сладкий гумус, мягкий гумус, нейтральный гумус
mildew плесень
mildewed заплесневелый
milfoil обыкновенный тысячелистник, *Achillea millefolium* L.
milk of lime известковое молоко
milk vetch сладколистный астрогал, *Astragalus glycyphyllos* L.
milk-cap, saffron рыжик, *Lactarius deliciosus* Fr.
mill мельница, комбинат
millet, wood развесистый бор, *Milium effusum* L.
milling, peripheral продольное фрезование
milling machine, peripheral продольно-фрезерный станок
millipede кивсяк
mimosa серебристая мимоза, серебристая акация, *Acacia dealbata* Link.
mine рудник
mine shaft столб
mineral ископаемый
mineral fertilizer минеральное удобрение
mineralization минерализация
mining timber крепежный лес, рудничный лес, пропсы (pl.)
Ministry of Forestry Минлесхоз
mink, European норка, *Mustela lutreola* L.
minor побочный
minor forest produce побочный продукт
mint мята, *Mentha* L.
mirror зеркало
mirror relascope зеркальный реласкоп
mist мгла, марь, туман
mistle-thrush деряба, *Turdus viscivorus* L.
mistletoe **1.** обыкновенная омела, белая омела, *Viscum album* L. **2.** европейский ремнецветник, *Loranthus europaeus* L.
mite клещ
mixed смешанный
mixed forest смешанный лес
mixed herbs разнотравный
mixed stand смешанное насаждение
mixer мешалка
mixing смешение
mixture смесь
mixture, Bordeaux бордосская жидкость
mobile подвижный
mock orange чубушник, *Philadelphus* L.
modal модальный
modal stand модальное насаждение
model макет, модель
moder модер

moderate умеренный
modulus модуль
modulus of elasticity модуль упругости
moist soil влажная почва
moistening увлажнение
moisture влага, влажность
moisture, bound связанная влага
moisture, free свободная влага
moisture absorption влагопоглощение
moisture capacity влагоемкость
moisture conduction влагопроводность
moisture content влажность, содержание влаги
moisture meter влагомер
molasses, wood древесная патока
mole крот, *Talpa europaea* L.
mole run кротовина
molehill кротовина
molluscs мягкотелые (pl.)
Mongolian oak монгольский дуб, *Quercus mongolica* Fisch.
monitoring мониторинг
monocotyledons однодольные (pl.)
monoculture монокультура
monoecious однодомный
montane горный
montane forest горный лес
montane soil горная почва
moor (wet) болото
mooring rope причал
mor кислый гумус, грубый гумус, сырой гумус, мор
moraine морена
mordant протрава, протравитель
morning frost утренник
morphology морфология
mortality гибель, отпад
mortise гнездо, долбеж, долбежка
mortise and tenon соединение шипом в гнездо
mortiser, chain цепнодолбежный станок
mortiser, slot сверлильно-пазовальный станок
mortising долбление
mortising machine долбежный станок
mosaic parquet мозаичный паркет, наборный паркет
mosaics (aerial photo) фотосхемы
moschatel мускатница, мускусница, *Adoxa moschatellina* L.
moss мох
moss, bog сфагнум, болотный мох, торфяной мох, *Sphagnum* L.
moss, hair обыкновенный кукушкин-лен, обыкновенный политрихум, *Polytrichum commune* Hedw.
moss, reindeer ягель, олений мох, *Cladonia rangiferina* (L.) Wigg.
moss bog моховое болото
moss fen моховое болото
moss tundra моховая тундра
moss-covered замшелый
mosses 1. мхи (pl.) 2. листостебельные мхи
mossiness мшистость
mossy мшистый, моховой
moth моль
moth, brown-tail златогузка, златогузка шелкопряд, *Euproctis chrysorrhoea* L.
moth, goat пахучий древоточец, ивовый древоточец, *Cossus cossus* L.
moth, green oak-roller дубовая зеленая листовертка, *Tortrix viridana* L.
moth, gypsy непарный шелкопряд, *Lymantria dispar* L.
moth, lackey кольчатый шелкопряд, *Malacosoma neustria* L.
moth, leopard въедливая древесница, *Zeuzera pyrina* L.
moth, mottled umber пяденица-обдирало, *Erannis defoliaria* L.
moth, nun монашенка, шелкопряд-монашенка, *Lymantria monacha* L.
moth, pine resin-gall побеговьюн-смолевщик, *Evetria resinella* L.
moth, pine-shoot листовертка-верхоед, зимующий побеговьюн, *Evetria buoliana* Schiff.
moth, winter зимняя пяденица, *Operophtera brumata* L.
mother материнский
mother tree материнское дерево, маточное дерево
motive power двигательная сила
motor двигатель
motor boat катер

mottled мраморный, пятнистый
mottled butt rot пестрая напенная гниль, корневая губка, подпар, *Fomes annosus* (Fr.) Cke
mottled rot пестрая гниль, мрамор
mottled rot of spruce пестрая гниль (ели), *Trametes abietis* Karst.
mottled umber moth пяденица-обдирало, *Erannis defoliaria* L.
mottling пятнистость
mould **1.** плесень **2.** форма **3.** (vb) калевать **4.** (vb) штамповать
mould fungi плесневые грибы
mouldboard отвал
moulder калевочная машина, калевочный станок
moulding **1.** калевка **2.** пластик **3.** багет
moulding cutter **1.** фреза **2.** шарошка
moulding machine калевочный станок, фрезерный станок, фрезер
moulding plane калевка, калевочник
mouldy заплесневелый
moulting линька, линяние
mound холмик, бугор
mound planting посадка в бугорки
mountain **1.** гора **2.** (adj.) горный
mountain ash обыкновенная рябина, *Sorbus aucuparia* L.
mountain hare заяц-беляк, беляк, *Lepus timidus* L.
mountain pine сосновый стланец, жереп, горная сосна, *Pinus mugo* Turra, (*Pinus montana* Mill.)
mountain range хребет
mountainous нагорный
mounted ranger объездчик
mouse мышь
mouse, harvest мышь-малютка, *Micromys minutus* Pall.
mouse, house домовая мышь, *Mus musculus* L.
mouse, long-tailed field лесная мышь, *Apodemus silvaticus* L.
mouse, yellow-necked field желтогорлая мышь, *Apodemus flavicollis* Melch.
mouse-like rodent (any small species of the Muridae) мышевидный, мышевидный грызун
movable передвижной
mower косилка
mowing обжинка, сенокос, скашивание
mowing machine косилка
mucilage слизь
Mucorales плесневые грибы
mud (and/or rock) flow сель, селевой поток, селевой вынос
muffler effect шуба
mugwort чернобыльник, обыкновенная полынь, *Artemisia vulgaris* L.
mulberry шелковица, тут, тутовик, тутовое дерево, *Morus* L.
mulberry, black черная шелковица, *Morus nigra* L.
mulberry, red красная шелковица, *Morus rubra* L.
mulberry, white белая шелковица, *Morus alba* L.
mulch мульча
mulching мульчирование
mull сладкий гумус, мягкий гумус, нейтральный гумус, мулль
mullein коровяк, *Verbascum* L.
multi-daylight press многоэтажный пресс
multi-function logging machinery агрегатная техника
multi-layered многослойный, многоярусный
multi-opening press многоэтажный пресс
multiple perforation множественная перфорация
multiple tending operation (e.g. thinning and fertilizing) полиуход
multiple top многовершинность
multiple use комплексное использование
multiplication размножение
multiply размножать
multi-ply многослойный
multi-storeyed многоярусный, сложный
multi-storeyed forest сложный лес
multi-storeyed stand многоярусное насаждение, сложное насаждение
Muridae мышиные (pl.)
Musci листостебельные мхи

mushroom гриб, шампиньон
musk deer кабарга, *Moschus moschiferus* L.
muskrat ондатра, *Ondatra zibethicus* L.
Mustelidae куньи
mutation мутация
mycelium мицелий
mycology микология
mycorrhiza грибокорень, микориза
mycotrophy микотрофия, микотрофность
myriapod многоножка
myrtle, common обыкновенный мирт, *Myrtus communis* L.

N

nail гвоздь
nail claw гвоздодер
nail pullers гвоздодер
nailing machine гвоздезабивный станок
naked голый
narrow board лата, драны, дранка, рейка
narrow buckler fern остистый щитовник, *Dryopteris spinulosa* Kuntze
narrow-gauge railway узкоколейная дорога, узкоколейка
narrow-ringed мелкослойный, узкослойный, тонкослойный
narrow-ringed wood узкослойная древесина, мелкослойная древесина
narrow-strip 1. (felling) каемчатый 2. (felling) кайма
narrow-strip felling каемчатая рубка
national park национальный парк
native туземный
natural естественный, природный
natural diameter class (expressed in decimal fractions of mean stand diam.) естественная ступень толщины
natural history природоведение
natural maturity (age when decline sets in) естественная спелость
natural monument памятник природы
natural origin естественное происхождение
natural pruning очистка ствола, очищаемость ствола, самоочищение ствола
natural reforestation самооблесение
natural regeneration естественное возобновление, естественное лесовозобновление, (from seed) самосев
natural root-grafting срастание корней
natural thinning естественное изреживание, самоизреживание
naturalization одичание
naturalized одичалый
nature натура, природа
nature conservation охрана природа
nature reserve заповедник
navigable river судоходная река
navigation навигация
navigation season навигационный период
neck шейка
necrosis некроз
nectar-bearing медоносный
nectar-bearing plant медонос
needle 1. игла 2. хвоинка
needle boom спицевая запань
needle-bearing twig лапка
needle cast опадение хвои
needle-eating insect хвоегрызущее насекомое
needle-foliage хвоя
needle-leaved хвойный, игролистный
nematode круглый червь, нематод
nest гнездо
nest planting гнездовая посадка
nest sowing гнездовой посев
Nesterov's fire-danger index комплексный гидротермический показатель
nesting place гнездовье
net сетка, сеть
net cost себестоимость
net income чистый доход
net return from forest land лесная рента
net thermal effect тепловый эффект
net weight чистый вес
nettle крапиво, *Urtica* L.
nettle, small жгучая крапива, *Urtica urens* L.
nettle, stinging двудомная крапива, *Urtica dioica* L.

nettle-tree, Caucasian кавказский каркас, *Celtis caucasica* Willd.
nettle-tree, southern южный каркас, *Celtis australis* L.
network сетка, сеть
network, road дорожная сеть
Neuroptera сетчатокрылые (pl.)
neutral нейтральный
New Year tree новогодная елка
newsprint газетная бумага
niche ниш
nightshade, deadly белладонна, *Atropa belladonna* L.
nightshade, woody сладко-горький паслень, *Solanum dulcamara* L.
nipplewort обыкновенный бородавник, *Lapsana communis* L.
nitragin (a preparation of *Rhizobium* used as a bacterial fertilizer) нитрагин
nitrification нитрификация
nitrogen азот
noctuid ночница, совка
noctuid, pine сосновая ночница, сосновая совка, *Panolis flammea* Schiff.
node узел
nodular желвачный
nodule клубень, желвак
nominal density условная плотность
non-forest soil нелесная почва
non-forest(ed) area нелесная площадь, непокрытая лесом площадь
non-resistant неустойчивый
non-standard timber некондиционная древесина
noose петля, силок
norm норм
normal forest нормальный лес
normal growing stock нормальный запас
normal increment нормальный прирост
normal (summer) water level межень
northern red-backed vole красная полевка, *Clethrionomys rutilus* Pall.
northern sea route Севморпуть
Norway maple остролистный клен, *Acer platanoides* L.
Norway spruce обыкновенная ель, европейская ель, *Picea abies* (L.) Karst.
nose bar прижимая линейка
'nosing' of logs затеска
notch выемка, зарубка, надруб, прорез
notch planting посадка в щель, щелевая посадка, посадка с защемлением
notched зазубренный
notching зарубание
notching spade клиновидная лопата, лопатка для посадки в щель
notodontid хохлатка
nourish питать
nucleolus ядрышко
nucleus ядро
nucleus (cell) клеточное ядро
number of head (of livestock) поголовье
numbering hammer нумератор
nun moth монашенка, шелкопряд-монашенка, *Lymantria monacha* L.
nurse подгон
nurse (accessory species) (encouraging straightness and self-pruning of crop trees) шуба
nursery питомник, лесной питомник, древесная школа
nut орех
nutcracker кедровка, *Nucifraga caryocatactes* L.
nuthatch поползень, *Sitta europaea* L.
nutlet орешок
nutrient питательное вещество
nutrient deficiency недостаточность питательных веществ
nutrient medium питательная среда
nutrient requirement трофность
nutrient solution питательный раствор
nutrient status трофность
nutrients, available усвояемые минеральные вещества
nutrition питание
nutritious питательный
nutshell скорпула ореха

O

oak **1.** дуб, *Quercus* L. **2.** дубовый

oak, Araks аракский дуб, араксинский дуб, *Quercus araxina* (Trautv.) Grossh.
oak, bog моренный дуб
oak, Caucasian крупнопыльниковый дуб, восточный дуб, *Quercus macranthera* F. et M.
oak, chestnut-leafed каштанолистный дуб, *Quercus castaneifolia* C.A.M.
oak, cork пробковый дуб, *Quercus suber* L.
oak, downy пушистый дуб, *Quercus pubescens* Willd.
oak, durmast сидячецветный дуб, скальный дуб, зимний дуб, *Quercus sessiliflora* Salisb. (*Quercus petraea* (Matt.) Liebl.)
oak, Georgian грузинский дуб, *Quercus iberica* Stev.
oak, Hartwiss дуб Гартвиса, *Quercus hartwissiana* Stev.
oak, holm каменный дуб, *Quercus ilex* L.
oak, Mongolian монгольский дуб, *Quercus mongolica* Fisch.
oak, pedunculate черешчатый дуб, стебельчатый дуб, летний дуб, *Quercus robur* L.
oak, Pontine понтиийский дуб, *Quercus pontica* Koch
oak, red красный дуб, *Quercus rubra* L. (*Quercus borealis* Michx.)
oak, sessile сидячецветный дуб, скальный дуб, зимний дуб, *Quercus sessiliflora* Salisb. (*Quercus petraea* (Matt.) Liebl.)
oak, willow иволистный дуб, *Quercus phellos*
oak bark-beetle дубовый заболонник, *Scolytus intricatus* Ratz.
oak forest дубняк, дубрава, *Quercetum*
oak mast дубовый плодокорм
oak powdery mildew (caused by *Microsphaera alphitoides* Gr.) мучнистая роса (дуба)
oak tanbark дубло
oaken дубовый
o.b. в/к, с/к, с корой, в коре
object of management хозяственная цель
occlude зарастать
occlusion зарастание
occurrence месторождение
ocular глазомерный
ocular estimation глазомер, глазомерная оценка
off-cut обрезок, срезок, отструг
offshoot отпрыск, отросток
offspring помет
oil масло
oil, essential эфирное масло
oil, fir пихтовое масло
oil, flotation флотационное масло
oil, light (flotation) всплывное масло
oil, resin смоляное масло, дегтярное масло
oil, tar дегтярное масло
oilstone оселок
oleaster 1. лох, *Elaeagnus* L. 2. джида, узколистный лох, *Elaeagnus angustifolia* L.
oleoresin живица
olive европейская маслина, *Olea europaea* L.
one year old годовалый, однолетний
one-year-old plant однолетнее растение
onion луковица
open 1. открытый 2. редкостойный 3. ажурный
open canker язвенный рак
open forest редкий лес, редколесье, светлый лес
open shelterbelt ажурная полоса
open stand редина, редкостой
open up освоить
open woodland редкий лес, редколесье
opening 1. отверстие 2. просвет, межлесье 3. размыкание 4. распускание
opening the canopy перерыв сомкнутости, нарушение сомкнутости, размыкание насаждения
opening up освоение
openness (e.g. of canopy or shelterbelt) ажурность
operating line тяговой канат
opposite супротивный
orchard (forest) лесосад, лесной сад

orchard (seed) семенная плантация, семенный участок
orchid, bird's nest настоящая гнездовка, *Neottia nidus-avis* Rich.
orchid, lesser butterfly двулистная любка, *Platanthera bifolia* (Cust.) Rich.
orchis ятрышник, *Orchis* L.
orchis, spotted кукушкины слезки (pl.), *Orchis maculata* L.
order **1.** отряд **2.** порядок **3.** приказ
order of fellings последовательность лесосек
ore руда
organization устройство
oribatid панцирный клещ
Oriental arbor-vitae восточная туя, *Thuja orientalis* L.
oriental beech восточный бук, кавказский бук, *Fagus orientalis* Lipsky
oriental plane восточный платан, *Platanus orientalis* L.
oriental spruce кавказская ель, восточная ель, *Picea orientalis* Carr.
oriented strand board структурно ориентированная доска, СОД
origin происхождение
origin, artificial искусственное происхождение
origin, natural естественное происхождение
origin, seed семенное происхождение
origin, vegetative вегетативное происхождение
original первобытный
ornamental plant декоративное растение
orthogonal cutting прямолинейное резание
Orthoptera прямокрылые (pl.)
ortstein ортштейн
osage orange маклюра, *Maclura aurantiaca* Nutt.
OSB структурно ориентированная доска, СОД
osier корзиночник, прут
osier, common лоза корзинная, корзиночная ива, прутовидная ива, *Salix viminalis* L.
osier, purple тал, талина, пурпурная ива, желтолозник, тальник, *Salix purpurea* L.
osier stand прутняк
otter выдра, *Lutra lutra* L.
outbreak (insects) вспышка
outer внешний
outer sapwood оболонь
outflow истечение
outgrowth вырост, нарост
outlay издержка, расходы
outlet **1.** водосток **2.** выпуск
outlier (forest) колок, колки (pl.), островный лес, колочный лес, остров
outline контур, абрис
output выработка, выпуск, добыча, продукция, производительность
output, annual комплексная выработка
outturn of timber отпуск леса
ouzel, ring белозобый дрозд, *Turdus torquatus* L.
ovary завязь
oven печь
over bark в/к, с/к, с корой, в коре
overall dimensions габарит
over-cut area переруб
over-cutting переруб
overflow водослив
overgrown переросший
overhang (in frame saw) перевешивание пилы, наклон пилы, уклон пилы, свес
overhead light верхний свет
overhead line подвесная дорога
overhead travelling crane мостовой кран
overlength припуск, допуск, прибавка
overload перегружать
overloading перегрузка, сверхнагрузка
overmature перестойный
overmature forest перестойиый лес
overmature stand перестой
overmature tree перезрелое дерево, переспелое дерево, перестой
overmeasure припуск, допуск, прибавка
over-population перенаселенность
overseer надзиратель
overstorey верхний ярус

overwintering 1. перезимовка 2. (adj.) зимующий
overwood вершняк
ovule семяпочка
owl сова, сыч
owl, barn сипуха, *Tyto alba* Scop.
owl, long-eared ушастая сова, *Asio otus* L.
owl, short-eared болотная сова, *Asio flammeus* Pont.
owl, snowy белая сова, *Nyctea scandiaca* L.
owner (forest) лесовладелец
ox бык
oxygen кислород

P

pace шаг
package пакет
packaged lumber пакет пиломатериалов
packaging (lumber) пакетование, пакетирование
packet пачка
packing 1. прокладка 2. тара 3. укладка 4. упаковка
packing piece прокладка, разлучка
pail ведро
paint краска, окраска
paint sprayer краскораспылитель
painting крашение, окраска
pair пара
paling тын, частокол
palisade частокол
palisade tissue столбчатая ткань, палисадная ткань
pallet поддон
palm пальма
palm, date финиконосный финик, *Phoenix dactylifera* L.
palmate пальчатый
pamphiliid ткач, пилильщик-ткач
panel 1. филенка 2. щит 3. палетка (e.g. of bark removed for counting insect populations)
panel parquet щитовый паркет
panel sizing контурная обработка (щитов)
panel-sizing machine форматный станок
panel-sizing saw форматная пила
panelled филенчатый
panelling филенка
panicle метелка
paper бумага
paper, abrasive шкурка
paper, bag мешочная бумага
paper, filter фильтровальная бумага
paper, grease-proof жиростойкая бумага
paper, litmus лакмусовая бумага
paper, sack мешочная бумага
paper, waste макулатура
paper, wet-strength влагопрочная бумага
paper, wrapping обверточная бумага
paper factory бумажная фабрика
paper industry бумажная промышленность
paper overlay текстурная бумага
Papilionaceae мотылковые (pl.)
parallel to grain вдоль волокон
parasite паразит, чужеядное растение
parcel 1. партия 2. пачка 3. парцелла (horizontal synusia)
parenchyma паренхима
parenchyma, aliform крыловидная паренхима
parenchyma, confluent сомкнутокрыловидная паренхима
parenchyma, diffuse диффузная паренхима
parenchyma, disjunctive разъединенная паренхима
parenchyma, phloem лубяная паренхима
parenchyma, ray лучевая паренхима
parenchyma, strand тяжевая паренхима
parenchyma, vasicentric вазицентрическая паренхима
parent производитель, родитель
parent material материнская порода
parent rock материнская порода
parent stand материнское насаждение

parent tree материнское дерево, маточное дерево
parent(al) материнский
park парк, лесопарк
parquet паркет
parquet, mosaic мозаичный паркет, наборный паркет
parquet, panel щитовый паркет
parquet block штучный паркет
parquet board паркетная доска, фриза
parquet strip фриза
parrot-crossbill клест-сосновик, *Loxia pytyopsittacus* Bork.
parthenogenesis партеногенез
partial enumeration частичный перечет, выборочный перечет
particle дробленка, стружка, частица, щепа, щепка
particleboards древесностружечные плиты, ДСП
partition перегородка
partridge серая куропатка, *Perdix perdix* L.
parts by weight в.ч., весовые части
party команда
pass билет
passability проходимость
passable проходимый
passage ход
passing пропускание
passing place (on road) разъезд
paste мазь
pasturage пастьба
pasture выгон, пастбище
patch площадка
patch, bald лысина, пролысина
patch, bare 1. голье 2. лысина, пролысина
patch barking лысение, пролыска
patch planting посадка местами, посадка площадками
patch sowing посев площадками, посев местами
path дорога, дорожка, путь, тропа
pathogen возбудитель
pathological disease патологическая болезнь
pattern 1. закономерность 2. шаблон
paulownia адамово дерево, *Paulownia tomentosa* (Thunb.) Steud.
pavement мостовая
pavement, wood-block торцовая мостовая
pavement, wooden деревянная мостовая
paving block торец
paw лапа
pay зарплата, плата, заработная плата
pay, hourly почасовая плата
pay, piece-work сдельная плата
pay-back period окупаемость, срок окупаемости
pea tree высокая карагана, желтая акация, *Caragana arborescens* Lam.
pea tree, dwarf кустарная карагана, чилига, *Caragana frutex* Koch
pea tree, pygmy карликовая карагана, *Caragana pygmaea* DC.
pea tree, spiny колючая карагана, *Caragana spinosa* DC.
peach персик, *Prunus persica* (L.) Batsch
pear 1. груша, *Pyrus* L. 2. обыкновенная груша, *Pyrus communis* L.
pear, common обыкновенная груша, *Pyrus communis* L.
pear, Ussuri уссурийская груша, *Pyrus ussuriensis* Max.
pear, willow-leafed иволистная груша, *Pyrus salicifolia* Pall.
peasant крестьянин
peat торф
peat bed торфяная залежь
peat bog торфяник, торфяное болото
peat deposit торфяная залежь
peat dust торфяная пыль
peat fire торфяной пожар
peat/mineral fertilizer ТМАУ, торфоминералное удобрение
peaty soil торфяная почва
peavy кондак, вертун, багровый крюк, поворотный крюк
pebble-bed галечник
pebbles галька
pecan пекан, *Carya pecan* E. et G.
pedicel плодоножка, цветонос, цветоножка
pedigree родовое дерево, родословие

pedology почвоведение
pedometer шагомер
peduncle плодоножка, цветонос, цветоножка
pedunculate oak черешчатый дуб, стебельчатый дуб, летний дуб, *Quercus robur* L.
peel 1. кожа 2. (vb) облупить, лупить
peeled veneer лущеная фанера, шпон
peeler log фанерный кряж
peeling (of veneer) лущение
peg 1. колок, колки (pl.) 2. шпилька, шкант, нагель
pencil карандаш
pencil slat карандашная дощечка
penduline tit ремез, *Parus pendulinus* L.
pendulum saw балансирная пила, висячая пила, маятниковая пила
penetrate проникать
perambulator дальномер
percent forest cover облесенность, лесистость, процент лесистости
percentage процент
perched water-table верховодка
peregrine falcon сапсан, сокол, *Falco peregrinus* Tunst.
perennial 1. многолетнее растение 2. (adj.) многолетний
perennial lupin многолетний люпин, *Lupinus polyphyllus* Lindl.
perforation перфорация
perforation, multiple множественная перфорация
perforation, simple простая перфорация
perforation plate перфорафионная пластинка
perforation rim перфорационный ободок
performance table таблица производительности
perianth околоцветник
pericarp околоплодник
periderm перидерма
perimeter периметр
periodic (annual) increment периодический прирост
periodic block периодная площадь
peripheral milling продольное фрезование
peripheral milling machine продольно-фрезерный станок
periwinkle, lesser малый барвинок, *Vinca minor* L.
permafrost вечная мерзлота
permanent nursery постоянный питомник
permanent sample plot пробная (постоянная) площадь
permanent study area стационар
permanent wilting point влажность устойчивого завядания
permeability проницаемость
permeability to gas газопроницаемость
permeability to water водопроницаемость
permeable shelterbelt продуваемая полоса
permeate проникать
permissible cut размер пользования, размер рубки, допустимая рубка
permit билет
permit (for felling) лесорубочный билет
perpendicular to the grain поперек волокон
Persian iron wood боккаут, *Parrotia persica* C.A.M.
persimmon хурма, *Diospyros* L.
persistent неопадающий
pest вредитель; (insect) энтомовредитель, вредное насекомое
pesticide пестицид
pesticide, contact контактный пестицид
pestle пестик
petal лепесток
petiolate черешчатый
petiole черешок
petrifaction окаменение
petrol бензин
petrol-driven power saw бензопила
phanerogam явнобрачное растение, цветочное растение
phase фаза
pheasant обыкновенный фазан, *Phasianus colchicus* L.
phellem феллем, пробка

phellogen пробковий камбий, феллоген
phenol-formaldehyde glue фенольно-формальдегидный клей
phenological form феноформ
phenomenon явление
phenotype фенотип
phloem проводящий луб, луб, флоэма
phloem, included включенный луб
phloem, internal внутренный луб
phloem, primary первичный луб
phloem, secondary вторичный луб
phloem fibre лубяное волокно
phloem parenchyma лубяная паренхима
phloem ray лубяный луч
phosphate фосфат
phosphorus фосфор
photogrammetry фотограмметрия
photograph снимок
photograph, aerial аэрофотоснимок
photography фотосъемка, фотография
photography, aerial аэрофотосъемка, аэрофотография
photosynthesis фотосинтез
phyllophage филлофаг
phyllotaxis листорасположение, размещение листьев
physical maturity физическая спелость
phytocoenosis фитоценоз
'phytolin' (substance excreted by a plant) фитолин
phytoncide фитонцид
pick-axe кирка
picket пикет
piece деталь
piece work сдельная работа, сдельщина
piece-work pay сдельная плата
pied flycatcher мухоловка-пеструшка, *Muscicapa hypoleuca* Pall.
pier пыжедержатель, пыжелом, устой, бык
pigeon голубь, *Columba* L.
pigeon, wood вяхирь, витютень, *Columba palumbus* L.
pigment краситель, красящее вещество
pika пищуха, *Ochotona* Link
pike pole багор
pile 1. костер, куча 2. свая (for construction in soft ground) 3. стопа, штабель
pileate fungi шляпочные грибы
piled fathom кубическая сажень
pile-up (of debris) залом
pileus шляпка
pillow подушка
pin заклепка, шпилька, шкант, нагель
pincers щипцы (pl.)
pinch off прищипывать
pinch out прищипывать
pinching (of saw) зажим
pinching off прищипка
pinching out прищипка
pine 1. сосна, *Pinus* L. 2. (adj.) сосновый
pine, Aleppo алепская сосна, *Pinus halepensis* Mill.
pine, Arolla европейский кедр, *Pinus cembra* L.
pine, black черная сосна, *Pinus nigra* Arn.
pine, Corsican калабрийская сосна, *Pinus brutia* Ten.
pine, Crimean крымская сосна, *Pinus nigra* var. *caramanica*
pine, Eldar эльдарская сосна, *Pinus eldarica* Medwed.
pine, Korean маньчжурская кедровая сосна, корейский кедр, маньчжурский кедр, корейская сосна, *Pinus koraiensis* Sieb. et Zucc.
pine, Macedonian румелийская сосна, *Pinus peuce* Grisob.
pine, maritime приморская сосна, *Pinus pinaster* Sol.
pine, mountain сосновый стланец, жереп, горная сосна, *Pinus mugo* Turra (*Pinus montana* Mill.)
pine, Pitsunda пицундская сосна, *Pinus pithyusa* Stev.
pine, Scots сосна, обыкновенная сосна, *Pinus sylvestris* L.
pine, Siberian 1. сибирский кедр, кедровая сибирская сосна, *Pinus sibirica* Mayr. 2. сибирская сосна, (the Siberian subspecies of *Pinus sylvestris*)

pine, umbrella итальянская сосна, пиния, *Pinus pinea* L.
pine, western yellow желтая сосна, *Pinus ponderosa* Dougl.
pine, Weymouth сосна веймутова, *Pinus strobus* L.
pine, white сосна веймутова, *Pinus strobus* L.
pine forest бор, боровой лес, сосняк, *Pinetum*
pine grosbeak щур, *Pinicola enucleator* L.
pine looper сосновая пяденица, *Bupalus piniarius* L.
pine marten лесная куница, *Martes martes* L.
pine noctuid сосновая ночница, сосновая совка, *Panolis flammea* Schiff.
pine 'nut' (kernel) (e.g. of *Pinus sibirica*) кедровый орех
pine resin сосновая смола
pine resin-gall moth побеговьюн-смолевщик, *Evetria resinella* L.
pine sawfly сосновый обыкновенный пилильщик, сосновый пилильщик, *Diprion pini* L.
pine-shoot beetle садовник
pine-shoot moth листовертка-верхоед, зимующий побеговьюн, *Evetria buoliana* Schiff.
pine tar сосновая смола
pine twisting rust сосновый вертун, сосново-осиновый ржавчинник, *Melampsora pinitorqua* Rost.
pine weevil большой сосновый долгоносик, *Hylobius abietis* L.
pinnate перистый
pinnate bladdernut перистая клекачка, *Staphylea pinnata* L.
pioneer пионер
pip fruits семячковые (pl.)
pipe труба, трубка
pistachio **1.** фисташка, *Pistacia* L. **2.** настоящая фисташка, *Pistacia vera* L.
pistachio woodland фисташник
pistil пестик
pit **1.** пора **2.** ров, рудник, шурф, яма
pit, blind слепая пора
pit, bordered окаймленная пора
pit, ramiform разветвленная пора
pit, simple простая пора
pit, vestured зубчатая пора, украшенная пора
pit aperture отверстие поры
pit border окаймление поры
pit cavity полость поры
pit chamber камера поры
pit channel канал поры
pit membrane пленка поры
pit-pair пара пор
pit planting посадка в ямки
pit-prop рудстойка, крепежная стойка, рудничная стойка
pit sowing посев в ямки
pitch вар, пек, лесохимический пек, древесный пек
pith сердцевина
pith, double двойная сердцевина
pith fleck сердцевинное повторение, сердцевинное пятно
Pitsunda pine пицундская сосна, *Pinus pithyusa* Stev.
pitted пористый
pitting поровость
pitting, alternate очередная поровость
pitting, scalariform лестничная поровость
pitting, sieve ситовидная поровость
pitting, unilaterally compound односторонне-сложная поровость
pitwood, long руддолготье
pivot стержень
plain равнина
plait коса
plan план, проект
plan, felling ведомость рубкам, план рубок
plan, five-year пятилетка
plan, intermediate перспективный план
plan, working план хозяйства
plane **1.** плоскость **2.** рубанок, струг, фуганок **3.** (vb) фуговать, выстрагивать, обстрогать
plane, grooving дорожник, фальцгобель
plane, jack шерхебель
plane, jointing фуганок
plane, moulding калевка, калевочник

plane, rebating фальцгобель
plane, rough шерхебель
plane off сострагивать
plane-table планшет
plane (tree) чинар, платан, *Platanus* L.
plane (tree), American западный платан, *Platanus occidentalis* L.
plane (tree), London кленолистный платан, *Platanus acerifolia* Willd.
plane (tree), oriental восточный платан, *Platanus orientalis* L.
planed выстроганный, строганный
planed board выстроганная доска
planed stock погонка, строганный погонаж
planed timber строганный товар, строганная древесина
planer строгальный станок, продольно-фрезерный станок, пригоночный станок, фуговальный станок, пригонно-строгальный станок
planer blade строгальный нож
planer saw строгальная пила
planing строгание, строжка, фуговка, фугование
plank толстая доска, планка, плансон, половица
plank road лежневая дорога
planking стлань, слань
planned (authorized) cut расчетная лесосека, сметная вырубка, сметная рубка
planned felling area сметная лесосека
plant 1. растение 2. оборудование 3. (vb) засаживать, посадить, сажать
plant, bare-rooted саженец с обнаженными корнями
plant, berry-bearing ягодник
plant, containerized саженец с закрытыми корнями
plant, flowering явнобрачное растение, цветочное растение
plant, fodder кормовое растение
plant, forest лесное растение
plant, herbaceous травянистое растение
plant, higher высшее растение
plant, host растение-хозяин
plant, indicator растение-индикатор
plant, leguminous бобовое растение
plant, medicinal лечебное растение, лекарственное растение
plant, nectar-bearing медонос
plant, one-year-old однолетнее растение
plant, ornamental декоративное растение
plant, poisonous ядовитое растение
plant, stump посадной пенек, пенек, торчок
plant, transplanted саженец, перешколенный саженец
plant, twining вьющееся растение
plant, two-year-old двулетнее растение
plant, vascular сосудистое растение
plant, woody древесное растение
plant bed посадная гряда
plant breeder растениевод
plant breeding растениеводство
plant growing растениеводство
plant out высаживать
plant sociology геоботаника
plant spacing размещение растений, шаг посадки
plantain подорожник, *Plantago* L.
plantation культура, рукотворный лес, плантация, посадка
plantation development развитие культуры
plantation establishment закладка культуры
planter сажальщик
planting посадка, засадка, обсадка
planting, amenity зеленое строительство
planting, ball посадка глыбами, посадка с глыбкой, посадка с комом
planting, bunch посадка пучками
planting, furrow бороздовая посадка
planting, hole посадка в ямки
planting, line рядовая посадка
planting, mound посадка в бугорки
planting, nest гнездовая посадка
planting, notch посадка в щель, щелевая посадка, посадка с защемлением
planting, patch посадка местами, посадка площадками
planting, pit посадка в ямки

planting, row рядовая посадка
planting, slant косая посадка
planting, slit посадка в щель, щелевая посадка, посадка с защемлением
planting, square квадратная посадка
planting, street обсадка дорог
planting, stump посадка на пень
planting cord посадный шнур
planting cuttings посадка черенками
planting dagger сажальный кинжал
planting hole посадочная лунка, посадочная яма
planting in advance предварительная культура
planting iron сажальный меч, сажальный нож
planting iron, Kolesov's меч Колесова
planting machine сажальная машина, сажалка
planting method способ посадки
planting of forest насаждение леса
planting of greenery (for amenity and decoration) озеленение
planting out высадка
planting spade сажальная лопата
planting stock культурный материал, посадочный материал
planting willow (to fix sands) шелюгование
plastic пластик, пластмасса
plastic flow пластическое течение
plasticity пластичность
plate пластина, плита
plateau плато, плоскогорье
platen (press) плита пресса
platform платформа, площадка, лабаз, помост
platypodid плоскоход
play (of saw) блуд, зазор, игра
pliers щипцы (pl.)
Plimsoll line грузовая марка
plinth плинтус
plot площадь, участок, делянка, земельный надел
plot, forest лесоучасток
plot, sample пробная площадь, пр. пл.
plot, temporary летучая площадь
plough **1.** плуг **2.** вспахивать, пахать
plough, deep плантажный плуг
plough, double-mouldboard двухотвальный плуг
plough, forest лесной плуг
plough, ridging двухотвальный плуг
plough, snow снегоочиститель
plough, trenching плантажный плуг
plough, wooden соха
plough beam грядиль
plough sole плужная подошва
plough (thoroughly) пропахивать
plough up распахивать
ploughed horizon пахотный горизонт, пахотный слой
ploughed layer пахотный слой
ploughing вспахивание, вспашка, опашка
ploughing, autumn зябь
ploughing, deep плантажная вспашка, плантаж
ploughing, shallow лущение
ploughing up распашка, распахивание
plough-land пахота, пашня
plough-share лемех, сошник
plum **1.** домашная слива, *Prunus domestica* L. **2.** алыча, *Prunus divaricata* Led.
plus-tree плюс-дерево
ply слой
plywood фанера-переклейка, фанера, клееная фанера, многослойная фанера
plywood factory фанерный завод, фанерная фабрика
pneumatic barking пневмоокорка
poacher браконьер
poaching браконьерство
pocket rot ямчатая гниль
pod коробочка, стручок
podzol подзол
podzolic подзолистый
podzolic soil подзолистая почва
podzolization подзолообразование
podzolized оподзоленный
podzolized brown soil подбур
podzolized soil оподзоленная почва
pointed заостренный
poison **1.** яд **2.** (vb) травить

poisoning отравление
poisonous ядовитый
poisonous plant ядовитое растение
poisonousness ядовитость
polar willow полярная ива, *Salix polaris* Wg.
pole жердь, столб, шест, штанга
pole caliper штангенциркуль
pole stand жердняк
polecat черный хорек, лесной хорек, *Mustela putorius* L.
policy (forest) лесная политика
polish 1. политура 2. полировать, шлифовать
polishing полировка, шлифование, шлифовка
pollard 1. кобла, безвершинник 2. (vb) обезвершинивать
pollard willow безвершинная ива
pollarded tree безвершинник
pollarding рубка на кобло, безвершинное хозяйство
pollen плодотворная пыль, пыльца
pollen grain пыльцевое зерно
pollen sac пыльцевой мешок
pollen shedding лет пыльцы
pollen tube пыльцевая трубочка
pollination опыление
pollution загрязнение
polygon полигон
polypody многоножка, блошник, сладкокорень, *Polypodium vulgare* L.
Polyporaceae трутовые грибы, трубчатые грибы
pomegranate гранат, гранатник, *Punica granatum* L.
pond пруд; (log pond) лесохранилище, бассейн, котлован
Pontine oak понтиийский дуб, *Quercus pontica* Koch
Pontine rhododendron понтийский рододендрон, *Rhododendron ponticum* L.
pontoon плашкаут, плашкот, понтон
pool (in bog) мочага, мочажина, мочежина
poor crop неурожай
poplar тополь, *Populus* L.
poplar, balsam бальзамический тополь, *Populus balsamifera* L.
poplar, black осокорь, черный тополь, *Populus nigra* L.
poplar, blue сизолистный тополь, сизый тополь, *Populus pruinosa* Schrenk
poplar, Bolle's тополь Болле, *Populus alba* var. *pyramidalis* L. (Bge.)
poplar, Canadian канадский тополь, *Populus deltoides* Marsh, but including also various hybrid black poplars
poplar, Chinese китайский тополь, *Populus simonii* Carr.
poplar, eastern balsam бальзамический тополь, *Populus tacamahaca* Mill.
poplar, fragrant душистый тополь, *Populus suaveolens* Fisch.
poplar, grey сереющий тополь, серый тополь, *Populus canescens* L.
poplar, laurel-leafed лавролистный тополь, *Populus laurifolia* Ldb.
poplar, Lombardy пирамидальный тополь, *Populus nigra* var. *italica* L. (Duroi), (*P. pyramidalis* Roz.)
poplar, Maximovich тополь Максимовича, *Populus maximowiczii* Henry
poplar, robusta "мощный" тополь, *Populus "Robusta"*
poplar, white белый тополь, серебристый тополь, *Populus alba* L.
poplar borer ольховый скрытохоботник, *Cryptorrhynchus lapathi* L.
poplar forest тополевник
population заселение, население, популяция
population build-up (of pests) градация
population density плотность населения
population dynamics динамика численности
porcupine дикобраз, *Hystrix leucura* Sykes
pore пора, просвет сосуда; (solitary pore) одиночный просвет
pore chain цепочка просветов
pore cluster группа просветов
pore multiple просветы сомкнутые (pl.)
pore space скважность
pore-filler порозаполнитель
porosity пористость, скважность

porous пористый, порозный
port гавань, порт
possible cut возможная рубка
post жердь, кол, стойка, столб, штанга
post-emergence повсходовый
post-glacial послеледниковый
pot trial вегетационный опыт
potash поташ, шадрик
potassium калий
powder порох, порошок
powdering опудривание
powder-post beetle древогрыз
powdery mildew мучнистая роса
powdery mildew (oak) мучнистая роса (дуба)
power saw мотопила, моторная пила
power saw (petrol-driven) бензопила
power source (forestry) ЛЭС, лесное энергетическое средство
power take-off вал отбора мощности, ВОМ
power unit энергосредство
power/weight ratio энергонасыщенность
power-line ride трасса
practicability (of terrain) проходимость
precipitate осадок
precipitation выпадение
precipitation (atmospheric) атмосферные осадки (pl.)
precipitation interception задержание осадков, перехватка осадков
precise точный
precision точность
predator хищник
predatory хищный
predominant (in numbers) преобладающий
pre-dominant tree сверхгосподствующее дерево
pre-drying (e.g. of trees before felling) подсушка
pre-emergence довсходовый
pre-emergence treatment довосходовая обработка
prefabricated house сборный дом
preparation подготовка, приготовление
preparatory (seeding) felling подготовительная рубка
preparatory stage подготовительная стадия
prepare заготовить
pre-pressing подпрессовка
prescribed burn управляемый огонь
prescribed burning выжигание
preservation 1. консервирование, хранение, сохранение 2. пропитывание, пропитка
preservative пропитывающее средство, антисептик
preservative-treated wood консервированная древесина
pre-skidding wood подача леса
pre-sowing предпосевной
press вайма, пресс
press, multi-daylight многоэтажный пресс
press, multi-opening многоэтажный пресс
press, single-daylight одноэтажный пресс
press, single-opening одноэтажный пресс
press platen плита пресса
pressed прессованный
pressing прессование, прижимание
Pressler increment borer бурав Пресслера
pressure давление
pressure bar контрнож, прижимая линейка
pressure-bar angle угол обжима
pressure-bar lead высота линейки
prevailing преобладающий
prevailing age господствующий возраст
prevention предохранение, борьба
price стоимость, цена
price, contract договорная цена
price, cost себестоимость
price, delivery сдаточная цена
price, fixed такса
price, market рыночная цена
price, selling отпускная цена
price, statutory такса

price, stumpage корневая плата, полная корневая стоимость, таксовая цена
price, transfer отпускная цена
price, wholesale оптовая цена
pricking out пикирование, пикировка
prickle колючка
prickly castor-oil tree калопанакс, диморфант, *Kalopanax pictus* Nakai (Thunb.)
prickly juniper красный можжевельник, *Juniperus oxycedrus* L.
primary conversion первичная обработка
primary forest девственный лес, первобытный лес, стихийный лес, пралес
primary phloem первичный луб
primary ray сердцевинный луч
primary wall первичная оболочка, первичная стенка
prime costs себестоимость
prime mover тягач, ЛЭС, лесное энергетическое средство
primeval первобытный
priming 1. грунт 2. грунтование
primula первоцвет, *Primula* L.
principal felling главное пользование, главная рубка, рубка главного пользования
principal species главная порода, основная порода
prism призма
private forest частновладельческий лес, частный лес
privet, common обыкновенная бирючина, *Ligustrum vulgare* L.
probe зонд, щуп
probing зондирование
process, alternating pressure циклический метод
process, batch периодический процесс
process, continuous непрерывный процесс
process, full-cell метод полного поглощения
process, intermittent периодический процесс
process, Lowry empty-cell метод полуограниченного поглощения (по Лауру)
process, Rueping empty-cell метод ограниченного поглощения (по Рюпингу)
processed обработанный
processing обработка, обрабатывание, переработка
procumbent ray cell лучевая лежачая клетка
procurement заготовка, лесозаготовка, добывание
producer производитель
producer gas генераторный газ
product изделие, продукт, произведение
product(s), forest лесной материал, лесоматериал(ы), лесопродукция
product, half-finished заготовка, болванка
product, manufactured фабрикат
product, semi-manufactured полуфабрикат
production 1. выработка, выпуск, добыча 2. продуктивность, продукция, произведение, производство
production cost себестоимость, стоимость производства
productive урожайный
productive capital производственные фонды
productiveness производительность
productivity производительность, производительная способность
profile профиль
profile cutter фасонная фреза
profiling профилировка
profit доход
profitability доходность
profitable рентабельный
profit-and-loss accountancy хозяйственный расчет, хозрасчет
progeny потомство
progressive kiln камера непрерывного действия, сушилка непрерывного действия
prohibition запрещение, запрет

project проект
proleptic shoot пролептический побег
proliferation израстание
prong зуб
prop подпор, подпорка
prop root опорный корень, ходульный корень
propagate размножать
propagation размножение
propagation, vegetative вегетативное размножение
propagation by division of clumps разделение куста
propagation by seed семенное размножение
property свойство
property, forest лесная собственность
proportion удельный вес
proportional limit предел пропорциональности
props (pit) крепежный лес, рудничный лес, пропсы (pl.)
prosenchyma прозенхима
prostrate (or shrubby) tree стланец, стлан(н)ик
protect защищать, охранять, предохранять
'protected ground' (i.e. greenhouse) защищенный грунт
protection защита, охрана, предохранение
protection costs стоимость охранения
protection forest защитный лес, охранный лес
protection of forest лесозащита, лесоохрана, лесоохранение, охрана леса
protective afforestation защитное лесоразведение
protective forest strip лесозащитная полоса
protective stand защитное насаждение
protective strip защитная полоса, охранная полоса
protective tree защитное дерево
protein 1. белок 2. (adj.) белковый
prototype пусковой образец
protuberance холмик, бугор
provenance климатип, происхождение
provenance of seed происхождение семян
provenance trial географическая посадка
provenance trial plantation географическая культура
provision снабжение
provisional условный
prune обрубать, подрезать
pruner сучкорез
pruning обрезка, обрезание, обрезка ветвей, обрезка сучьев, обрубка сучьев, опилка сучьев, очистка ствола, подрезка
pruning, dry обрезка засохших ветвей, обрезка сухих ветвей
pruning, green обрезка зеленых ветвей
pruning, natural очистка ствола, очищаемость ствола, самоочищение ствола
pruning saw пила для обрезки сучьев
pseudocarp ложный плод
psychrometer психрометр
psyllid листоблошка
ptarmigan тундровая куропатка, *Lagopus mutus* L.
Pteridophyta папоротникообразные (pl.)
pubescent пушистый
pull тяга
pulley шкив, блок
pulley block полиспаст
pulling down (trees) стаскивание
pulling through (a tree through a debranching unit) протяжка
pulp 1. масса 2. мякоть
pulp, chemical целлюлоза
pulp, chemi-thermomechanical ХТММ
pulp, ground-wood древесная масса
pulp, mechanical древесная масса
pulp, semi-chemical полуцеллюлоза
pulp, soda натронная целлюлоза
pulp, sulphate сульфатная целлюлоза
pulp, sulphite сульфитная целлюлоза
pulp, thermo-mechanical ТММ
pulp industry целлюлозная промышленность
pulpwood балансы (pl.)

pulvinus листовая подушечка
punch штамп
punk knot табачный сучок, крапивный сучок
pupa куколка
pupation окукление
pure crop монокультура
pure plantation монокультура
pure stand чистое насаждение
purity чистота
purple loosestrife иволистный дербенник, *Lythrum salicaria* L.
purple osier тал, талина, пурпурная ива, желтолозник, тальник, *Salix purpurea* L.
purple smallreed ланцетный вейник, *Calamagrostis lanceolata* Roth
purple willow тал, талина, пурпурная ива, желтолозник, тальник, *Salix purpurea* L.
push pole (in felling) валочная вилка
putty замазка, шпатлевка
pyrolysis пиролиз

Q

quail обыкновенный перепел, *Coturnix coturnix* L.
quaking bog трясина, зыбунное болото
qualitative maturity (yielding max. price per unit of wood volume) качественная спелость
quality качество, добротность
quality (site) добротность места произрастания
quality (stand) добротность насаждения
quality class класс добротности
quantitative maturity (yielding max. volume production) количественная спелость
quantity количество
quarry камнеломня, каменоломня
quarter log четвертина, четвертинник
quarter sawing радиальная распиловка
quay пристань, водоподпорная стена
quebracho квебрахо, *Aspidosperma quebracho-blanco* Schlecht.
quicklime жженая известь, едкая известь
quince айва, *Cydonia oblonga* Mill.
quincunx конверт
quickthorn однопестичный боярышник, *Crataegus monogyna* Jacq.
quotient коэффициент, частное

R

rabbit кролик, *Oryctolagus cuniculus* L.
race раса
rack полка
radial section радиальный разрез
radiation излучение
radicle зародышевый корень
radio-frequency field поле высокой частоты
radius радиус
raft плот, кошма
raft assembly сплотка
raft maker сплотчик
raft section челен
rafter стропило
rafting **1.** плотовой сплав **2.** (adj.) сплоточный
ragged robin кукушкин-цвет, *Lychnis flos-cuculi* L.
rail рельс
railing перила
rails (of chain-saw guide bar) направляющие гребни
railside прирельсовый
railway железная дорога, рельсовая дорога
railway, narrow-gauge узкоколейная дорога
railway line колея
railway shelterbelt железнодорожная полоса
railway spur лесовозный ус
railway system железнодорожная сеть
railway track рельсовый путь
railway wagon, open платформа
rain дождь
rain forest дождевой лес
rain gauge дождемер

raise выращивать, разводить
raised bed клумба
raised bog выпуклое болото, рям
raised platform (for shooting game) лабаз, помост
raised water table подтопление
raising воспитание, разведение
raising by cuttings разведение черенками
raising by layers разведение отводками
rake грабли (pl.)
raker tooth скоблящий зуб, очижающий зуб, подрезающий зуб, режущий зуб
raking litter сгребание подстилки
ram таран
ramiform pit разветвленная пора
ramp платформа, эстакада
ramsons черемша, *Allium ursinum* L.
random sample случайная выборка
range 1. диапазон 2. распространение 3. ареал 4. (mountain) хребет
range finder дальномер
ranger лесничий, лесник, объездчик
rank ранг
rapids пороги (pl.)
rapprochement (vegetative) вегетативное сближение
rare редкий
rasp рашпиль
raspberry малина, *Rubus idaeus* L.
rat крыса
rat, black черная крыса, *Rattus rattus* L.
rat, brown серая крыса, пасюк, *Rattus norvegicus* Berk.
rating plate (of a machine) паспорт
ratio of age classes соотношение классов возраста
raven ворон, *Corvus corax* L.
ravine ложбина, лощина, овраг, ущелье, балка, лог, байрак
ravine forest байрачный лес
raw сырой
raw humus кислый гумус, грубый гумус, сырой гумус, мор
raw material сырой материал, сырье, сырец
ray луч
ray, aggregate сложный луч
ray, phloem лубяный луч
ray, primary сердцевинный луч
ray, wood древесный луч
ray, xylem древесный луч
ray parenchyma лучевая паренхима
ray tracheid лучевая трахеида
reach (of a river) бьеф
reach pole дышло
reaction взаимодействие, реакция
realizability ликвидность
realizable ликвидный
'realizable' volume (the exploitable volume after deduction of reserved species and stands of less than 30 m^3/ha) ликвидный запас, ликвидная масса
'realizable' wood ликвидная древесина
rebate фальц
rebating фальцевание
rebating plane фальцгобель
reburning (in shifting cultivation) опрятывание, прятание
receptacle вместилище, приемник, цветоложе
reciprocal обратный
reciprocal crossing обратное скрещивание
reclamation мелиорация, рекультивация
reconnaissance рекогносцировка
recovered stock скоп
recreation отдых
recreational impact рекреационная нагрузка
recreational loading рскрсационная нагрузка
red dead-nettle пурпуровая яснотка, *Lamium purpureum* L.
red deer благородный олень, *Cervus elaphus* L.
red earth краснозем
red elder красная бузина, *Sambucus racemosa* L.
red heart(wood) красное ядро
red lime красная липа, *Tilia rubra* DC.
red mulberry красная шелковица, *Morus rubra* L.
red oak красный дуб, *Quercus rubra* L. (*Quercus borealis* Michx.)

red rot (of pine and larch) сосновая губка, красная гниль (сосны), *Trametes pini* Fr.

red squirrel векша, обыкновенная белка, *Sciurus vulgaris* L.

red stain краснина, краснополосица, красная полосатость

red streak краснополосица, красная полосатость

red whortleberry брусника, *Vaccinium vitis-idaea* L.

'red wood' (usually mahogany) красное дерево

redcurrant красная смородина, *Ribes rubrum* L.

red-footed falcon кобчик, *Falco vespertinus* L.

red-rattle болотный мытник, *Pedicularis palustris* L.

redstart горихвостка-лысушка, *Phoenicurus phoenicurus* L.

redstart, black горихвостка-чернушка, *Phoenicurus ochruros* Gmel.

reel катушка, шпулька

reforest облесить, залесить, олесить, разводить лес

reforestation разведение леса, залесение, лесоразведение, облесение

reforestation, natural самооблесение

refrigerator ледник

refuse отброс

regenerate возобновлять, облесить, залесить, олесить, разводить лес

regenerated облесенный

regeneration лесовозобновление, возобновление, лесовосстановление, восстановление, демутация, перерождение, разведение леса, залесение, лесоразведение, облесение, регенерация

regeneration, artificial искусственное возобновление, искусственное лесовозобновление

regeneration, coppice порослевое возобновление

regeneration, group куртиное возобновление

regeneration, natural естественное возобновление, естественное лесовозобновление, самосев

regeneration, strip каемчатое возобновление

regeneration, sucker порослевое возобновление

regeneration, supplementary дополнительное возобновление

regeneration, vegetative вегетативное возобновление

regeneration by seed семенное возобновление

regeneration felling возобновительная рубка

regeneration interval период возобновления

regeneration period период возобновления

regrowth разрастание

regularity закономерность

regulation регулирование

rehabilitation реконструкция

reindeer северный олень, *Rangifer tarandus* L.

reindeer moss ягель, олений мох, *Cladonia rangiferina* (L.) Wigg.

reinforced concrete железобетон

reject браковать

rejected бракованный

rejected goods бракованный товар, брак

rejection браковка

rejects брак, бракованный товар

rejuvenate омолодить

rejuvenation омолаживание, омоложение

relascope полнотомер Биттерлиха, реласкоп

relascope, mirror зеркальный реласкоп

relationship отношение

relative humidity относительная влажность

release выставить на свободу, освобождать

releasing изоляция, изолирование, освобождение

relict реликт

relief рельеф

remote дистанционный
remote crossing отдаленное скрещивание
remoteness дальность
removal выбор, уборка
removal (delivery) of timber отпуск леса
rendzina рендзина
renewable возобновляемый
renewable resources возобновляемые ресурсы
renewal лесовозобновление, возобновление, лесовосстановление
rent аренда
repair 1. поправка, ремонт 2. (vb) чинить, исправлять
repeat streak повторная подновка
repeating повторение
repel отпугивать
repellant отпугивающее вещество
repetition повторение
replacement замена
replacement of species смена пород
replacement shoot замещающий побег
replant подсаживать
replanting подсадка
replication повторение опыта
report ведомость, отчет, сообщение
reproduce размножать
reproduction воспроизведение, воспроизводство, размножение
reproductive генеративный
requirement потребленость
requirement, nutrient трофность
re-saw делительная пила, ребровый станок, прирезный станок
re-saw frame рама второго ряда, распускная рама
re-saw machine делительный станок
re-saw shop (for producing dimension stock) раскройный цех
re-sawing распиловка, раскрой
reserve заказник, заповедник
reserved forest заказник
reserved seed stand заказник
reserved tree оставленное дерево, маяк
reservoir водохранилище
residual tree оставленное дерево
residue остаток
residues from logging лесной хлам, лесосечные остатки, порубочные остатки, лесорубочные остатки (pl.)
resin живица, смола
resin acid смоляная кислота
resin bleeding смолотечение
resin canal смоляной ход
resin collection собирание живицы
resin content смолистость
resin duct смоляной ход
resin flow смолотечение
resin gall смоляной желвак
resin oil смоляное масло, дегтярное масло
resin pocket смоляной карман, смоляной кармашек, просмолка, серница
resin productivity смолопродуктивность
resin streak засмолок, засмол, просмолок, серница
'resin top' серянка, смоляной рак
resiniferous смолоносный
resiniferous cavity смоляное вместилище, смоловместилище
resinosis осмоление, засмоление, просмоление
resinous смолистый
resinous splay knot роговой сучок
resinous stemwood стволовой осмол
resinous stumpwood пневой осмол
resinous wood (usually Scots pine) осмол, смольняк
resinousness смолистость
resin-soaked засмоленный
resin-tapper вздымщик
resin-tapping подсочка
resistance выносливость, прочность, сопротивление, стойкость, устойчивость
resistance strain gauge датчик сопротивления
resistant выносливый, устойчивый
resonance wood резонансовая древесина, резонансовый лес
resources ресурсы
resowing подсев
respiration дыхание
rest покой

restoration восстановление
result итог
reticulate willow сетчатая ива, *Salix reticulata* L.
retort казан
retort tar камерная смола
return line возвратный канат
revenue доход
reverse обратный
reversible обратимый
revision ревизия
revision period ревизионный период
revolution 1. оборот, вращение 2. обращение
rhachis ось соцветия, плодонос
rhizoid ризоид
rhizome корневище
rhododendron рододендрон, *Rhododendron* L.
rhododendron, Caucasian кавказский рододендрон, *Rhododendron caucasicum* Pall.
rhododendron, Dahurian даурский рододендрон, *Rhododendron dauricum* L.
rhododendron, Kamchatka камчатский рододендрон, *Rhododendron camtschaticum* Pall.
rhododendron, Pontine понтийский рододендрон, *Rhododendron ponticum* L.
rhododendron, yellow желтый рододендрон, *Rhododendron luteum* Sweet
rib 1. жилка 2. ребро 3. (of boat) шпангоут
ribbon spreading (glue) линейчатый способ
ribwort ланцетный подорожник, *Plantago lanceolata* L.
ride коридор, прорубка, просека
ride between compartments квартальная просека
ridge гряда, кряж, хребет
ridging plough двухотвальный плуг
rigging оснастка
right право
right (forest) лесное право
right-of-way трасса
rigidity жесткость
rill erosion ручейная эрозия
rim обод
rind кожура
rind graft прививка за кору
ring кольцо, хомут
ring (annual) годичное кольцо, годичный слой
ring ouzel белозобый дрозд, *Turdus torquatus* L.
ring shake отсталь, отлуп, луночка, облуп, отслойка, кольцевая трещина, отлупная трещина
ringing кольцевание, окольцовывание
ring-porous кольцепоровый, кольцесосудистый
ring-porous wood кольцепоровая древесина
ring-shaped кольцеобразный
rip saw продольная пила, разрезная пила
riparian приречный
riparian woodland урема, тугайный лес, тугай
ripe созрелый, спелый
ripe for felling спелый к рубке
ripen спеть
ripeness зрелость, спелость
ripening вызревание, дозревание, назревание, созревание
ripewood спелая древесина
ripper рыхлитель
rip-sawing продольное пиление, попутное пиление, продольная распиловка
river река
river bed ложе реки
river drive сплавной попуск
river driving гонка
river raft речной плот
riverside приречный
riverside landing катище
rivet заклепка
riving knife расклинивающий нож
road дорога
road bed полотно
road building дорожное строительство
road cross-section профиль дороги
road network дорожная сеть

road system дорожная сеть
road train автопоезд
roading освоение
roads рейд
roadside придорожный
robin зарянка, *Erithacus rubecula* L.
robusta poplar "мощный" тополь, *Populus "Robusta"*
rock порода, горная порода, скала
rock flow селевой вынос
rod жезл, прут, стержень, шест
rodent грызун
rod-maker круглопалочный станок
roe deer козуля, косуля, *Capreolus capreolus* L.
roll over перекатывать
roller 1. вал, валок, каток, цилиндр 2. сизоворонка, *Coracias garrulus* L.
roller drier роликовая сушилка
rolling (e.g. saws) вальцовка
rollway рольганг
rollway (for logs) бревнокатка
roof крыша, кровля
roofing board кровельная доска
roofing shingle кровельная плитка
rook грач, *Corvus frugilegus* L.
root 1. корень 2. (vb) укореняться
root, absorbing сосательный корень, сосущий корень, всасывающий корень
root, adventitious придаточный корень, адвентивный корень
root, aerial воздушный корень
root, anchor якорный корень
root, buttress досковидный корень
root, conducting проводящий корень
root, extension ростовый корень
root, horizontal поверхностный корень
root, main главный корень, скелетный корень
root, prop опорный корень, ходульный корень
root, skeleton скелетный корень
root, stilt ходульный корень
root, supporting опорный корень
root, surface поверхностный корень
root, tap стержневой корень
root cap корневой чехлик
root channel корневина
root collar корневая шейка
root-collar scorch опал шейки
root crop корнеплод
root cutting корневый черенок
root-grafting (natural) срастание корней
root hair корневой волосок
root nodule клубень
root occupation (of soil) корненаселенность, корненасыщенность
root pruner корнерез, нож-корнерез
root rot корневая гниль
root spur лапа, корневая лапа
root-spur wood (for shipbuilding) кокорный лес
root sucker корневой побег, отпрыск, корневой отпрыск
root suckers поросль от корня
root swelling комлеватость, закомелистость, закомлеватость, корневый наплыв, бедра дерева
root system корневая система
rooting окоренение, укоренение
rooting out корчевка, корчевание, выкорчевывание
rootlet корешок
rootstock подвой, корневище
rope канат, снасть, трос
Rosaceae розоцветные (pl.)
rose роза, шиповник, *Rosa* L.
rose chafer золотистая бронзовка, *Cetonia aurata* L.
rosebay willow-herb Иван-чай, узколистный кипрей, *Chamaenerion angustifolium* (L.) Scop.
rosette розетка
rosin канифоль, гарпиус
rot 1. гниение, загнивание 2. гниль 3. (vb) сгнивать
rot, brown бурая гниль, деструктивная гниль
rot, butt комлевая гниль
rot, cubical призматическая гниль
rot, heart сердцевинная гниль, центральная гниль
rot, honeycomb ситовина
rot, mottled пестрая гниль, мрамор
rot, pocket ямчатая гниль

rot, root корневая гниль
rot, sap периферическая гниль, заболонная гниль
rot, soft плесневая гниль, умеренная гниль, мягкая гниль
rot, stem стволовая гниль
rot, top вершинная гниль
rot, trunk стволовая гниль
rot, white коррозионная гниль, белая гниль
rotary cultivating фрезовка, фрезерование
rotary cultivator фреза, почвенная фреза
rotary cutting фрезовка, фрезерование
rotary planer пригоночный станок, фуговальный станок, пригонно-строгальный станок
rotatable поворотный
rotating поворотный
rotation 1. оборот, вращение, круговорот 2. оборот рубки
rotation age оборотный возраст, возраст рубка, оборот рубки
rotation period период оборота рубки
rotten гнилой, загнивший, прелый, трухлявый
rotten wood гнилая древесина, прелая древесина, труха
rotting выпревание, гниение, загнивание, истлевание, истление
rough необработанный, суровый, юркий
rough hewing отеска
rough plan абрис
rough plane шерхебель
rough-sawn stock черновая заготовка
rough spindle tree бородавчатый бересклет, *Euonymus verrucosus* Scop.
roughness (of wood surface) мшистость, шероховатость
round обход
round knot круглый сучок
round sleeper круглая шпала
rounding (off) округление
roundwood кругляк, круглый лес
roundwood (small) подтоварник, топорник
route маршрут, трасса
route location трассирование
route survey маршрут
route-line трасса
router пазовальный станок
row ряд
row(s), distance between расстояние между рядами
row(s), space between междурядие
row, spacing within расстояние в ряду
row planting рядовая посадка
row sowing рядовой посев
row thinning линейная рубка
rowan 1. рябина, *Sorbus* L. 2. обыкновенная рябина, *Sorbus aucuparia* L.
rubber каучук
rubble щебень
rudiment зачаток
Rueping empty-cell process метод ограниченного поглощения (по Рюпингу)
rule 1. закономерность 2. линейка
ruler линейка
rules устав
ruling gradient (road) руководящий подъем
running fire беглый пожар
running mean скользяшее среднее
running metre погонный метр
running skyline тяговонесущий канат
running wild 1. (adj.) одичалый 2. (n.) одичание
run-off сток
run-off, subsurface подземный сток, внутрипочвенный сток
run-off, surface поверхностный сток
rupture 1. (vb) разрушать 2. (n.) разрушение, разрыв
rural сельский
rush ситник, *Juncus* L.
rush, soft расходящийся ситник, развесистый ситник, *Juncus effusus* L.
rush, wood ожика, *Luzula* DC.
Russian Academy of Sciences РАН
Russian broom русский ракитник, *Cytisus ruthenicus* Fisch.
Russian olive лох, oleaster, *Elaeagnus* L.
Russian Wood Exchange Рослесбиржа

Russian willow русская ива, *Salix rossica* Nas.
rust ржавчина
rust fungi ржавчинные грибы
rust fungus ржавчинник
rut 1. (zool.) гон, течка 2. колея
rye-grass плевел, *Lolium* L.

S

sable соболь, *Martes zibellina* L.
sable hunting соболевание
sac мешок
sac, pollen пыльцевой мешок
Saccharomycetes дрожжевые грибы
sack мешок
sack paper мешочная бумага
saddle graft прививка седлом
safety безопасность
safety belt предохранительный пояс
saffron milk-cap рыжик, *Lactarius deliciosus* Fr.
sag прогиб, стрела прогиба
sage шалфей, *Salvia* L.
saiga сайга, сайгак, *Saiga tatarica* L.
Sakhalin fir сахалинская пихта, *Abies sachalinensis* Mast.
Sakhalin spruce ель Глена, *Picea glehnii* Mast.
salary оклад
sale продажа, сбыт
sale standing продажа леса на корню
salinified засоленный
salinity засоленность
salinization засоление, осолонение
sallow козья ива, ракита, бредина, *Salix caprea* L.
salt соль
salt lick лизалка, лизальный камень
salt resistance солестойкость
salt tree чингиль, *Halimodendron halodendron* Voss
salvage felling санитарная рубка
samara крылатка
sample 1. образец, проба 2. (adj.) пробный
sample, random случайная выборка
sample area эталон
sample plot пробная площадь, пр. пл.
sample tree модельное дерево, модель
sand 1. песок 2. шлифовать
sandbank мель, отмель, перекат
sander шлифовальный станок
sanding шлифование, шлифовка
sandpaper шкурка, стеклянная шкурка
sandstone песчаник
sandy песчаный
sandy loam 1. супесчаный 2. супесь
sandy soil песчаная почва
sandy terrace арена
sanicle европейский подлесник, *Sanicula europaea* L.
sanitation felling санитарная рубка
sap пасок, сок
sap bleeding сокоистечение
sap rot периферическая гниль, заболонная гниль
sapling гейстер, деревцо, саженец
sappy сочный
saprophyte сапрофит
sapwood заболонь, оболонь, болонь
sapwood, included внутренняя заболонь, двойная заболонь
sapwood species заболонная порода
sapwood tree заболонная порода
saturated насыщенный
saturation насыщение
saturation point (fibre) предел насыщения волокна, точка насыщения волокна
savanna саванна
savin казацкий можжевельник, *Juniperus sabina* L.
savings скоп
saw 1. пила 2. пилить, опиливать, распиливать
saw, band ленточнопильный станок, ленточная пила
saw, bow лучковая пила
saw, chain цепная пила
saw, circular дисковая пила, круглая пила, круглопильный станок, циркулярка
saw, concave шаровая пила
saw, cross-cut поперечная пила

saw, cylinder кольцевая пила, цилиндрическая пила
saw, dimension прирезный станок
saw, docking торцовочная пила, торцовочный станок
saw, drunken пьяная пила
saw, electric power электропила
saw, end-trimming торцовочный станок
saw, frame лесорама, рамная пила, рама, лесопильная рама, пильная рама
saw, fret лобзик, ажурная пила, прорезная пила, лобзиковый станок
saw, hack ножевка, ножовка, ножевая пила
saw, hand ручная пила
saw, head головная пила, головной древнопильный станок
saw, hollow-ground строгальная пила
saw, joinery столярная пила
saw, panel-sizing форматная пила
saw, pendulum балансирная пила, висячая пила, маятниковая пила
saw, planer строгальная пила
saw, power мотопила, моторная пила
saw, pruning пила для обрезки сучьев
saw, reciprocating cross-cut штанговая пила
saw, reciprocating log cross-cut лисий хвост, пила лисий хвост, станок лисий хвост
saw, rip продольная пила, разрезная пила
saw, swage двухстороннеконическая пила, правоконическая пила, коническая пила, одностороннеконическая пила
saw, taper двухстороннеконическая пила, правоконическая пила, коническая пила, одностороннеконическая пила
saw, travelling летучая пила
saw, veneer фанерная пила
saw blade полотнище пилы, полотно
saw block козлы (pl.)
saw carriage захватывающая тележка
saw damage (to wood) запил
saw doctor направщик пил, пилоправ, пилостав
saw off спиливать
saw set разводной ключ, разводка, разводной шаблон
saw-shift упряг
saw up распиливать
saw wobble блуждание пилы
sawdust опилки (pl.)
sawfly пилильщик
sawfly, fox-coloured сосновый рыжий пилильщик, сосновый пилильщик, *Diprion sertifer* Geoffr.
sawfly, large larch лиственничный большой пилильщик, лиственничный пилильщик, *Lygaeonematus erichsoni* Htg.
sawfly, pine сосновый обыкновенный пилильщик, сосновый пилильщик, *Diprion pini* L.
sawing выпиливание, опиловка, пиление, пилка, распиловка, распил
sawing chain зубчатый венец
sawing horse козлы (pl.)
sawing off опиловка, отпиливание
sawing out выпиливание
sawing pattern постав
sawing schedule постав
sawing up распил
sawlog пиловочный кряж, пиловочник, пиловочное бревно
sawmill лесопильный завод, лесопильня, лесопилка
sawmilling лесопиление
sawmilling industry лесопильная промышленность
sawn опиленный, пиленый
sawn timber пиломатериалы (pl.), пиленый лес, пиленый материал, пилопродукция, пиленый товар
sawn veneer пиленая фанера
sawtimber пильный лес, пиловочник
sawyer пильщик
saxaul **1.** саксаул, *Haloxylon* Bge. **2.** зайсанский саксаул, *Haloxylon ammodendron* Bge.
saxaul, black черный саксаул, *Haloxylon aphyllum* Iljin

saxaul, white белый саксаул, песчаный саксаул, *Haloxylon persicum* Bge.
scab парша
scaffolding подмостки (pl.)
scalariform лестничный
scalariform pitting лестничная поровость
scald ожог
scale 1. масштаб, скала, шкала 2. чешуя, чешуйка 3. щитовка
scale insect червец
scales весы (pl.)
scalper дерносним, дерноснимaтель
scalping сдирание
scantling(s) брусок, скантлинги (pl.)
scar рубец
scar (fire) подсушина, пожарная подсушина, шрам
scarabaeid пластинчатоус
scarf joint соединение вполдерева, соединение на ус, усовое соединение
scarification поранение почвы, скарификация
scarifier взрыхлитель, покровосдиратель, рыхлитель
schist сланец
Schrenk's spruce тянь-шанская ель, ель Шренка, *Picea schrenkiana* F. et M.
scion прививок, привой, черенок
sclerenchyma склеренхима
scolytid заболонник
scorch нагар, ожог
scorch mark нагар
scorched опаленный
score врезать
Scots pine сосна, обыкновенная сосна, *Pinus sylvestris* L.
scour промоина
scrape (in resin tapping) засохшая живица, твердая живица, серка, баррас
scraper скребок
scraper blade цикля
scraping циклевание
screen 1. заслон, щит 2. решето, сито
screening сортировка
screes гольцы
screw винт
scribe рейсмус, скрейбер
scrub коряжник, криволесье, кустарник, мелколесье, стланец, стлан(н)ик
scrub forest кустарный лес
scrub tree корявое дерево
scythe коса
sea buckthorn облепиха, *Hippophae rhamnoides* L.
sea-going raft морской плот
seasoned высушенный
seasoned wood высушенный лес
seasoning сушение, сушка
seasoning check сухобойна
secateurs секатор
secondary felling вторичная рубка, осветительная рубка
secondary forest вторичный лес
secondary growth вторичный рост
secondary phloem вторичный луб
secondary species второстепенная порода
secondary stage осветительная стадия
secondary wall вторичная оболочка, вторичная стенка
secondary xylem вторичная древесина
secretion выделение
section звено, отдел, отделение, участок
sedge осока, *Carex* L.
sediment осадок, гуща
sedimentary rock осадочная порода
seed посевной материал, семя, семечко
seed bed семенная гряда
seed certification сертификация семян
seed coat семенная оболочка
seed collection сбор семян, семенозаготовка
seed disinfection протравливание семян
seed dispersal разлет семян, распространение семян
seed dormancy покой семян
seed drill 1. сеялка 2. посевная бороздка
seed extraction plant лущильня, семеносушильня, семяносушильня
seed fall опад семян
seed fullness полнозернистость семян
seed gauge семеномер
seed harvest урожай семян
seed harvesting семенозаготовка

seed kiln семеносушильня, семяносушильня

seed orchard семенная плантация, семенный участок

seed origin семенное происхождение

seed production семеноводство, семеношение

seed propagation семенное размножение

seed provenance происхождение семян

seed regeneration семенное возобновление

seed reserve лесосеменный заказник, ЛСЗ

seed-spot лунка

seed stand семенный участок

seed trap семяноуловитель, семеномер

seed tree семенное дерево, обсеменитель, семенник

seed weight (in g/l) натура семян

seed year семенной год, урожайный год

seed yield урожайность

seedbearer семенник

seeding 1. обсеменение, семеношение 2. (adj.) обсеменительный

seeding felling обсеменительная лесосека, обсеменительная рубка, семенная рубка

seeding stage обсеменительная стадия

seedling (usually up to age 2) сеянец

seedling section (of nursery) посевное отделение

seedling (young) всход

seepage фильтрация

segment отрезок

segregation 1. (genetics) выщепление 2. расщепление

selection выбор, отбор, подбор, селекция

selection differential селекционная разница

selection felling выборочная рубка, выборочная добровольная рубка, добровольно-выборочная рубка

selection forest выборочный лес

selection system выборочное хозяйство

selective выборочный, избирательный

selective logging выборочная рубка, выборочная добровольная рубка, добровольно-выборочная рубка, приисковая рубка, выборочная рубка на прииск, рубка на прииск, рубка с вершка, подневольная рубка, выборочная рубка с вершка, выборочная подневольная рубка

selects отборные (pl.)

self-ignition самовоспламенение

self-pollination самоопыление

self-pruning очищаемость ствола, самоочищение ствола

self-sowing самосев

selling price отпускная цена, продажная цена

semi-chemical pulp полуцеллюлоза

semi-circular planting spade полукруглая лопата

semi-coke полукокс

semi-desert полупустыня

semi-digger body полувинтовый отвал

semi-digger plough body полувинтовый плуг

semi-hard board полутвердая плита

semi-manufactured product полуфабрикат

semi-trailer полуприцеп

semiglobose juniper саур-арча, полушаровидный можжевельник, *Juniperus semiglobosa* Rgl.

sepal чашелистник

septate перегородчатый

septate parenchyma cell паренхимная перегородчатая клетка

sequence порядок

sequence of felling очередь рубки, примыкание лесосек, последовательность лесосек

sequential sampling последовательный учет

sequoia секвоя, *Sequoia* Endl.

sequoia, giant мамонтовое дерево, гигантская секвоя, *Sequoia gigantea* Decne.

Seravshan juniper кара-арча, зеравшанский можжевельник, *Juniperus seravschanica* Kom.

serozem серозем

serrate(d) пильчатый, зубчатый
service moisture content эксплуатационная влажность
service tree, true крупноплодная рябина, садовая рябина, *Sorbus domestica* L.
service tree, wild рябина глоговина, глоговина, лечебная берека, *Sorbus torminalis* Crantz
sessile сидячий
sessile oak сидячецветный дуб, скальный дуб, зимний дуб, *Quercus sessiliflora* Salisb. (*Quercus petraea* (Matt.) Liebl.)
set 1. (of seed or fruit) завязываемость, завязывание 2. комплект, набор, пакет
set of sliced veneers кноль
set of veneers or plies пакет шпона
set-maker сборочный станок
sett кол, прут
setting gauge разводной шаблон
setting of standards (esp. standard times) нормирование, нормировка
setting pliers разводка, разводные щипцы
settled tar отстойная смола
settlement 1. поселок, селение 2. (in new places) расселение
severe суровый
severing срезание, срезывание
sex пол
sexual specialization сексуализация
shackle вертлюг
shade 1. тень 2. отенение 3. (vb) затенивать
shade clump зеленый зонт, древесный зонт
shade tolerance теневыносливость
shade-bearing теневыносливый
shade-bearing species теневыносливая порода
shaded теневой
shade-loving тенелюбивый
shade-tolerant теневыносливый
shading затенение, отенение, притенение
shadow line (in sawmilling) теневая линия
shady теневой
shaft 1. вал 2. дышло 3. оглобля
shaft wood оглобельник
shake 1. трещина 2. (visible at butt end only) ветренница, ветреница
shake, heart метик, метиковая трещина, сердцевинная трещина
shake, internal внутренняя трещина
shake, lightning громобойна
shake, ring отсталь, отлуп, луночка, облуп, отслойка, кольцевая трещина, отлупная трещина
shake, star звездчатый метик, лучистый метик
shaking off (cones) отряхивание (шишек)
shale сланец
shale tar сланцевая смола
shallow ploughing лущение
shallow water мелководье, отмель
'shapeliness' (height/d.b.h.) стройность
shaping 1. формирование 2. (of saw teeth) профилировка
sharpen натачивать, точить
sharpened точеный
sharpening заточка, точение, точка
sharpening stone точильный камень
sharpness angle угол заострения, угол заточки
shaving стружка, витая стружка
shear сдбиг
shear parallel to grain скалывание
shear perpendicular to grain перерезание, перерезывание, срезывание, срез, срезание
shear strength прочность на сдбиг, сопротивление сдвигу
shear strength parallel to grain сопротивление скалыванию
shear strength perpendicular to grain прочность на срез, прочность на срезывание, сопротивление срезу
shearing force срезывающая сила
shears ножницы
sheath влагалище, кожух
sheath cell кроющая клетка

sheathing обшивка
sheave каток
shed навес, сарай
sheep's bit горный букашник, *Jasione montana* L.
sheep's fescue типчак, овечья овсяница, *Festuca ovina* L.
sheep's sorrel малый щавель, *Rumex acetosella* agg.
sheet **1.** лист **2.** слой
sheet erosion смыв, смывание, плоскостная эрозия, поверхностная эрозия
shellac шеллак
shelling лущение
shelter **1.** защита **2.** защищать
shelter tree защитное дерево
shelterbelt полезащитное насаждение, защитная полоса, охранная полоса
shelterbelt establishment полезащитное лесоразведение
shelterbelt, field полезащитная полоса
shelterbelt, impermeable непродуваемая полоса
shelterbelt, open ажурная полоса
shelterbelt, permeable продуваемая полоса
shelterbelt, railway железнодорожная полоса
shelterwood felling постепенная рубка
shelterwood system семенно-лесосечное хозяйство
shepherd's purse пастушья сумка, *Capsella bursa-pastoris* Med.
shield **1.** заслон **2.** щит
shield budding окулировка щитком
shield fern многорядник, *Polystichum* Roth.
shift **1.** перевод **2.** сдвиг **3.** смена
shifting cultivation огневое земледелие, паловое земледелие, лесное земледелие, подсечное земледелие
shingle гонт, кровельная дранка, кровельная плитка
shingle roof кровельная крыша
ship судно
ship's hold трюм
shipbuilding timber корабельный лес
shipping-dry condition транспортная влажность
shoal мелководье
shock удар
shock resistance сопротивление удару
shoe **1.** ползун **2.** башмак
shook (box) тара
shoot отросток, побег (usually one year's growth)
shoot, annual годовой побег
shoot, coppice порослевина, побег от пня, пневая поросль, поросль от пня
shoot, epicormic волчок, водяной побег, стволовой побег
shoot, Lammas Иван-побег, летний побег, побег Иванов, вторичный прирост (в высоту)
shoot, lateral боковой побег
shoot, proleptic пролептический побег
shoot, replacement замещающий побег
shoot, short коротыш, укороченный побег
shoot, side боковой побег
shoot, sylleptic силлептический побег
shoot, terminal верхушечный побег, вершинный побег, конечный побег
shoot-borer, larch лиственничная побеговая моль, *Argyresthia laevigatella* H.S.
shoot cutting побеговый черенок
shoot distortion искривление побегов
shoot-forming capacity побегопроизводительность
shore берег
short log коротье
short measure недомер
short shoot коротыш, укороченный побег
short split плаха
short timber коротье
shortage недостаток
short-eared owl болотная сова, *Asio flammeus* Pont.
short-toed eagle змееяд, *Circaetus gallicus* Gmel.
shower (heavy) ливень
shred измочаливать
shredding сдирание

shrew, common лесная землеройка, *Sorex araneus*
shrike сорокопут, *Lanius* L.
shrink ссыхаться, усыхать
shrinkage усадка, усушка
shrinkage check трещина усушки
shrinking усыхание
shrub кустарник, куст
shrub layer кустарниковый ярус
shrub tundra кустарниковая тундра
shrubbery кустарник
shrubby кустовой
shutter ставень
shuttle челнок
Siberian apricot сибирский абрикос, *Prunus sibirica* L.
Siberian chipmunk бурундук, *Eutamias sibiricus* Laxm.
Siberian elm 1. приземистый вяз, карагач, ильмовник, *Ulmus pumila* L. 2. перистоветвистый вяз, мелколистный вяз, *Ulmus pumila* L. var. *arborea* Litv. 3. карагач, перистоветвистый вяз, *Ulmus pinnato-ramosa* Dieck.
Siberian fir сибирская пихта, *Abies sibirica* Ledeb.
Siberian juniper сибирский можжевельник, *Juniperus sibirica* Burgsd.
Siberian larch сибирская лиственница, *Larix sibirica* Led.
Siberian linden сибирская липа, *Tilia sibirica* Bayer
Siberian marmot тарбаган, *Marmota sibirica* Radde
Siberian pine 1. сибирский кедр, кедровая сибирская сосна, *Pinus sibirica* Mayr. 2. сибирская сосна (the Siberian subspecies of *Pinus sylvestris*)
Siberian red deer марал, *Cervus elaphus sibiricus* Sev.
Siberian spruce сибирская ель, *Picea obovata* Ldb.
Siberian weasel колонок, *Mustela sibirica* Pall.
Siberian willow сибирская ива, *Salix sibirica* Pall.
Siberian yellow birch ребристая береза, желтая (далневосточная) береза, *Betula costata* Traut.
sickle серп
side drain кювет
side graft прививка сбоку
side shoot боковой побег
sierozem серозем
sieve решето, сито
sieve area ситовидное поле
sieve cell ситовидная клетка
sieve field ситовидное поле
sieve pitting ситовидная поровость
sieve plate ситовидная пластинка
sieve tube ситовидная трубка
sieve-tube element ситовидный элемент
sight 1. визировать 2. прицел
sight line визир, визирная линия
sighting bar 1. (modification of Bitterlich's angle gauge) таксационный прицел 2. (for measuring height) мерная скоба
sighting line визир, визирная линия
sign марка, знак, признак
signal tower маяк
sika deer пятистая олень, *Cervus nippon* Temm.
silica кремнезем
silicon кремний
silk шелк
silk tree шелковая акация, *Albizzia julibrissin* Dur.
silkworm шелковичный червь
silo силосная башня
silt пыль
silt deposition кольматаж
silting заиление
silver birch белая береза, бородавчатая береза, повислая береза, *Betula pendula* Roth. (*Betula verrucosa* Ehrh.)
silver fir европейская пихта, белая пихта, гребеньчатая пихта, *Abies alba* Mill.
silver linden войлочная липа, серебристая липа, *Tilia tomentosa* Moench

silver wattle, серебристая мимоза, серебристая акация, *Acacia dealbata* Link.
silvics лесоведение
silvicultural лесоводственный
silvicultural character лесоводственное свойство
silvicultural system способ хозяйства
silviculture лесоводство
silviculturist лесохозяйственник, лесовод
simple простой
simple perforation простая перфорация
simple pit простая пора
single единичный
single tree солитер
single-entry table таблица с одним входом
single-storeyed одноэтажный, одноярусный, простой
single-storeyed forest простой лес
single-storeyed stand одноярусное насаждение, простое насаждение
single-tree assortment table сортиментная таблица
sinker топняк, топляк, утоп
siricid рогохвост
site местонахождение, местообитание, местопроизрастание, месторождение, стация, лесной участок
site class класс бонитета, бонитет
site class determination бонитирование, бонитировка
site conditions условия произрастания
site index показатель местообитания
site quality добротность места произрастания
situation местоположение
skeletal content (of soil) скелет
skeleton 1. каркас 2. остов
skeleton root скелетный корень
sketch абрис
ski лыжа
ski log assortment лыжный кряж
skid трелевать, стрелевать
skidder трелевочная машина, подтрелевщик, скиддер, трелевочный трактор, трелевщик, тягач
skidding подвозка леса, подтаскивание, подтрелевка, трелевка
skidding, aerial подвесная трелевка
skidding, cable лебедочная трелевка
skidding, ground наземная трелевка
skidding, tractor тракторная трелевка
skidding, winch лебедочная трелевка
skidding cap трелевочный колпак
skidding out вытрелевка
skidding pan пэн
skidding sledge юмпар
skidding track волок
skidding tractor движитель, трелевочный трактор, трелевщик
skidding trail волок
skilled workers кадры (pl.)
skilled workman мастер
skim coulter дерносним, дерноснимател ь
skimming лущение, сдирание
skin шкура, кожа, кожица
skirting board плинтус
skull-cap обыкновенный шлемник, *Scutellaria galericulata* L.
skyline 1. ВТУ, воздушная трелевочная установка, подвесная установка, канатно-подвесная дорога, воздушная канатная дорога 2. несущий канат
slab 1. горбина, горбыль 2. обапол, абапол 3. плита
slab re-saw ребровый станок
slabbing брусовка, раскрой с брусовкой
slacking skyline несущий опускающий канат
slag, basic томасшлак
slag, Thomas томасшлак
slaked lime гашенная известь
slant planting косая посадка
slash лесной хлам, лесосечные остатки, порубочные остатки, лесорубочные остатки, лесосечные отходы
slash and burn agriculture огневое земледелие, паловое земледелие
slash burning огневая очистка
slash-collector подборщик сучьев, подборщик, сучкоподборщик
slasher слешер

slat дощечка, лата, дрань, дранка, рейка
slate сланец
slats слетинги (pl.)
sledge сани (pl.)
sledge runner санный полоз
sledge trailer санный прицеп
sledge-hammer кувалда
sledge-way скат для саней
sleeper шпала
sleeper, crossing переводный брус
sleeper, flat (slabbed) пластинчатая шпала
sleeper, round круглая шпала
sleeper, square-edged обрезная шпала
sleeper, waney-edged брусковая шпала
sleeper block (baulk of 10 x 10 in (= *c.* 25 x 25 cm) cross-section, sawn to give two sleepers) слипер
sleeper length тюлька
sleeper log шпальный кряж
sleeper timber шпальный лес, шпальник, тюлька
sleeve гильза, втулка
'slenderness' (height/d.b.h.) стройность
slice срез
sliced строганный
sliced veneer ножевая фанера, строганная фанера
sliced veneers (set of) кноль
slicer строгальщик
slicer (veneer) строгальный станок, фанерострогальный станок
slicing строгание
slicing blade строгальный нож
slide древоспуск, лесоскат, лесоспуск, скат, спуск
slide block ползун
slide rule счетная линейка
slider ползун
sliding скольжение
sliding caliper штангенциркуль
slime слизь
slime flux слизетечение
sling петля, строп
slip скольжение
slip line линия скольжения
slip plane плоскость скольжения
slipping скольжение
slit planting посадка в щель, щелевая посадка, посадка с защемлением
sloe терн, терновик, колючая слива, *Prunus spinosa* L.
slope косогор, уклон, наклон, склон, откос, отлог, скат, спуск
slope of grain наклон волокон
sloping покатый
slot долбеж, долбежка, паз, прорез
slot mortiser сверлильно-пазовальный станок
slot mortising пазовое сверление
slot-mortising bit пазовое сверло, пазовая фреза
slotter пазовальный станок
slotting долбление
sloven расщеп
sluice шлюз
sluice gate шандора
small nettle жгучая крапива, *Urtica urens* L.
small-leafed linden мелколистная липа, *Tilia cordata* Mill.
small-leafed maple мелколистный клен, *Acer mono* Maxim.
smallreed, purple ланцетный вейник, *Calamagrostis lanceolata* Roth
smallwood мелкотоварная древесина, мелкотоварный лес
smoke дым
smoke-jumper пожарный парашютист, пожарный десантник
smoke-tree желтинник, скумпия, *Cotinus coggygria* Scop.
smooth bark зеркальная кора
smooth elm берест, лиственный вяз, листоватый вяз, *Ulmus carpinifolia* Gled. (*U. foliacea* Gilib.)
smoothing шлихтование
smoothly начисто
smoothness чистота
smothered заглушенный
smothering заглушение
smothering of seedlings (e.g. by *Thelephora terrestris* Ehr.) удушье сеянцев
snag карча, коряга, кокорник, кокорье, кокорвяга, остолоп, сора
snail улитка

snake змея
snake-root горлец, рачьи шейки (pl.), *Polygonum bistorta* L.
snedding обрезка, обрезание
snipe, jack гаршнеп, *Lymnocryptes minimus* Brünn.
snow снег
snow blight (caused by *Phacidium infestans* Karst.) снежное шютте
snow breakage снеголом
snow clearer снегоочиститель
snow cover снежный покров
snow pack снежный покров
snow plough снегоочиститель
snow retention снегозадержание
snow road снежная дорога
snow roller снежный каток
snow storm метелица, метель
snowberry снежеягодник, снежная ягода, *Symphoricarpos rivularis* Suks.
snowdrop белый подснежник, *Galanthus nivalis* L.
snowfall снегопад
snowy owl белая сова, *Nyctea scandiaca* L.
soaking замачивание, вымачивание, мочение, мочка, намачивание
social vole общественная полевка, *Microtus socialis* Pall.
society общество
socle цоколь
sod дерн, дернина
soda pulp натронная целлюлоза
soddy soil дерновая почва
soft rot плесневая гниль, умеренная гниль, мягкая гниль
soft rush расходящийся ситник, развесистый ситник, *Juncus effusus* L.
soft-grass, creeping мягкий бухарник, *Holcus mollis* L.
softwood хвойная древесина, хвойная порода
soft-wooded broadleaved мягколиственный
soft-wooded broadleaved species мягколиственная порода
soil 1. почва 2. почвенный
soil, acid кислая почва
soil, alkaline щелочная почва
soil, alluvial наносная почва
soil, bog болотная почва
soil, boggy болотистая почва
soil, brown бурая почва
soil, calcareous известковая почва, карбонатная почва
soil, chalky меловая почва
soil, chestnut каштановая почва
soil, cinnamon-brown коричневая почва
soil, clay глинозем
soil, clayey илистая почва, глинистая почва
soil, clay-loam суглинистая почва
soil, coarse крупнозем
soil, ferruginous железистая почва
soil, floodplain пойменная почва
soil, forest лесная почва
soil, fresh (i.e. fairly moist) свежая почва
soil, gleyed оглееная почва
soil, grey серая почва
soil, grey desert серозем
soil, leached палевая почва
soil, light легкая почва
soil, loess лессовая почва
soil, loose рыхлая почва
soil, meadow луговая почва
soil, moist влажная почва
soil, montane горная почва
soil, non-forest нелесная почва
soil, peaty торфяная почва
soil, podzolic подзолистая почва
soil, podzolized оподзоленная почва
soil, podzolized brown подбур
soil, sandy песчаная почва
soil, skeletal content of скелет
soil, soddy дерновая почва
soil, virgin целинная почва, целина
soil, wet мокрая почва, сырая почва
soil, yellowish палевая почва
soil and subsoil почво-грунт
soil auger почвенный бурав
soil cohesion связность почвы
soil cultivation обработка почвы, уход за почвы
soil-degrading почвоухудшающий
soil density плотность почвы
soil disinfection протравливание почвы

soil exhaustion утомление почвы
soil fauna эдафон
soil-forming почвообразовательный
soil horizon горизонт почвы
soil-improving почвоулучшающий
soil loosener взрыхлитель
soil loosening разрыхление почвы, рыхление почвы
soil mantle почвенный покров
soil moisture влажность почвы
soil-pit шурф
soil probe почвенная трость
soil-protective почвозащитный
soil science почвоведение
soil structure структура почвы
soil-thrower (fire-fighting) грунтомет
soil type тип почвы
soil with a bleached horizon подбел
soil wounding поранение почвы
sole подошва
sole, plough плужная подошва
solid cubic metre плотный кубометр, плотный метр, фестметр
solid measure мера плотнои массы
solid volume плотная масса
solid-volume/stacked-volume ratio коэффициент полнодревесности, полнодревесность
solid wood цельная древесина
solitary единичный
solitary pore одиночный просвет
solod солодь
solonchak солончак
solonetz солонец
solution раствор
song thrush певчий дрозд, *Turdus philomelos* L.
sorrel обыкновенный щавель, *Rumex acetosa* L.
sorrel, sheep's малый щавель, *Rumex acetosella* agg.
sorrel, wood кислица, *Oxalis acetosella* L.
sort 1. вид 2. род 3. сортировать
sort out выбраковать
sorter бракер
sorting сортировка, разбивка по сортам
sorting installation сортировочная сеть
sound knot здоровый сучок, плотный сучок
sound seed полноценная семя
sound wood здоровая древесина
sour cherry обыкновенная вишня, *Prunus cerasus* L.
sour felling (leaving the crown intact on felled trees) подвяливание после повала
source 1. источник 2. родник
souslik суслик, *Citellus* L.
south юг
southern nettle-tree южный каркас, *Celtis australis* L.
sovkhoz совхоз
sovkhoz forest совхозный лес
sow сеять
sowing посев, обсейвание, обсеменение
sowing, additional подсев
sowing, aerial аэросев
sowing, broadcast посев вразброс, сплошной посев
sowing, drill бороздковый посев
sowing, furrow бороздовый посев
sowing, hole посев в ямки
sowing, line рядовой посев
sowing, nest гнездовой посев
sowing, patch посев площадками, посев местами
sowing, pit посев в ямки
sowing, row рядовой посев
sowing, strip посев полосами
sowing, winter озимый посев
sowing lath сеяльная доска, сеяльный брусок
sowing machine сеялка
space пространство
space between rows междурядие
space between strips межполосное пространство
spacing размещение, распоряжение, расположение, расстояние
spacing, plant размещение растений, шаг посадки
spacing in the row расстояние в ряду
spade заступ, лопата
spade, notching клиновидная лопата, лопатка для посадки в щель

spade, planting сажальная лопата
spade, semi-circular (planting) полукруглая лопата
spade's depth штык
spadix початок
span пролет
spar мачта, шпат
spar, head головая мачта
spar, tail тыловая мачта
sparrow воробей
sparrow, house домовый воробей, *Passer domesticus* L.
sparrow, tree полевой воробей, *Passer montanus* L.
sparrow-hawk ястреб-перепелятник, перепелятник, *Accipiter nisus* L.
sparse редкостойный
sparse forest редкий лес, редколесье
sparse stand редкое насаждение, редина, редкостой
specialists кадры (pl.)
specialization, sexual сексуализация
species 1. вид 2. порода
species, accessory подгон
species, associate сопутствующая порода
species, broadleaved лиственная порода
species, coniferous хвойная порода
species, dominant господствующая порода
species, exotic экзот
species, forest-forming лесообразователь
species, hard-wooded твердая порода
species, hard-wooded broadleaved твердолиственная порода
species, light-demanding светолюбивая порода
species, principal главная порода, основная порода
species, sapwood заболонная порода
species, secondary второстепенная порода
species, shade-bearing теневыносливая порода
species, soft-wooded broadleaved мягколиственная порода
species, subordinate подчиненная порода
species, tree древесная порода
species, undesirable нежелательный порода
species, valuable ценная порода
species, woody древесная порода
species choice выбор пород
species distribution распределение пород
species diversity видовое разнообразие
species mixture смешение пород
species replacement смена пород
species succession смена пород
specific удельный
specific gravity удельный вес
specific heat теплоемкость
specific work удельная работа
specimen проба
specimen tree солитер
speckled alder белая ольха, серая ольха, *Alnus incana* Moench
specks (in pulp) сорность
speedwell, common лекарственная вероника, *Veronica officinalis* L.
spent liquor барда
sphagnum peat верховой торф
sphingid бражник
spider паук
spillway водослив
spindle веретено
spindle moulder фрезерный станок, фрезер
spindle moulding фрезовка, фрезерование
spindle tree бересклет, *Euonymus* L.
spindle tree, European европейский бересклет, *Euonymus europaeus* L.
spindle tree, rough бородавчатый бересклет, *Euonymus verrucosus* Scop.
spindle-mould фрезовать
spine 1. колючка 2. хребет
spiny almond колючейший миндаль, *Prunus spinosissima* Franch.
spiny pea tree колючая карагана, *Caragana spinosa* DC.
spiral grain косослой, косослойность
spiral heart shake несогласный метик

spiral-grained wood косослойная древесина
splay knot продолговатый сучок, сшивный сучок
spleenwort костенец, *Asplenium* L.
splice grafting копулировка, прививка в пенек, простая копулировка
splinter лучина, заноза, задорина, осколок
splintering расщеп
split **1.** трещина, щель, раскол **2.** (vb) колоть, раскалывать **3.** половинник, пластина, плаха
splitting колка, расколка, раскалывание, растрескивание, расщепление
spoke спица
spoke boom спицевая запань
spontaneous combustion самовоспламенение
spool катушка, шпулька
spoon bit ложечное сверло
spore спора
spot пятно
spot fire пятнистый пожар
spotted пятнистый
spotted dead-nettle крапчатая яснотка, *Lamium maculatum* L.
spotted flycatcher серая мухоловка, *Muscicapa striata* Pall.
spotted orchis кукушкины слезки (pl.), *Orchis maculata* L.
spottiness пятнистость
sprawling раскидистый
spray gun краскораспылитель
sprayer опрыскиватель, форсунка
spraying опрыскивание, распыление
spreading **1.** разброс, разбрасывание, распространение **2.** (adj.) раскидистый
spreading crown развесистая крона
spring frosts весенние заморозки
spring-set **1.** (saw) разведенный **2.** (vb.) разводить
spring-set tooth разведенный зуб
spring-setting разведение, разводка, развод пилы
spring wood весенняя древесина, ранняя древесина
sprinkler irrigation дождевание
sprinkling дождевание
sprocket звездочка
sprout **1.** росток, поросль **2.** (vb) наклюнуться, наклевываться, произрастать
sprouting произрастание
spruce **1.** ель, *Picea* Dietrich **2.** (adj.) еловый
spruce, blue колючая ель, *Picea pungens* Engel
spruce, Norway обыкновенная ель, европейская ель, *Picea abies* (L.) Karst.
spruce, oriental кавказская ель, восточная ель, *Picea orientalis* Carr.
spruce, Sakhalin ель Глена, *Picea glehnii* Mast.
spruce, Schrenk's тянь-шанская ель, ель Шренка, *Picea schrenkiana* F. et M.
spruce, Siberian сибирская ель, *Picea obovata* Ldb.
spruce, Yezo аянская ель, *Picea jezoensis* Carr.
spruce forest ельник, *Piceetum*
spur **1.** ус **2.** отрог, укороченный побег, шпора
spur railway лесовозный ус
spur road лесовозный ус
spurge молочай, *Euphorbia* L.
spurge, wood миндалевидный молочай, *Euphorbia amygdaloides* L.
squad команда
square брус, четырехкантный брус
square metre квадратный метр
square planting квадратная посадка
squared timber кантовый лес
square-edged sleeper обрезная шпала
square-sawn timber призма
squaring off log ends торце-выравнивание
squirrel белка, *Sciurus* L.
squirrel, red векша, обыкновенная белка, *Sciurus vulgaris* L.
squirrel hunting белкование
S-shaped cleat эс
St John's wort зверобой, *Hypericum* L.
stability устойчивость

stabilization закрепление, укрепление
stable устойчивый
stack **1.** костер, поленница, стопа, штабель **2.** укладывать
stacked cubic metre складочный кубометр, объемный метр, складочный метр
stacked measure складочная мера
stacked volume складочный объем
stacker укладчик, штабелер
stacking кладка, складывание, укладка
stacking loose рыхлая кладка
stacking tight плотная кладка
stag-beetle, common жук-олень, *Lucanus cervus* L.
staggered вразбежку
stagheaded суховершинный
stagheaded tree суховершинное дерево
stagheadedness суховершинность
stagnant застойный
stagnate застаиваться
stagnation застой
stain **1.** окраска **2.** пятно **3.** травить
staining морение, окраска, протравливание, травление
stake кол, тычина
stalk стебель
stalk (flower) цветонос, цветоножка
stalk (fruit) плодоножка
stalking выслеживание
stamen тычинка
stand насаждение, лесостой, древонасаждение, древостой
stand, amenity зеленое насаждение
stand, closed сомкнутое насаждение
stand, coppice низкоствольник
stand, even-aged одновозрастное насаждение
stand, forest лесонасаждение, лесное насаждение
stand, fully stocked полное насаждение
stand, gappy насаждение с просветами
stand, indicator указательное насаждение
stand, main главное насаждение
stand, mature спелое насаждение
stand, mixed смешанное насаждение
stand, modal модальное насаждение
stand, multi-storeyed многоярусное насаждение, сложное насаждение
stand, open редина, редкостой
stand, overmature перестой
stand, parent материнское насаждение
stand, pole-stage жердняк
stand, protective защитное насаждение
stand, pure чистое насаждение
stand, pure even-aged элемент леса
stand, seed семенный участок
stand, single-storeyed одноярусное насаждение, простое насаждение
stand, sparse редкое насаждение, редина, редкостой
stand, thinned изреженное насаждение, разреженное насаждение
stand, trial пробное насаждение
stand, two-storeyed двухярусное насаждение
stand, uneven-aged разновозрастное насаждение
stand, young молодняк, молодое насаждение
stand approaching maturity приспевающее насаждение
stand assortment table товарная таблица
stand canopy полог насаждения
stand canopy density сомкнутость насаждения
stand canopy opening размыкание насаждения
stand composition состав насаждения
stand condition состояние насаждения
stand constitution форма насаждения, форма леса
stand density плотность насаждения, густота насаждения, полнота леса, степень полноты
stand development образование насаждения
stand formation образование насаждения
stand height высота насаждения
stand increment прирост насаждения
stand maturity спелость насаждения
stand merchantability товарность древостоя

stand quality добротность насаждения
stand storey structure форма насаждения, форма леса
stand structure структура насаждения
stand volume запас насаждения
standard 1. маяк 2. норм, стандарт, эталон 3. (hort.) штамбовый
standard deviation среднеквадратическая ошибка
standard plywood рядовая фанера
standard timber кондиционная древесина
standard time норм времени, норм выработки
standard tree оставленное дерево
standardization нормирование, нормировка, стандартизация
standing dead tree сухостой
standing timber древесина на корню, лес на корню
standing volume запас, запас леса на корню
standing volume (total) общий запас
standstill простой
star shake звездчатый метик, лучистый метик
starch крахмал
starling обыкновенный скворец, *Sturnus vulgaris* L.
state farm совхоз
state forest государственный лес
state forest enterprise Гослесхоз
state forest inventory ГИЛФ
state forest resource (the total area of all land and water in the charge or control of the state forestry service) Гослесфонд
state plan Госплан
state reserve Госзаповедник
state standard ГОСТ
static bending статический изгиб
station местонахождение, местопроизрастание, станция, стация
statistics (forest) лесная статистика
stave клепка, бочарная клепка
stave, dry сухотарная клепка
stave, wet заливная клепка
stave bolt клепочный кряж
stave wood лад
staves (for casks) боковник
steam 1. пар 2. пропаривать
steam engine паровоз
steamed wood пропаренная древесина
steamer пароход
steaming пропаривание, пропарка, распаривание
steaming pit парильная яма
steaming tank паровой котел
steeping замачивание, вымачивание, мочение, мочка
stellite стлеллит
stem ствол, стебель, штамб
stem analysis анализ ствола
stem canker рак ствола
stem disk кружок от ствола
stem flow сток по стволам
stem rot стволовая гниль
stem volume объем ствола
stem volume formula формула обьема ствола
stemwood стволовая древесина
stemwood, resinous стволовой осмол
stencil шаблон
steppe степь
stepped ступенчатый
sterile стерильный
sterilize обесплодить
stevedore стивидор
stick палка, трость
stick, measuring мерная рейка, трость таксатора
sticker (in stack) прокладка
sticker bluestain подлежка, подпарина, прокладная синева
stigma рыльце
stilt ходуля
stilt root ходульный корень
stimulate стимулировать
stinging nettle двудомная крапиво, *Urtica dioica* L.
stinking juniper вонючий можжевельник, тяжелопахутий можжевельник, *Juniperus foetidissima* Willd.
stint обход
stipule прилистник

stitchwort, wood дубравная звездчатка, *Stellaria nemorum* L.
stoat горностай, *Mustela erminea* L.
stock **1.** заготовка **2.** запас **3.** (for grafting) подвой **4.** фонд
stock dove клинтух, *Columba oenas* L.
stock map план лесонасаждений
stocking (on a basal-area basis) плотность насаждения, густота насаждения, полнота леса
stocking (basic) элементарный запас
stoma устьице
stone камень
stone birch каменная береза *Betula ermanii* Cham.
stone cell каменистая клетка
stone fruit косточковый плод, костянка
stone marten каменная куница, белодушка, *Martes foina* Erx.
stoniness каменистость
stool **1.** пень **2.** напенный
stool (coppice) пневая поросль, поросль от пня
stool bed маточная плантация
stool shoots пневая поросль, поросль от пня
stool-bed маточный сад
stopwatch секундомер
storage хранение
storage basin подпруда
storage boom задерживающий бон
storage place (wood) **1.** лесобиржа, биржа **2.** лесохранилище, лесосклад
storage pond подпруда
store **1.** заготовка **2.** лабаз **3.** склад
stored запасный
storehouse амбар
storey этаж, ярус
storey, lower нижний ярус
storey, top верхний ярус
storey structure (stand) форма насаждения, форма леса
storeyed ярусный
storeyed cambium ярусный камбий
storeyed structure ярусность
storing складывание
storm ураганъ, буря
stowage штивка
straddle carrier автолесовоз, портальный автолесовоз
straight-grained прямослойный
straight-stemmed прямоствольный
strain деформация
strand тяж
strand board, oriented структурно ориентированная доска, СОД
strand tracheid тяжевая трахеида
strand parenchyma тяжевая паренхима
stranded tree кокша
stranding застрявание
strap ремень
stratification стратификация
stratum **1.** пласт, залежь, гряда **2.** слой **3.** этаж, ярус
straw солома
strawberry, hautbois клубника, *Fragaria moschata* Duch.
strawberry, wild лесная земляника, *Fragaria vesca* L.
strawberry tree земляничник, земляничное дерево, *Arbutus* L.
streak (in resin tapping) вздымка, рана-подсочка, подновка
streak angle угол подновки
streak interval шаг подновки
streaking (in resin tapping) вздымка, рана-подсочка, подновка
stream flow русловый сток
street planting обсадка дорог
strength прочность, сопротивление, крепость
strength, bending прочность на изгиб, сопротивление изгибу
strength, compression прочность на сжатие, сопротивление сжатию
strength, crushing прочность на сжатие, сопротивление сжатию
strength grading прочностная сортировка
strength, impact прочность на удар, сопротивление удару
strength, shear прочность на сдбиг, сопротивление сдвигу
strength, tensile прочность на растяжение, сопротивление растяжению, сопротивление разрыву

strength, torsional сопротивление скручиванию
strengthening закрепление, укрепление
stress напряжение, напряженность
stress, allowable допускаемое напряжение
stress grading силовая сортировка
stress, internal внутреннее напряжение
stress, working допускаемое напряжение
string шнур
strip 1. коридор, лента, полоса 2. багет 3. (vb) облупить, лупить
strip barking лысение, пролыска
strip enumeration ленточный перечет
strip felling каемчатая лесосека, полосная рубка
strip plantation кулисная культура
strip regeneration каемчатое возобновление
strip road колейная дорога
strip sowing посев полосами
stripe полоса
striped field-mouse полевая мышь, *Apodemus agrarius* Pall.
stripping сдирание
stripping (bark) погрыз коры, сдирание коры
strips стрипсы (pl.)
stroke (of frame saw) ход пилы
structure сложение, строение, структура
struggle борьба
stub 1. остолоп 2. пень, пенек
stub, branch пенек от сучка, торчок
stubble стерня
stump 1. пень, пенек 2. (adj.) пневой, напенный
stump clearance корчевание пней
stump grubber корчевальная машина
stump grubbing раскорчевка пней
stump plant посадной пенек, пенек, торчок
stump planting посадка на пень
stump removal раскорчевка пней
stumpage price корневая плата, полная корневая стоимость, таксовая цена
stumpage price (fixed) попенная плата, лесная такса
stumpage value корневая себестоимость, попенная цена, корневая стоимость, таксовая стоимость, корневая цена
stump-grubber корчеватель
stump-puller, tripod треножный корчеватель
stumpwood пневой материал, пневая древесина
stumpwood (resinous) пневой осмол
stunted growth корявый рост
stunted tree корявое дерево
style (of flower) столбик
subalpine субальпийский
subalpine meadow полонина
subcompartment выдел, комплексный участок, таксационный участок, элементарный участок
subdivision подразделение
suberization опробкование
subfloor накат
subgenus подрод
submerged stump коряга, кокорник, кокорье, кокорвяга
subordinate подчиненный
subordinate species подчиненная порода
subordinate tree подчиненное дерево
sub-shrub полукустарник
subsidiary побочный
subsoil грунт, подпочва
subsoiler подпочвенник, почвоуглубитель
sub-species подвид
substitution замена
subsurface run-off подземный сток, внутрипочвенный сток
subterranean подземный
suburban forest пригородный лес
subzone подзона
succession демутация, смена
succession of species смена пород
successive (regeneration) felling последовательная рубка
succulent сочный

sucker корнеотпрыск, корневой побег, отпрыск, корневой отпрыск
sucker regeneration порослевое возобновление
suckers (root) поросль от корня
suction всасывание
suffrutex полукустарник
suitable годный
suite свита
Sukachev's larch лиственница Сукачева, *Larix sukaczewii* Dyl.
sulphate pulp сульфатная целлюлоза
sulphite pulp сульфитная целлюлоза
sulphur сера
sulphuric acid серная кислота
sum итог
sumac сумах, *Rhus* L.
sumac, tanners' дубильный сумах, *Rhus coriaria* L.
summer лето
summer chafer нехрущ, июньский хрущ, *Amphimallon solstitialis* L.
summer wood летняя древесина
summit верхушка, вершина, вершок
sun crack солнечная трещина, солнцепечная трещина
sun scorch солнцепек
sundew круглолистная росянка, *Drosera rotundifolia* L.
sunfleck блик
sunflower подсольнечник, *Helianthus annuus* L.
sunken log коряга, кокорник, кокорье, кокорвяга
sunken timber топняк, топляк, утоп
sunken tree кокша, сора
sunken wood топняк, топляк, утоп
sunscald ожог коры, солнечный ожог
superhard board сверхтвердая плита
superheated перегретый
superstructure надстройка
supplementary regeneration дополнительное возобновление
support опора, подпор, подпорка, стойка
supporting root опорный корень
suppress угнетать
suppressed заглушенный, угнетенный
suppressed tree заглушенное дерево, угнетенное дерево
suppression заглушение, угнетение, угнетенность
surface **1.** плоскость, площадь, поверхность **2.** (vb) пригонять
surface check поверхностная трещина
surface fire наземный пожар, низовой пожар
surface fungal disease (spots, mildew, etc.) грибный налет
surface root поверхностный корень
surface run-off поверхностный сток
surplus избыток
survey засъемка, обзор, съемка, учет
survey line визир, визирная линия
surveying землемерное дело, межевание
surveying pole веха
surveying rod веха
surveyor землемер, межевик, съемщик
survival выживание, приживаемость
survival capacity выживаемость
susceptibility чувствительность, восприимчивость
susceptibility to rot загниваемость
suslik суслик, *Citellus* L.
suspension velocity скорость витания
sustainable устойчивый
sustainable management устойчивое управление
sustained постоянный
sustained management непрерывное хозяйство, постоянное хозяйство
sustained use постоянное пользование
sustained yield постоянство пользования
sustained-yield forest непрерывнопроизводительный лес
swage saw двухстороннеконическая пила, правоконическая пила, коническая пила, одностороннеконическая пила
swaged плющенный
swaged tooth плющенный зуб, лопаточка
swaging плющение, расплющивание, сплющивание
swaging saw teeth расклепка зубьев

swage-set tooth плющенный зуб
swamp топь, трясина, зыбунное болото, болото
swamp cypress болотный кипарис, *Taxodium distichum* Rich.
swamp formation заболачивание
swampy заболоченный, болотистый, топяный
sward дерн, дернина, мурава
swarm of bees рой пчел
sweep кривизна
sweep-back angle (of chipper knives) угол поворота
sweet chestnut сьедобный каштан, благородный каштан, посевной каштан, *Castanea sativa* Mill.
sweet gale обыкновенный восковник, *Myrica gale* L.
sweet woodruff пахучий ясменник, *Asperula odorata* L.
swell разбухать
swelling набухание, разбухание **2.** опухоль
swelling (at butt) комлеватость, закомелистость, закомлеватость, корневый наплыв, бедра дерева
swollen butt комлеватость, закомелистость, закомлеватость, корневый наплыв
sword меч
sycamore maple явор, клен явор, сикомор, *Acer pseudoplatanus* L.
sylleptic shoot силлептический побег
symbiosis симбиоз, сожительство
syncarp сборный плод
systematic (i.e. non-selective) thinning схематическая рубка
systemic системный

T

table **1.** стол **2.** таблица
table, double-entry таблица с двумя входами
table, height таблица высот
table, local volume местная массовая таблица
table, log volume таблица объемов бревен
table, performance таблица производительности
table, single-entry таблица с одним входом
table, single-tree assortment сортиментная таблица
table, stand assortment товарная таблица
table, taper таблица сбега, таблица сбежистости
table, tariff тариф
table, volume массовая таблица, объемная таблица
table, yield опытная таблица, таблица хода роста
tachinid ежемуха, тахина
tackle снасть, такелаж
tagged меченый
taiga **1.** тайга, черневый лес **2.** (adj.) таежный, черневый
tail хвост
tail-plate щит
tail spar тыловая мачта
take root укореняться
taking root укоренение
tall oil rosin талловая канифоль
tall tree бревенник
tally **1.** паспорт пакета **2.** учет **3.** бирка
tally (log) учет по числу бревен, штучный учет
tallyman присяжный счетник
talus осыпь
tamarisk тамарикс, гребенщик, *Tamarix* L.
tamarisk, branchy ветвистый гребенщик, *Tamarix ramosissima* Ldb.
tamarisk, graceful изящный гребенщик, *Tamarix gracilis* Willd.
tamarisk, Kashgar жестковолистный гребенщик, опушенный гребенщик, *Tamarix hispida* Willd.
tamarisk, limp рыхлый гребенщик, *Tamarix laxa* Willd.
tamarisk, tetrandrous четырехтычинковый гребенщик, *Tamarix tetrandra* Pall.

tan waste одубина
tanbark корье, дубильная кора
tanbark (oak) дубло
tangential section тангенциальный разрез
tank чан, бак
tanners' sumac дубильный сумах, *Rhus coriaria* L.
tannery дубильня
tannin таннид, таннин
tannin content таннидоносность
tanning дубление
tanning material дубитель, дубильное вещество
tanstuff дубитель, дубильное вещество
tansy обыкновенная пижма, *Tanacetum vulgare* L.
tap **1.** кран **2.** (resin) подсачивать
tap root стержневой корень, стержень
tape лента
tape, measuring мерная лента, рулетка, мерная тесьма, тесьмовый метр
taper сбег, сбежистость
taper saw двухстороннеконическая пила, правоконическая пила, коническая пила, одностороннеконическая пила
taper table таблица сбега, таблица сбежистости
tapering сбежистый
tapped подсоченный
tapping подсочка
tar деготь, смола
tar, coal каменноугольная смола
tar, generator генераторная смола
tar, pine сосновая смола
tar, retort камерная смола
tar, settled отстойная смола
tar, shale сланцевая смола
tar, wood древесная смола
tar distillation перегонка смолы, смолокурение
tar/earth (for stabilized earth roads) смологрунт
tar oil смоляное масло, дегтярное масло
tar pit майдан
tar separator дегтеотделитель
tar water подсмольная вода
tar works дегтярня
tare тара
target canker ступенчатый рак
target growing-stock целевой запас
tariff тариф
tariff table тариф
tarring осмоление, смоление
tarry смолистый
Tatarian dogwood татарский дерен, *Cornus tatarica* Mill.
Tatarian honeysuckle татарская жимолость, *Lonicera tartarica* L.
Tatarian maple татарский клен, неклен, черноклен, *Acer tataricum* L.
tax налог, пошлина, оклад
taxonomy таксономия
tea-chest чайный ящик
teak тиковое дерево, тик, *Tectona grandis* L.
team бригада
team leader бригадир
technical maturity (yielding the max. of required assortment(s)) техническая спелость
technology техника, технология
telegraph pole телеграфный столб
temperate умеренный
tempering закаливание, закалка
temporary nursery временный питомник
temporary plot летучая площадь
temporary sample plot пробная (временная) площадь
tending воспитание, пестование, уход (usually by felling)
tending a plantation уход за культурой
tending felling рубка ухода
tending of the growing stock уход за запасом, пестование запаса
tending the forest уход за лесом
tendril ус, усик
tenon шип
tenon joint шиповое соединение
tenoner шипорезный станок
tensile force сила растяжения
tensile strength прочность на растяжение, сопротивление растяжению, сопротивление разрыву

tensility растяжимость
tension 1. (of saw blades) натяжение 2. растяжение
tension member растяжка
tension wood тяговая древесина
tensioning (e.g. saws) вальцовка, натяжение, правка, проковка, рихтовка
tenthredinid пилильщик
terminal boom запань
terminal bud верхушечная почка
terminal shoot верхушечный побег, вершинный побег, конечный побег
termite термит
terpene терпен
terrace терраса
terracing террасирование
terrain местность
terrestrial наземный
tertiary wall третичная оболочка, третичная стенка
test испытание, опыт, проба
testing испытание
testing area полигон
tetrandrous tamarisk четырехтычинковый гребенщик, *Tamarix tetrandra* Pall.
Tetraonidae (grouse, ptarmigan, capercaillie, hazel hen) тетеревиные (pl.), боровая дичь
texture текстура
thallus слоевище
thaw 1. таяние 2. таять
thaw water талая вода
thawed талый
thawing оттаивание, таяние
Thelephoraceae пленчатые грибы
thermal conductivity теплопроводимость, теплопроводность
thermolysis термолиз
thermophilous теплолюбивый
thermoplastic glue термореактивный клей, термопластичный клей
thermo-setting glue термореактивный клей, термопластичный клей
thickening утолщение
thicket глушь, дебри, заросль, чаща, гуща
thickness толщина, мощность, диаметр
thickness (of board) высота пропила
thickness gauge толщиномер
thicknesser рейсмус, пропускной станок, реймусовый станок
thin изреживать, прореживать
thin pole дрючок, тычина
thinned изреженный
thinned stand изреженное насаждение, разреженное насаждение
thinning 1. (including natural thinning by suppression) изреживание, разреживание 2. (in younger stands, usually of age 21-40 years) рубка прореживания, прореживание 3. (in older stands, usually over 40 years) проходная рубка
thinning, crown верховой метод ухода, верховое прореживание, проходная верховая рубка
thinning, late heavy простор, рубка простора
thinning, line линейная рубка
thinning, low низовой метод ухода, низовое прореживание, проходная низовая рубка
thinning, natural естественное изреживание, самоизреживание
thinning, row линейная рубка
thinning, systematic (i.e. non-selective) схематическая рубка
thinning cycle повторяемость рубок ухода
thinning from above верховое прореживание
thinning from below низовое прореживание
thinning grade степень прореживания
thinning in younger stands уход за формы ствола и кроны
thinning out прореживание, разреживание
thistle 1. чертополох, *Carduus* L. 2. бодяк, *Cirsium* Mill.
Thomas slag томасшлак
thorn шип
thorn-apple дурман, *Datura stramonium* L.
thorny тернистый
thread (of screw) нарезка

three-storeyed трехэтажный, трехярусный
threshold порог
threshold allowable concentration пдк, кларк
through and through (in sawing) вразвал, раскрой вразвал
throwing (into water) сбрасывание, сброска
thrown down поваленный
thrush дрозд, *Turdus* L.
thrush, song певчий дрозд, *Turdus philomelos* L.
thrust отжим
thunder storm гроза
thyme чабрец, тимьян, *Thymus* L.
thyme, wild обыкновенный тимьян, *Thymus serpyllum* L.
tick клещ
ticket билет
tight knot плотный сучок
tight stacking плотная кладка
tile cell черепитчатая клетка
tillage вспахивание, вспашка, запашка, обработка почвы, уход за почвы, пахота
tillering кущение
timber 1. древесина, лес 2. (all assortments except fuelwood) поделочный лес, деловая древесина, деловой лес, поделочная древесина 3. лесной материал, лесоматериал(ы)
timber, aircraft авиалес
timber, boatbuilding судостройлес
timber, building строевой лес, строительный лес, стройлес
timber, cleft колотый лес
timber, constructional строевой лес, строительный лес, стройлес
timber, defective фаутный лес
timber, felled срубленный лес
timber, floated сплавной лес
timber, hewn обтесанный лес, тесанный лес
timber, knotty сучковатый лес
timber, large толстомерный лес, крупномер, крупная древесина, крупный лес, толстомер
timber, long долготье, длиномерный лес
timber, loose-floated молевой лес, сплавляемый россыпью лес
timber, mast мачтовый лес
timber, mining крепежный лес, рудничный лес, пропсы (pl.)
timber, non-standard некондиционная древесина
timber, planed строганный товар
timber, sawn пиломатериалы (pl.), пиленый лес, пиленый материал, пилопродукция, пиленый товар
timber, shipbuilding корабельный лес
timber, short коротье
timber, sleeper шпальный лес, шпальник, тюлька
timber, small тонкомерный лес
timber, squared кантовый лес
timber, square-sawn призма
timber, standard кондиционная древесина
timber, standing древесина на корню, лес на корню
timber, sunken топняк, топляк, утоп
timber carrier автолесовоз
timber concession закрепление лесосырьевой базы
timber drier древосушка, лесосушилка
timber floater сплавщик
timber floating лесосплав
timber-floating channel сплавной канал
timber harvest урожай древесины
timber impregnation plant деревопропиточная установка
timber lorry лесовоз
timber sale продажа леса, сбыт леса
timber sledge лесовозные сани
timber thief порубщик
timber trade лесная торговля
timber transport road лесовозная дорога
timber tree деловое дерево, бревенник
timber trust лесотрест
timber yard лесобиржа, биржа
time study хронометраж
timothy тимофеевка, *Phleum pratense* L.
tinder трут
tinder fungus огневка

tine зуб
tip **1.** верхушка, вершина, вершок, кончик **2.** (adj.) верхушечный
tipper опрокидыватель
tirr очес
tissue ткань
tit, blue лазоревка, *Parus caeruleus* L.
tit, coal московка, *Parus ater* L.
tit, crested хохлатая синица, *Parus cristatus* L.
tit, great большая синица, *Parus major* L.
tit, long-tailed длиннохвостая синица, *Aegithalos caudatus* L.
tit, penduline ремез, *Parus pendulinus* L.
toad жаба, *Bufo* L.
toadflax обыкновенная льнянка, *Linaria vulgaris* Mill.
toadstool поганка, *Amanita* Pers.
tolerance припуск, допуск, прибавка
tomentum войлок
ton (metric) тонна
tongs клещи, щипцы (pl.)
tongue and groove дорожка и шпонка, шпунт
tongue-and-groove joint соединение гребнем в паз, шпунтовое соединение, соединение шпунтом в фальц
tonguing and grooving шпунтование
tonguing cutter шпунтовая фреза
tooth зуб
tooth, chipper Г-образный зуб, режкообразный зуб, фуговочный зуб
tooth, cutter резающий зуб
tooth, gouge-type Г-образный зуб, режкообразный зуб, фуговочный зуб
tooth, inserted вставной зуб
tooth, raker скоблящий зуб, очижающий зуб, подрезающий зуб, режущий зуб
tooth, spring-set разведенный зуб
tooth, swaged плющенный зуб
tooth, swage-set плющенный зуб
tooth, triangular треугольный зуб
tooth pitch шаг зуба
tooth shape форма зуба
toothed пильчатый, зубчатый
toothwort чешуйчатый крест петров, *Lathraea squamaria* L.
top **1.** верхушка, вершина, вершок **2.** обезвершинивать
top diameter диаметр верхнего отруба, верхний отруб
top end верхний отруб
top log вершинник
top rot вершинная гниль
top storey верхний ярус
top-dressing подкормка
topiary фигурная стижка
topping обезвершинивание
tormentil, common узик, лесная лапчатка, прямостоящая лапчатка, *Potentilla erecta* (L.) Räusch.
torn grain вырванное волокно
torsion кручение
torsional strength сопротивление скручиванию
tortricid плодожорка, листовертка
tortuous извилистый, корявый
torus торус
total итог
total forest area лесной фонд
total standing volume общий запас
touch-me-not желтая недотрога, бальзамин-недотрога, *Impatiens noli-tangere* L.
toughness вязкость
tower вышка, башня
tower, fire пожарная вышка, пожарная каланча
tower, look-out сторожевая башня, вышка
tower, water водонапорная башня
tower crane башенный кран
towing буксировка
towing boom кошель
toxicity токсичность
trace element микроэлемент
tracer меченый атом
trachea сосуд, трахей
tracheary element трахеальный элемент
tracheid трахеида
tracheid, disjunctive разъединенная трахеида
tracheid, fibre волокнистая трахеида

tracheid, ray лучевая трахеида
tracheid, strand тяжевая трахеида
tracheid, vascular сосудистая трахеида
tracheid, vasicentric вазицентрическая трахеида
track дорожка, колея, путь, тропа, волок
track width ширина колеи
tracking выслеживание
traction тяга
tractor трактор, тягач
tractor, amphibious плавающий трактор
tractor, crawler гусеничный трактор
tractor, skidding движитель, трелевочный трактор, трелевщик
tractor, wheeled колесный трактор
tractor driver тракторист
tractor skidding тракторная трелевка
tractor sprayer тракторный опрыскиватель
tractor winch тракторная лебедка
trade промысел, торговля
trade, foreign внешняя торговля
trade, home внутренняя торговля
trade, timber лесная торговля
trade mark товарный знак
traffic load грузооборот
trafficability проходимость
trail, skidding волок
trail, winter зимник
trailer прицеп, роспуск
trailer, sledge санный прицеп
trailer, wheeled колесный прицеп
train поезд
training воспитание, образование
training forest учебная дача
trample вытаптывать
trampling вытаптывание, растаптывание
transfer 1. перевод, передача 2. (from one method of transport to another) перевалка
transfer price отпускная цена
transformation превращение, преобразование
transitional bog переходное болото
translocation передвижение
transmission 1. передача 2. (e.g. of light) пропускание
transom переплет
transparency (of canopy) сквозистость
transpiration транспирация
transplant 1. саженец, перешколенный саженец 2. (vb) перешколивать, пересаживать
transplant nursery древесная школа, лесная школа
transplanted plant саженец, перешколенный саженец
transplanting перешколивание, пересадка, пересаживание, пикирование, пикировка
transport возка, перевозка, подвоз, транспорт
transverse 1. поперек 2. поперечный
transverse boom поперечная запань, поперечный бон
transverse order (of floating logs) поперечная щеть
transverse section поперечный разрез, отруб
trap капкан, ловушка, силок
trap boom запань, поперечная запань
trap ditch ловчяя канава
trap ring ловчее кольцо
trap tree ловчее дерево
travelling saw летучая пила
traverse траверса, траверс
treatment переработка
tree 1. дерево, древо (poetical) 2. (adj.) древесный
tree, border окраинное дерево, краевое дерево
tree, boundary пограничное дерево
tree, broadleaved лиственное дерево
tree, broken полом
tree, Christmas рождественская елка
tree, co-dominant согосподствующее дерево
tree, coniferous хвойное дерево
tree, coppice низкоствольник
tree, dead standing сухостойное дерево, подстой, подсушина
tree, defective фаутное дерево
tree, dominant господствующее дерево

tree, dwarf карликовое дерево
tree, dying отмирающее дерево
tree, evergreen вечнозеленое дерево
tree, family родовое дерево
tree, final crop дерево будущего, избранник
tree, firewood дровяное дерево
tree, forest лесное дерево
tree, forked двойчатка, развилка
tree, freshly felled свежесрубленное дерево
tree, fruit плодовое дерево
tree, individual отдельное дерево
tree, lodged зависшее дерево
tree, lopped подсечное дерево
tree, marginal окраинное дерево, краевое дерево
tree, marked заклейменное дерево
tree, mean среднее дерево
tree, mother материнское дерево, маточное дерево
tree, New Year новогодная елка
tree, overmature перезрелое дерево, переспелое дерево, перестой
tree, parent материнское дерево, маточное дерево
tree, pollarded безвершинник
tree, pre-dominant сверхгосподствующее дерево
tree, prostrate стланец, стлан(н)ик
tree, protective защитное дерево
tree, reserved оставленное дерево
tree, residual оставленное дерево
tree, sample модельное дерево, модель
tree, sapwood заболонная порода
tree, scrub корявое дерево
tree, seed семенное дерево, обсеменитель, семенник
tree, shelter защитное дерево
tree, single солитер
tree, small деревцо
tree, specimen солитер
tree, stagheaded суховершинное дерево
tree, standard оставленное дерево
tree, standing dead сухостой
tree, stranded кокша
tree, stunted корявое дерево
tree, subordinate подчиненное дерево
tree, sunken кокша, сора
tree, suppressed заглушенное дерево, угнетенное дерево
tree, tall бревенник
tree, timber деловое дерево, бревенник
tree, trap ловчее дерево
tree, wind-broken ветролом, бурелом
tree, windthrown валежник, ветровал
tree, wolf волк, разросшееся дерево
tree climber верхолаз
tree cutter лесорез, древовал
tree heath древовидный вереск, *Erica arborea* L.
tree-injector древесный инжектор
tree layer древесный ярус
tree-length log хлыст
tree-like древовидный
tree line верхняя граница древесной растительности
tree measurement дендрометрия
tree of heaven айлант, *Ailanthus altissima*
tree-pipit лесной конек, *Anthus trivialis* L.
tree-ring chronology дендрошкала
tree sparrow полевой воробей, *Passer montanus* L.
tree species древесная порода
tree vegetation древесная растительность
treecreeper обыкновенная пищуха, *Certhia familiaris* L.
treeless безлесный
trefoil трилистник, клевер, *Trifolium* L.
trench канава, ров, траншея
trenching plough плантажный плуг
trestle подмостки (pl.), эстакада
trial испытание, опыт, проба
trial field опытное поле
trial stand пробное насаждение
triangular tooth треугольный зуб
tributary приток
trim обрубать, подрезать, подстригать
trimmer сучкорез, сучкорезная машина
trimming **1.** обрезка, обрезание **2.** обрезка ветвей, обрезка сучьев, обрубка сучьев, очистка ствола **3.** обрезок, срезок, отструг

trip рейс
tripod тренога, треножник
tripod stump-puller треножный корчеватель
trolley вагонетка, тележка
trough корыто, ванна, желоб
truck грузовик
truck-tractor тягач
true service tree крупноплодная рябина, садовая рябина, *Sorbus domestica* L.
truing up log ends торце-выравнивание
trunk ствол
trunk rot стволовая гниль
trust трест
tub кадка
tube труба, трубка
tube, latex млечник, млечная трубка
tube, pollen пыльцевая трубочка
tube, sieve ситовидная трубка
tuber клубень
tufted vetch мышиный горошек, *Vicia cracca* L.
tug boat буксирный катер, буксир
tulip tree тюльпанное дерево, *Liriodendron tulipifera* L.
tumour желвак, опухоль
tundra тундра
tundra, hummocky кочковатая тундра
tundra, lichen ягельная тундра
tundra, moss моховая тундра
tundra, shrub кустарниковая тундра
tung тунг, *Aleurites* Forst.
tunnel (of barkbeetle) ход короеда
tunnel kiln туннельная сушилка
Turcoman juniper туркменский можжевельник, *Juniperus turcomanica* B.Fed.
turf дерн, дернина
turf cutter дернорез
turf formation задернение, одерновение
turfiness задернелость
turf-remover дерносним, дерноснимател ь
turfy задернелый, задерненный
turgor тургор
Turkestan ash туркестанский ясень, *Fraxinus potamophila* Herd.
Turkestan juniper туркестанский можжевельник, тасва-арча, урюк-арча, балх-арча, *Juniperus turkestanica* Kom.
Turkish hazel медвежий орех, древовидная лещина, медвежий орешник, *Corylus colurna* L.
turn точить, обтачивать
turn over перекатывать
turnbuckle талреп, тальреп
turned точеный
turner токарь
turnery wood токарный лес
turning обточка, точение
turning lathe токарный станок
turning place место поворота
turning the log (in sawing) разворачивание бревна
turnover оборот
turnover, annual годовой оборот
turnover, freight грузооборот
turpentine скипидар
turpentine, gum живичный скипидар
turpentine, wood древесный скипидар
turtle dove горлица, *Streptopelia turtur* L.
tussock кочка
twayblade тайник, *Listera* R.Br.
twig ветвь, ветка
twig (needle-bearing) лапка
twining вьющийся
twining plant вьющееся растение
twist искривление, винтовая покоробленность
twist bit спиральное сверло, винтовое сверло
twisted перекошенный
twisting кручение
two-storeyed двухэтажный, двухярусный
two-storeyed stand двухярусное насаждение
two-year двухлетний
two-year-old plant двулетнее растение
tylosis тилла
tylosoid тиллоид
tyre шина

U

u.b. б/к, без коры
ultrasonics ультразвук
ultrasound ультразвук
ultraviolet radiation ультрафиолетовая радиация
umbel зонтик
Umbelliferae зонтичные
umbrella pine итальянская сосна, пиния, *Pinus pinea* L.
unbranched несучковатый
undemanding нетребовательный
under bark б/к, без коры
undercut 1. надруб, подпил, подруб 2. недоруб
undercutting 1. зарубание 2. подкопка
underground подземный
undergrowth заросль, подлесок
underlay подкладка
undermining (of stream banks, etc.) подмыв
underplant подсаживать
underplanted layer искусственный ярус
underplanting подсадка, предварительная культура
understorey подрост, нижний ярус
undertaking предприятие
underwood подлесок, подрост
underwood, auxiliary выполняющий подлесок
undesirable species нежелательный порода
undressed (of lumber) обезличенный
unedged необрезный
unedged board необрезная доска
uneven-aged разновозрастный
uneven-aged stand разновозрастное насаждение
unexploited forest неосвоенный лес, резервный лес
unfinished необработанный
unfloatable несплавный
ungulate копытный
uniform равномерный
uniform high forest лесосечный лес
uniform shelterwood felling семенно-лесосечная рубка
uniform (shelterwood) system равномерная система
unilaterally compound pitting одностороние-сложная поровость
uninterrupted непрерывный
unit (of management) хозяйственная единица, хозяйство
unite срастаться
uniting сращивание, срастание
unload отгружать, разгружать
unloading выгрузка, разгрузка
unmarketable нетоварный
unmerchantable нетоварный
unproductive непроизводительный
unproductive area непроизводительная площадь
unripe недозрелый, незрелый, неспелый
unripeness недозрелость
unsorted нерассортированный, бессортный
unsorted goods бессортный товар
unsound knot гнилой сучок, загнивший сучок
unstable неустойчивый
untreated необработанный
untying (log bundles) размолевание, размолевка
upland нагорный
upland bog верховое болото
upper landing верхний склад
upper storey верхний ярус, вершняк
upright ray cell лучевая стоячая клетка
uproot выкорчевывать
uprooting корчевка, корчевание, выкорчевывание
upstroke (of saw) подъем пилы
uptake поглощение
urban forest городской лес
urea мочевина
urea glue карбамидный клей
urea resin карбамидная смола
urea-formaldehyde glue мочевино-формальдегидный клей
urea-formaldehyde resin КФС, мочевино-формальдегидная смола
Uredinales ржавчинные грибы
useful полезный
useful insect полезное насекомое

Ussuri pear уссурийская груша, *Pyrus ussuriensis* Max.

utilization (of forest) использование, освоение, пользование, пользование лесом, эксплуатация

utilize 1. использовать 2. освоить

V

valerian лекарственная валериана, *Valeriana officinalis* L.

valley долина

valley line тальвег

valuable species ценная порода

valuation оценка, стоимостная оценка

value стоимость, цена, ценность

value increment прирост по ценности

variance дисперсия

variegated пестрый

variety разновидность, сортообразец, сорт

variety testing сортоиспытание

varnish лак

varnish tree лаковое дерево, *Rhus verniciflua* Stokes

varnishing лакировка

vascular сосудистый

vascular bundle сосудистый пучок

vascular plant сосудистое растение

vascular tracheid сосудистая трахеида

vasicentric parenchyma вазицентрическая паренхима

vasicentric tracheid вазицентрическая трахеида

vat корыто, ванна, кадка, чан

vegetable растительный

vegetable kingdom растительное царство

vegetation растительность

vegetation, tree древесная растительность

vegetation, woody древесная растительность

vegetation cover растительный покров

vegetative origin вегетативное происхождение

vegetative propagation вегетативное размножение

vegetative rapprochement (in Michurinist theory) вегетативное сближение

vegetative regeneration вегетативное возобновление

vehicle повозка

vehicle, all-terrain вездеход

vein жилка

velocity скорость

velvet maple бархатный клен, *Acer velutinum* Boiss.

venation жилкование

veneer фанера, шпон

veneer, face облицовочная фанера

veneer, peeled лущеная фанера, шпон

veneer, sawn пиленая фанера

veneer, sliced ножевая фанера, строганная фанера

veneer, ultra-thin микрошпон

veneer bolt чурак

veneer factory фанерный завод, фанерная фабрика

veneer lathe лущильный станок

veneer lay-up укладка листов шпона

veneer log фанерный кряж

veneer peeler (rotary) лущильный станок

veneer saw фанерная пила

veneer scraps шпон-рванина

veneer slicer строгальный станок, фанерострогальный станок

veneer take-off съем листов шпона

ventilating kiln вентиляционная сушилка

vernalization яровизация

vernier нонуис

vertical frame saw вертикальная рама

vertical interface (in resin tapping) межкарровая перемычка

vertical light верхний свет

vervain лекарственная вербена, *Verbena officinalis* L.

vesicle пузырек

vessel сосуд, трахей

vestured pit зубчатая пора, украшенная пора

vetch горошек, вика, *Vicia* L.

vetch, crown разноцветный вязель, *Coronilla varia* L.
vetch, milk сладколистный астрогал, *Astragalus glycyphyllos* L.
vetch, tufted мышиный горошек, *Vicia cracca* L.
VHF kiln СВЧ-камера
viability всхожесть, жизнеспособность
viable всхожный, всхожий
vice тиски (pl.)
violet фиалка, *Viola* L.
violet, water болотная турча, *Hottonia palustris* L.
virgin девственный
virgin forest девственный лес, первобытный лес, стихийный лес, пралес
virgin land целина
virgin soil целинная почва, целина
virus вирус
viscose вискоза
viscosity вязкость
vise клещи, тиски (pl.)
visible knot открытый сучок
vision, field of поле зрения
vitamin flour (from conifer needles) витаминная мука, хвойно-витаминная мука
volatile летающий, летучий
vole полевка
vole, bank рыжая лесная полевка, *Clethrionomys glareolus* Schr.
vole, common обыкновенная полевка, *Microtus arvalis* Pall.
vole, northern red-backed красная полевка, *Clethrionomys rutilus* Pall.
vole, social общественная полевка, *Microtus socialis* Pall.
vole, water водяная крыса, *Arvicola terrestris* L.
voltage напряжение
volume кубатура, масса, объем, вместимость
volume, exploitable эксплуатационный запас
volume, 'realizable' ликвидный запас, ликвидная масса
volume, solid плотная масса
volume, stacked складочный объем
volume, stand запас насаждения
volume, standing запас, запас леса на корню
volume, stem объем ствола
volume calculator (automatic) автокубатурник
volume increment массовый прирост, прирост по массе, прирост по объеме
volume measurement учет по обьему
volume of timber (usually over 7 cm diam.) объем крупной древесины
volume table массовая таблица, объемная таблица
Vorratspflege уход за запасом, пестование запаса

W

wage, day поденная плата
wages зарплата, плата, заработная плата
wagon вагон, повозка, телега
wagtail трясогузка, *Motacilla* L.
wainscot billet ванчес
wall стена, стенка, оболочка, перегородка
wall, cell оболочка клетки, стенка клетки
wall, primary первичная оболочка, первичная стенка
wall, secondary вторичная оболочка, вторичная стенка
wall, tertiary третичная оболочка, третичная стенка
wallpaper обойная бумага
walnut **1.** орех, *Juglans* L. **2.** грецкий орех, *Juglans regia* L.
walnut, black черный орех, *Juglans nigra* L.
walnut, Japanese орех Зибольда *Juglans sieboldiana* Maxim.
walnut, Manchurian маньчжурский орех, *Juglans mandshurica* Maxim.
wane обзол, обливина
waney обзольный
waney balk брус с обзолом
waney board обзольная доска

waney-edged sleeper брусковая шпала
warden, forest лесничий, лесник
warehouse амбар
warmth тепло, теплота
warp коробиться
warped покоробленный
warping 1. (of land) кольматаж 2. коробление, покоробленность
washer прокладка
washer (for logs) обмыватель бревен
washing out вымывание
wasp оса
waste отброс, отходы (pl.)
waste land пустошь, пустырь
waste liquor щелок
waste paper макулатура
watch вахта
watch tower вышка, сторожевая башня
water 1. вода 2. (vb) поливать
water absorption водопоглощение
water avens речной гравилат, *Geum rivale* L.
water bombing (aerial fire fighting) водослив
water catchment водосбор
water level уровень воды
water table зеркало грунтовых вод
water table, perched верховодка
water table, raised подтопление
water tower водонапорная башня
water violet болотная турча, *Hottonia palustris* L.
water vole водяная крыса, *Arvicola terrestris* L.
water-conservation forest водоохранный лес
water-conservation zone водоохранная зона
watercore водослой, мокрослой
watercress жеруха, жерушник, *Nasturtium officinale* R. Br.
watering поливка, полив
watering place водопой
waterlogged заболоченный, болотистый
waterlogging заболачивание
waterproof водонепроницаемый
water-regulating forest водорегулирующий лес
water-repellent treatment гидрофобизация
watershed водораздельная линия, водораздел
waterside landing пристань
watertight водонепроницаемый
wattle, silver серебристая мимоза, серебристая акация, *Acacia dealbata* Link.
wattle fence плетень
wattled плетеный
wave-breaker волнолом
waviness волнистость
wavy волнистый, свилеватый
wavy grain свиль, свилеватость
wavy-fronted felling бухтообразная рубка
wax воск
wax, grafting древесный воск, садовая смола
waxwing свиристель, *Bombycilla garrulus* L.
way дорога, путь
wayfaring tree гордовина, *Viburnum lantana* L.
weakly чахлый
wear износ, истирание
wear resistance сопротивление износу
weasel ласка, *Mustela nivalis* L.
weasel, Siberian колонок, *Mustela sibirica* Pall.
weather погода
weather forecast предсказание погоды
weather service метеослужба
weathering выветривание
wedge клин
wedge, hydraulic felling гидроклин
wedge graft прививка в вырез
wedge grafting инкрустация
wedging заклинивание
wedge-shaped клиновидный
weed 1. сорное растение, сорняк, сорная трава 2. (vb) пропалывать, полоть
weeder осветлитель
weeder, chain-flail цепной осветлитель
weediness засоренность
weeding выпалывание, полка, прополка

weeding unit осветлитель
weeds бурьян
weeping willow плакучая ива, бабилонская ива, *Salix babylonica* L.
weevil долгоносик, слоник, смолевка
weighed portion (or sample) навеска
weight вес, тяжесть
weight, air-dry воздушно-сухой вес
weight, measurement by учет по весу
weight, net чистый вес
weight, parts by весовые части, в.ч.
weight, seed натура семян
weight loss потеря веса
weir плотина
well колодезь, колодец
western yellow pine желтая сосна, *Pinus ponderosa* Dougl.
wet мокрый, сырой
wet barrel заливная бочка
wet moor болото
wet soil мокрая почва, сырая почва
wet stave заливная клепка
wet-strength paper влагопрочная бумага
wettability смачиваемость
wetting замачивание, вымачивание, мочение, мочка, намачивание
wetwood водослой, мокрослой
Weymouth pine сосна веймутова, *Pinus strobus* L.
wharf пристань
wheel колесо
wheelbarrow тачка
wheeled skidder колесный тягач
wheeled tractor колесный трактор
wheeled trailer колесный прицеп
whetstone оселок
whip **1.** охлестивающее дерево, охлестыватель **2.** (vb) охлестывать
whip grafting простая копулировка
whip-and-tongue grafting копулировка с язычком, улучшенная копулировка
white ash американский ясень, *Fraxinus americana* L.
white dead-nettle глухая крапиво, белая яснотка, *Lamium album* L.
white groundwood белая древесная масса
white mulberry белая шелковица, *Morus alba* L.
white pine сосна веймутова, *Pinus strobus* L.
white poplar белый тополь, серебристый тополь, *Populus alba* L.
white rot коррозионная гниль, белая гниль
white saxaul белый саксаул, песчаный саксаул, *Haloxylon persicum* Bge.
white willow серебристая ива, белая ива, ветла, *Salix alba* L.
white-backed woodpecker белоспинный дятел, *Dendrocopos leucotos* Bech.
whitebeam **1.** рябина, *Sorbus* L. **2.** круглолистная рябина, *Sorbus aria* Crantz
white-collared flycatcher мухоловка-белошейка, *Muscicapa albicollis* Temm.
whiteness белизна
wholesale price оптовая цена
whorl мутовка, розетка сучьев
whorled мутовчатый
whortleberry черника, *Vaccinium myrtillus* L.
whortleberry, bog голубика, *Vaccinium uliginosum* L.
whortleberry, red брусника, *Vaccinium vitis-idaea* L.
wicker плетеный
widely spaced изреженный
wide-ringed крупнослойный, широкослойный, мендовый, мяндовый
wide-ringed wood крупнослойная древесина
width ширина
width (of annual ring) ширина годичного слоя
width (of track) ширина колеи
wild дикий, дикорастущий, одичалый
wild angelica дудник, *Angelica sylvestris* L.
wild animal зверь
wild basil обыкновенный душевик, пахучка, *Clinopodium vulgare* L.
wild boar кабан, вепрь, дикая свинья, *Sus scrofa* L.

wild cherry черешня, птичья вишня, *Prunus avium* L.

wild date palm лесной финик, *Phoenix sylvestris* Roxb.

wild radish полевая редька, *Raphanus raphanistrum* L.

wild service tree рябина глоговина, глоговина, лечебная берека, *Sorbus torminalis* Crantz

wild strawberry лесная земляника, *Fragaria vesca* L.

wild thyme обыкновенный тимьян, *Thymus serpyllum* L.

wildcat дикая кошка, *Felix silvestris* Schreb.

wilderness глушь, дичь

wilding дичок, дикарь

wildlife дичь

willow **1.** ива, *Salix* L. **2.** белотал, белолоз, верболоз, *Salix viminalis* L. **3.** краснотал, красная верба, красная шелюга, *Salix acutifolia* Willd. **4.** верба, желтая шелюга, *Salix daphnoides* Vill.

willow, almond миндальнолистная ива, *Salix amygdalina* L.

willow, bay пятитычинковая ива, чернолоз, чернотал, *Salix pentandra* L.

willow, blue сизая ива, *Salix glauca* L.

willow, Caspian каспийская ива, *Salix caspica* Pall.

willow, crack ломкая ива, *Salix fragilis* L.

willow, creeping ползучая ива, *Salix reptans* Rupr.

willow, dwarf травянистая ива, *Salix herbacea* L.

willow, goat козья ива, ракита, бредина, *Salix caprea* L.

willow, grey серая ива, пепельная ива, *Salix cinerea* L.

willow, hastate копьевидная ива, *Salix hastata* L.

willow, polar полярная ива, *Salix polaris* Wg.

willow, pollard безвершинная ива

willow, purple тал, талина, пурпурная ива, желтолозник, тальник, *Salix purpurea* L.

willow, reticulate сетчатая ива, *Salix reticulata* L.

willow, Russian русская ива, *Salix rossica* Nas.

willow, Siberian сибирская ива, *Salix sibirica* Pall.

willow, weeping плакучая ива, бабилонская ива, *Salix babylonica* L.

willow, white серебристая ива, белая ива, ветла, *Salix alba* L.

willow, woolly мохнатая ива, *Salix lanata* L.

willow grouse белая куропатка, *Lagopus lagopus* L.

willow-herb кипрей, *Epilobium* L.

willow-herb, rosebay Иван-чай, узколистный кипрей, *Chamaenerion angustifolium* (L.) Scop.

willow-leafed pear иволистная груша, *Pyrus salicifolia* Pall.

willow oak иволистный дуб, *Quercus phellos*

willow planting шелюгование

willow stand ветлянник, вербняк

willow wood ивняк, *Salicetum*

wilt **1.** увядание **2.** усыхать

wilting увядание

wilting point (permanent) влажность устойчивого завядания

winch лебедка, ворот

winch, hoisting подъемная лебедка

winch, tractor тракторная лебедка

winch skidding лебедочная трелевка

wind breakage ветролом, бурелом

wind-broken trees ветролом, бурелом

wind dispersal распространение ветром

wind erosion дефляция, выдувание

wind-pollinated ветроопыляемый

wind pollination ветроопыляемость

wind-resistant ветроупорный, ветроустойчивый

wind-throw ветровал, повал, буревал

wind-thrown tree(s) ветровал, валежник

windbreak затишка, защитная полоса, ветроупорная опушка

windfall валежник, падалица

window окно

window frame оконный блок

window jamb оконный косяк
windowsill подоконник
wing крыло
wing boom рея, рей, реевный бон
wing knot лапчатый сучок
winged крылатый
wing-nut, Caucasian кавказская лапина, *Pterocarya caucasica* C.A.M.
winter 1. зима 2. межвегетационный период
winter crop озимь
winter drought зимняя засуха, зимнее высушивание
winter drying зимняя засуха, зимнее высушивание
winter moth зимняя пяденица, *Operophtera brumata* L.
winter road зимняя дорога
winter sowing озимый посев
wintergreen грушанка, *Pyrola* L.
wintergreen, chickweed европейский седмичник, *Trientalis europaea* L.
wintergreen, common малая грушанка, *Pyrola minor* L.
wintergreen, larger круглолистная грушанка, *Pyrola rotundifolia* L.
winter-hardy зимнестойкий
wintering 1. зимование 2. (adj.) зимующий
wire проволока
wire, barbed колючая проволока
wire rope проволочный канат
wire slide проволочный лесоспуск
wire-stitching machine проволокосшивный станок
wireworm проволочник, проволочный червь
witches' broom ведьмина метла, метла ведьмина
wither сохнуть
withering засыхание
withy прут
wobble (saw) блуждание пилы
wolf волк, *Canis lupus* L.
wolf tree волк, разросшееся дерево
wood 1. древесина, лес, лесок, лесной материал, лесоматериал(ы) 2. (adj.) древесный, деревянный, лесной
wood, air-dry воздушно-сухая древесина
wood, barked окоренная древесина
wood, bluestained посиневшая древесина
wood, box ящичный кряж, тарный кряж
wood, cleft плашник, плашняк
wood, comminuted измельченная древесина
wood, compressed прессованная древесина, ДП, спрессованная древесина
wood, compression крень, креневая древесина, кремень, кремнина, креноватость
wood, cooperage бочарный лес
wood, dead мертвый лес
wood, defective фаутная древесина, порочная древесина
wood, diffuse-porous рассеяннопоровая древесина
wood, early весенняя древесина, ранняя древесина
wood, faggot фашинник
wood, fascine фашинник
wood, feeding подача древесины
wood, fencing штакетик
wood, figured узорчатая древесина
wood, green сырая древесина
wood, hard твердая древесина
wood, impregnated пропитанная древесина
wood, joinery столярный лес
wood, late летняя древесина, осенняя древесина, поздняя древесина
wood, loose-floated молевая древесина, россыпь
wood, mature спелая древесина
wood, merchantable ликвидная древесина
wood, narrow-ringed узкослойная древесина, мелкослойная древесина
wood, planed строганная древесина
wood, preservative-treated консервированная древесина
wood, resinous осмол, смольняк
wood, resonance резонансовая древесина, резонансовый лес

wood, ring-porous кольцепоровая древесина
wood, root-spur кокорный лес
wood, rotten гнилая древесина, прелая древесина, труха
wood, seasoned высушенный лес
wood, shaft оглобельник
wood, small 1. мелкотоварная древесина, мелкотоварный лес 2. перелесок
wood, solid цельная древесина
wood, sound здоровая древесина
wood, spiral-grained косослойная древесина
wood, spring весенняя древесина, ранняя древесина
wood, stave лад
wood, steamed пропаренная древесина
wood, stump пневая древесина
wood, summer летняя древесина
wood, sunken топняк, топляк, утоп
wood, tension тяговая древесина
wood, turnery токарный лес
wood, wide-ringed крупнослойная древесина
wood alcohol древесный спирт
wood anemone белая ветреница, *Anemone nemorosa* L.
wood ant рыжий лесной муравей, *Formica rufa* L.
wood ash шадрик
wood avens городской гравилат, *Geum urbanum*
wood-based materials древесный материал
wood-block pavement торцовая мостовая
wood-carving резьба по дереву
wood cement древесный цемент
wood charring обугливание дерева
wood chemistry (including chemical utilization, resin tapping etc.) лесная химия, лесохимия
wood concrete деревобетон, арболит, древесный бетон
wood conversion (e.g. furniture manufacture, hydrolysis, paper making) деревопереработка, разработка леса
wood cudweed лесной бессмертник, *Gnaphalium sylvaticum* L.
wood density плотность древесины
wood-destroying fungi дереворазрушающие грибы
wood distillation перегонка дерева
wood drying plant древосушка, лесосушилка
wood durability сохраняемость древесины
wood exchange лесная биржа
wood fibre древесное волокно
wood fibreboard древесноволокнистая плита
wood floater сгонщик
wood flour древесная мука
wood launcher скатчик
wood lemming лесной лемминг, *Myopus schisticolor* Lill.
wood measurement древоизмерение
wood melick одноцветная перловка, *Melica uniflora* Retz.
wood millet развесистый бор, *Milium effusum* L.
wood molasses древесная патока
wood particleboard древесностружечная плита
wood particles (compressed into mouldings with resin) древпресскрошки, ДПК
wood pigeon вяхирь, витютень, *Columba palumbus* L.
wood pile поленница
wood pitch лесохимический пек, древесный пек
wood/plastic composite древопласт, древопластик
wood ray древесный луч
wood rush ожика, *Luzula* DC.
wood science древесиноведение, древоведение
wood sorrel кислица, *Oxalis acetosella* L.
wood spurge миндалевидный молочай, *Euphorbia amygdaloides* L.
wood-staining fungi деревоокрашивающие грибы

wood stitchwort дубравная звездчатка, *Stellaria nemorum* L.
wood storage place лесобиржа, биржа, лесохранилище
wood structure строение древесины
wood sugar древесный сахар
wood tar древесная смола
wood-tar oil древесносмоляное масло
wood turpentine древесный скипидар
wood volume древесная масса, объем древесины
wood wool древесная шерсть, стружка
woodcock вальдшнеп, *Scolopax rusticola* L.
woodcutter рубщик
wooded лесистый, облесенный
wooded district полесье
wooded ravine байрак
wooden деревянный
wooden pavement деревянная мостовая
wooden plough соха
wooden slide деревянный лесоспуск
wood(land) лес
woodland, open редкий лес, редколесье
woodland, riparian тугайный лес, тугай, урема
woodlark лесной жаворонок, *Lullula arborea* L.
woodlouse мокрица, *Oniscus* L.
woodpecker дятел
woodpecker, black желна, черный дятел, *Dryobates martius* L.
woodpecker, greater spotted большой пестрый дятел, *Dryobates major* Brehm
woodpecker, green зеленый дятел, *Picus viridis* L.
woodpecker, lesser spotted малый пестрый дятел, *Dryobates minor* L.
woodpecker, white-backed белоспинный дятел, *Dendrocopos leucotos* Bech.
woodpecker hole дятловина
woodruff, sweet пахучий ясменник, *Asperula odorata* L.
woods camp вахта
woodwasp рогохвост
woodwasp, alder ольховый рогохвост, *Xiphydria camelus* L.
woodworking (e.g. sawmilling, planing, plywood manufacture) деревообработка
woodworking industry деревообрабатывающая промышленность
woody деревянистый, деревянный, древесный, древовидный
woody cutting деревянистый черенок
woody nightshade сладко-горький паслень, *Solanum dulcamara* L.
woody plant древесное растение
woody species древесная порода
woody vegetation древесная растительность
wool (wood) древесная шерсть
woolliness (of grain) ворсистость
woolly willow мохнатая ива, *Salix lanata* L.
work **1.** работа, труд **2.** (vb) работать, разрабатывать
work, day поденная работа
work, hand ручная работа
work, impact ударная работа
work, piece сдельная работа; сдельщина
work, specific удельная работа
work-day (unit of work) трудодень
work safety охрана труда
work to failure работа разрушения, работа на излом
workbench верстак
worked обработанный
worker работник, рабочий
worker, boom запанщик
workers, gang of артель рабочих
workers, skilled кадры (pl.)
working разработка
working circle хозяйственная секция, хозсекция
working clothes спецодежда
working day рабочий день
working plan план хозяйства
working-plan officer лесоустроитель, таксатор
working stress допускаемое напряжение
workman, skilled мастер
workshop мастерская, цех

worm червь, червяк
worm-eaten червоточный
worm-hole червоточина
wormwood горькая полынь, *Artemisia absinthium* L.
wound рана
wounding ранение
wrapping paper оберточная бумага
wren крапивник, *Troglodytes troglodytes* L.
wrench ключ
wrinkled holly морщинистый падуб, *Ilex rugosa* Fr. Schmidt
wryneck вертишейка, *Jynx torquilla* L.
wych elm горный ильмь, шершавый вяз, *Ulmus glabra* Huds. (*U. scabra* Mill.)

X

xerophyte ксерофит
xylan ксилан
xylem ксилема
xylem, secondary вторичная древесина
xylem ray древесный луч
xylene ксилол
xylitol ксилит
xylometer ксилометр
xylose ксилоза

Y

yarrow обыкновенный тысячелистник, *Achillea millefolium* L.
yearbook ежегодник
yearling **1.** годовик **2.** (adj.) годовалый, однолетний
yearly однолетний, годичный
yeasts дрожжевые грибы
yeasts, fodder кормовые дрожжи (pl.)
yellow archangel желтый зеленчук, *Gabeobdolon luteum* Huds.
yellow bird's-nest волосистый подъельник, *Monotropa hypopithys* agg.
yellow loosestrife обыкновенный вербейник, *Lysimachia vulgaris* L.
yellow wood anemone лютиковая ветреница, *Anemone ranunculoides* L.
yellowing пожелтение
yellowish soil палевая почва
yellow-necked field mouse желтогорлая мышь, *Apodemus flavicollis* Melch.
yew тисс, *Taxus* L., usually *Taxus baccata* L.
yew, common ягодный тисс, европейский тисс, *Taxus baccata* L.
yew, Japanese восточный тисс, *Taxus cuspidata* Sieb. et Zucc.
Yezo spruce аянская ель, *Picea jezoensis* Carr.
yield выход, добыча, пользование, продуктивность, производительность, размер пользования, размер рубки, допустимая рубка, урожайность, ход роста
yield, final главное пользование
yield, seed урожайность
yield, sustained постоянство пользования
yield capacity производительная способность
yield determination определение размера пользования
yield regulation by area площадный метод, периодно-площадный метод
yield regulation by volume массовый метод, периодно-массовый метод
yield table опытная таблица, таблица хода роста
yoke коромысло
Yorkshire fog шерстистый бухарник, *Holcus lanatus* L.
young plantation культура
young seedling (usually first-year) всход
young stand (before thicket stage) молодняк, молодое насаждение
youth юность

Z

zone зона, полоса, пояс
zygote зигота

A

Abies Mill. пихта, fir

Abies alba Mill. европейская пихта, белая пихта, гребеньчатая пихта, silver fir

Abies gracilis Kom. тонкая пихта, камчатская пихта, грациозная пихта, Kamchatka fir

Abies holophylla Maxim. цельнолистная пихта, Manchurian fir

Abies mayriana M. et K. пихта Майра

Abies nephrolepsis Maxim. белокорая пихта, East Siberian fir

Abies nordmanniana Spach. пихта Нордманна, кавказская пихта, Caucasian fir

Abies sachalinensis Mast. сахалинская пихта, Sakhalin fir

Abies sibirica Ledeb. сибирская пихта, Siberian fir

Acacia акация, acacia

Acacia dealbata Link. серебристая мимоза, серебристая акация, silver wattle, mimosa

Acer L. клен, maple

Acer campestre L. полевой клен, паклен, field maple

Acer ginnala Maxim. приречный клен, Amur maple

Acer laetum C.A.M. яркий клен, Cappadocian maple

Acer mandshuricum Maxim. маньчжурский клен, Manchurian maple

Acer mono Maxim. мелколистный клен, small-leafed maple

Acer negundo L. американский клен, ясенелистный клен, box-elder

Acer platanoides L. остролистный клен, Norway maple

Acer pseudoplatanus L. явор, клен явор, сикомор, sycamore maple

Acer tataricum L. татарский клен, неклен, черноклен, Tatarian maple

Acer tegmentosum Maxim. зеленокорый клен, greenbark maple

Acer velutinum Boiss. бархатный клен, velvet maple

Aesculus hippocastanum L. конский каштан, horse chestnut

Ailanthus altissima Swingle железистый ясень, айлант, tree of heaven

Albizzia julibrissin Dur. шелковая акация, silk tree

Aleurites Forst. тунг, tung

Alnus Mill. ольха, alder

Alnus barbata C.A.M. бородатая ольха, bearded alder

Alnus fruticosa Rupr. кустарникова ольха, dwarf alder

Alnus glutinosa Gaertn. клейкая ольха, черная ольха, common alder

Alnus incana Moench белая ольха, серая ольха, speckled alder

Alnus kamtschatica Kom. камчатская ольха, Kamchatka alder

Alnus subcordata C.A.M. сердцелистная ольха

Amelanchier ovalis Med. коринка, обыкновенная ирга, service-berry

Amorpha fruticosa L. крутик, кустовидная аморфа

Aralia mandshurica Rupr. et Max. маньчжурская аралия, Manchurian aralia

Arbutus L. земляничник, земляничное дерево, strawberry tree

B

Berberis L. барбарис, barberry

Berberis vulgaris L. обыкновенный барбарис, common barberry

Betula L. береза, birch

Betula costata Traut. ребристая береза, желтая (далневосточная) береза, Siberian yellow birch

Betula dahurica Pall. даурская береза, черная береза, black birch

Betula ermanii Cham. каменная береза, stone birch

Betula fruticosa Pall. кустарниковая береза, dwarf birch

Betula humilis Schrank. низкая береза

Betula nana L. карликовая береза, березовый стланец, dwarf birch

Betula pendula повислая береза, silver birch

Betula pubescens Ehrh. пушистая береза, downy birch

Betula schmidtii Reg. железная береза, береза Шмидта, iron birch

Betula verrucosa Ehrh. бородавчатая береза, silver birch

Betula verrucosa f. ***carelica*** карельская береза, Karelian birch

Buxus sempervirens L. кавказская пальма, самшит, box

C

Caragana Lam. карагана

Caragana arborescens Lam. высокая карагана, желтая акация, pea tree

Caragana frutex Koch кустарная карагана, чилига, dwarf pea tree

Caragana jubata Poir. хвост верблюжий

Caragana pygmaea DC. карликовая карагана, pygmy pea tree

Caragana spinosa DC. колючая карагана, spiny pea tree

Carpinus L. граб, hornbeam

Carpinus betulus L. обыкновенный граб, hornbeam

Carpinus caucasica Grossh. кавказский граб, Caucasian hornbeam

Carpinus cordata Blume сердцелистный граб

Carpinus orientalis Mill. грабинник, восточный граб, eastern hornbeam

Carya pecan E. et G. пекан, pecan

Castanea sativa Mill. съедобный каштан, благородный каштан, посевной каштан, sweet chestnut

Cedrus atlantica Man. атласский кедр, Atlas cedar

Cedrus deodara (Roxb.) Loud. гималайский кедр, deodar

Cedrus libani Loud. ливанский кедр, cedar of Lebanon

Celtis L. каркас, hackberry

Celtis australis L. южный каркас, southern nettle-tree

Celtis caucasica Willd. кавказский каркас, Caucasian nettle-tree

Celtis glabrata Stev. голый каркас

Ceratonia siliqua L. рожечник, рожковое дерево, carob

Cercidiphyllum japonicum Sieb. et Zucc. багрянник, katsura tree

Cercis siliquastrum L. иудино дерево, Judas tree

Chamaecyparis Spach. кипарисовик, cypress

Chosenia macrolepis Kom. крупночешуйчатая чозения, chosenia

Colutea L. пузырник, bladder-senna

Colutea arborescens L. древовидный пузырник, bladder-senna

Colutea orientalis Mill. восточный пузырник, eastern bladder-senna

Cornus mas L. кизил, Cornelian cherry

Cornus sanguinea L. красный дерен, глог, свидина, dogwood

Cornus tatarica Mill. татарский дерен, Tatarian dogwood

Corylus L. лещина, орешник, hazel

Corylus avellana L. обыкновенная лещина, common hazel

Corylus colurna L. медвежий орех, древовидная лещина, медвежий орешник, Turkish hazel

Corylus heterophylla Fisch. разнолистная лещина, hazel

Corylus mandshurica Max. маньчжурская лещина, Manchurian hazel

Corylus maxima Mill. крупная лещина, фундук, filbert

Corylus sieboldiana Bl. лещина Зибольда, Japanese hazel

Cotinus coggygria Scop. желтинник, скумпия, smoke-tree

Cotoneaster Ehrh. кизильник, cotoneaster

Cotoneaster integerrima Med. обыкновенный кизильник

Cotoneaster melanocarpa Lodd. черноплодный кизильник

Crataegus L. боярышник, hawthorn

Crataegus monogyna Jacq. однопестичный боярышник, hawthorn, quickthorn

Crataegus orientalis Pall. восточный боярышник, eastern hawthorn
Crataegus oxyacantha L. обыкновенный боярышник, колючий боярышник, Midland hawthorn
Crataegus pentagyna W.K. пятипестичный боярышник, hawthorn
Cupressus Tourn. кипарис, cypress
Cupressus sempervirens L. обыкновенный кипарис, Mediterranean cypress
Cydonia oblonga Mill. айва, quince
Cytisus L. ракитник, broom
Cytisus borysthenicus Grun. днепроский ракитник, Dnieper broom
Cytisus nigricans L. чернеющий ракитник, black broom
Cytisus ruthenicus Fisch. русский ракитник, Russian broom

D

Diospyros L. хурма, persimmon
Diospyros ebenum König черное дерево, эбеновое дерево, ebony
Diospyros lotus L. обыкновенная хурма, кавказская хурма, date plum

E

Elaeagnus L. лох, oleaster, Russian olive
Elaeagnus angustifolia L. джида, узколистный лох, oleaster
Erythrophloeum fordii Oliv. лим
Eucalyptus L'Herit. эвкалипт, eucalypt
Eucommia ulmoides Oliv. гуттаперчевое дерево, gutta-percha tree
Euonymus L. бересклет, spindle tree
Euonymus europaeus L. европейский бересклет, European spindle tree
Euonymus verrucosus Scop. бородавчатый бересклет, rough spindle tree

F

Fagus L. бук, beech
Fagus orientalis Lipsky восточный бук, кавказский бук, oriental beech
Fagus sylvatica L. европейский бук, лесной бук, beech
Fagus taurica Popl. крымский бук, Crimean beech
Ficus carica L. инжир, фига, смоковница, fig
Frangula alnus Mill. ломкая крушина, alder buckthorn
Fraxinus L. ясень, ash
Fraxinus americana L. американский ясень, white ash
Fraxinus excelsior L. обыкновенный ясень, common ash
Fraxinus mandshurica Rupr. маньчжурский ясень, Manchurian ash
Fraxinus pennsylvanica Marsh. зеленый ясень, пенсильванский ясень, green ash
Fraxinus potamophila Herd. туркестанский ясень, Turkestan ash

G

Ginkgo biloba L. гинкго, maidenhair tree
Gleditsia L. гледичия, honey-locust
Gleditsia caspica L. каспийская гледичия, Caspian honey-locust
Gleditsia triacanthos L. обыкновенная гледичия, honey-locust
Guaiacum officinale L. бакаутовое дерево, бакаут, lignum vitae

H

Halimodendron halodendron Voss чингиль, salt tree
Haloxylon Bge. саксаул, saxaul
Haloxylon ammodendron Bge. зайсканский саксаул, saxaul

Haloxylon aphyllum Iljin черный саксаул, black saxaul
Haloxylon persicum Bge. белый саксаул, песчаный саксаул, white saxaul

I

Ilex L. падуб, holly
Ilex aquifolium L. остролистный падуб, holly
Ilex rugosa Fr. Schmidt морщинистый падуб, wrinkled holly

J

Juglans L. орех, walnut
Juglans cinerea L. масляный орех, серый орех, butternut
Juglans mandshurica Maxim. маньчжурский орех, Manchurian walnut
Juglans nigra L. черный орех, black walnut
Juglans regia L. грецкий орех, walnut
Juglans sieboldiana Maxim. орех Зибольда, Japanese walnut
Juniperus L. можжевельник, juniper
Juniperus communis L. обыкновенный можжевельник, common juniper
Juniperus dahurica Pall. даурский можжевельник, Dahurian juniper
Juniperus excelsa Bieb. высокий можжевельник, Grecian juniper
Juniperus foetidissima Willd. вонючий можжевельник, тяжелопахутий можжевельник, stinking juniper
Juniperus oxycedrus L. красный можжевельник, prickly juniper
Juniperus polycarpos Koch кара-арча, многоплодный можжевельник, black juniper
Juniperus sabina L. казацкий можжевельник, savin
Juniperus semiglobosa Rgl. саур-арча, полушаровидный можжевельник, semiglobose juniper
Juniperus seravschanica Kom. кара-арча, зеравшанский можжевельник, Seravshan juniper
Juniperus sibirica Burgsd. сибирский можжевельник, Siberian juniper
Juniperus turcomanica B.Fed. туркменский можжевельник, Turcoman juniper
Juniperus turkestanica Kom. туркестанский можжевельник, тасва-арча, урюк-арча, балх-арча, Turkestan juniper

K

Kalopanax pictus (Thunb.) Nakai калопанакс, диморфант, prickly castor-oil tree

L

Laburnum anagyroides Med. золотой дождь, golden-chain
Larix Mill. лиственница, larch
Larix czekanowskii Szaf. чекановская лиственница, Czekanow's larch
Larix decidua Mill. европейская лиственница, European larch
Larix gmelinii Litv. даурская лиственница, Dahurian larch
Larix gmelinii var. ***japonica*** Pilg. курильская лиственница, Kurile larch
Larix leptolepis Murr. тонкочешуйчатая лиственница, японская лиственница, Japanese larch
Larix sibirica Led. сибирская лиственница, Siberian larch
Larix sukaczewii Dyl. лиственница Сукачева, Sukachev's larch
Laurus nobilis L. благородный лавр, laurel
Libocedrus decurrens Torr. речной кедр, incense cedar
Ligustrum vulgare L. обыкновенная бирючина, common privet

Liriodendron tulipifera L. тюльпанное дерево, tulip tree
Lonicera L. жимолость, honeysuckle
Lonicera caprifolium L. каприфоль, жимолость-каприфоль, goat-leaf honeysuckle
Lonicera coerulea L. синяя жимолость, blue honeysuckle
Lonicera iberica M.B. грузинская жимолость, Georgian honeysuckle
Lonicera orientalis Lam. кавказская жимолость, eastern honeysuckle
Lonicera tartarica L. татарская жимолость, Tatarian honeysuckle
Lonicera xylosteum L. обыкновенная жимолость, лесная жимолость, fly honeysuckle

M

Maackia amurensis Rupr. акатник, амурская акация, Amur acacia
Maclura aurantiaca Nutt. маклюра, osage orange
Magnolia L. магнолия, magnolia
Malus sylvestris Mill. лесная яблоня, дикая яблоня, crab apple
Mespilus germanica L. мушмула, medlar
Morus L. шелковица, тут, тутовик, тутовое дерево, mulberry
Morus alba L. белая шелковица, white mulberry
Morus nigra L. черная шелковица, black mulberry
Morus rubra L. красная шелковица, red mulberry
Myrica gale L. обыкновенный восковник, sweet gale
Myrtus communis L. обыкновенный мирт, common myrtle

O

Olea europaea L. европейская маслина, olive
Ostrya Scop. хмелеграб, hop-hornbeam
Ostrya carpinifolia Scop. обыкновенный хмелеграб, European hop-hornbeam

P

Paliurus spina-christi Mill. держидерево, Christ-thorn
Parrotia persica C.A.M. железное дерево, боккаут, Persian iron wood
Phellodendron amurense Rupr. бархатное дерево, амурский бархат, Amur cork tree
Philadelphus L. чубушник, mock orange
Phoenix L. финиковая пальма, финик
Phoenix dactylifera L. финиконосный финик, date palm
Phoenix sylvestris Roxb. лесной финик, wild date palm
Picea Dietrich ель, spruce
Picea abies (L.) Karst. обыкновенная ель, европейская ель, Norway spruce
Picea glehnii Mast. ель Глена, Sakhalin spruce
Picea jezoensis Carr. аянская ель, Yezo spruce
Picea obovata Ldb. сибирская ель, Siberian spruce
Picea orientalis Carr. кавказская ель, восточная ель, oriental spruce
Picea pungens Engel. колючая ель, blue spruce
Picea schrenkiana F. et M. тянь-шанская ель, ель Шренка, Schrenk's spruce
Pinus L. сосна, pine
Pinus brutia Ten. калабрийская сосна, Corsican pine
Pinus cembra L. европейский кедр, Arolla pine
Pinus eldarica Medwed. эльдарская сосна, Eldar pine
Pinus halepensis Mill. алепская сосна, Aleppo pine
Pinus koraiensis Sieb. et Zucc. маньчшурская кедровая сосна, корейский кедр, маньчжурский кедр, корейская сосна, Korean pine

Pinus montana Mill. сосновый стланец, жереп, горная сосна, mountain pine
Pinus mugo Turra сосновый стланец, жереп, горная сосна, mountain pine
Pinus nigra var. ***caramanica*** крымская сосна, Crimean pine
Pinus nigra Arn. черная сосна, black pine
Pinus peuce Grisob. румелийская сосна, Macedonian pine
Pinus pinaster Sol. приморская сосна, maritime pine
Pinus pinea L. итальянская сосна, пиния, umbrella pine
Pinus pithyusa Stev. пицундская сосна, Pitsunda pine
Pinus ponderosa Dougl. желтая сосна, western yellow pine
Pinus sibirica Rupr. сибирский кедр, кедровая сибирская сосна, Siberian pine
Pinus strobus L. сосна веймутова, white pine, Weymouth pine
Pinus sylvestris L. сосна, обыкновенная сосна, Scots pine
Pistacia L. фисташка, pistachio
Pistacia lentiscus L. мастиковое дерево, mastic
Pistacia mutica F. et M. кевевое дерево, туполистная фисташка
Pistacia vera L. настоящая фисташка, pistachio
Platanus L. чинар, платан, plane
Platanus acerifolia Willd. кленолистный платан, London plane
Platanus occidentalis L. западный платан, American plane
Platanus orientalis L. восточный платан, oriental plane
Populus L. тополь, poplar
Populus alba L. белый тополь, серебристый тополь, white poplar
Populus alba var. ***pyramidalis*** L. (Bge.) тополь Болле, Bolle's poplar
Populus balsamifera L. бальзамический тополь, balsam poplar
Populus canescens L. сереющий тополь, серый тополь, grey poplar
Populus deltoides Marsh канадский тополь, 'Canadian poplar'
Populus diversifolia Schrenk разнолистный тополь
Populus laurifolia Ldb. лавролистный тополь, laurel-leafed poplar
Populus maximowiczii Henry тополь Максимовича, Maximovich poplar
Populus nigra L. осокорь, черный тополь, black poplar
Populus nigra var. ***italica*** L. (Duroi) пирамидальный тополь, Lombardy poplar
Populus pruinosa Schrenk сизолистный тополь, сизый тополь, blue poplar
Populus pyramidalis Roz. пирамидальный тополь, Lombardy poplar
Populus "Robusta" "мощный" тополь, robusta poplar
Populus simonii Carr. китайский тополь, Chinese poplar
Populus suaveolens Fisch. душистый тополь, fragrant poplar
Populus tacamahaca Mill. бальзамический тополь, eastern balsam poplar
Populus tremula L. осина, aspen
Populus tremula var. ***davidiana*** осина Давида, David's aspen
Prunus amygdalus Batsch обыкновенный миндаль, almond
Prunus armeniaca L. абрикос, обыкновенный абрикос, apricot
Prunus avium L. черешня, птичья вишня, wild cherry
Prunus cerasus L. обыкновенная вишня, sour cherry
Prunus divaricata Led. алыча, plum, damson
Prunus domestica L. домашная слива, plum
Prunus fruticosa Pall. вишарник, кустарниковая вишня, степная вишня, ground cherry
Prunus institia L. тернослив, bullace
Prunus laurocerasus L. лавровишня, cherry-laurel
Prunus mahaleb L. кучина, антипка, mahaleb cherry

Prunus mandshurica (Maxim.) Koehne маньчжурский абрикос, Manchurian apricot
Prunus padus L. черемуха, bird-cherry
Prunus persica (L.) Batsch персик, peach
Prunus sibirica L. сибирский абрикос, Siberian apricot
Prunus spinosa L. терн, терновик, колючая слива, blackthorn, sloe
Prunus spinosissima Franch. колючейший миндаль, spiny almond
Prunus tenella Batsch степной миндаль, низкий миндаль, dwarf Russian almond
Pseudotsuga menziesii Mirb. тиссолистная лжетсуга, дугласия, дугласовая пихта, Douglas fir
Pseudotsuga menziesii var. ***caesia*** серая лжетсуга, grey Douglas fir
Pseudotsuga menziesii var. ***glauca*** сизая лжетсуга, blue Douglas fir
Pseudotsuga menziesii var. ***viridis*** зеленая лжетсуга, green Douglas fir
Pterocarya caucasica C.A.M. кавказская лапина, Caucasian wing-nut
Punica granatum L. гранат, гранатник, pomegranate
Pyrus L. груша, pear
Pyrus communis L. обыкновенная груша, common pear
Pyrus elaeagnifolia Pall. похолистная груша
Pyrus salicifolia Pall. иволистная груша, willow-leafed pear
Pyrus ussuriensis Max. уссурийская груша, Ussuri pear

Q

Quercus L. дуб, oak
Quercus araxina (Trautv.) Grossh. аракский дуб, араксинский дуб, Araks oak
Quercus borealis Michx. красный дуб, red oak
Quercus castaneifolia C.A.M. каштанолистный дуб, chestnut-leafed oak
Quercus hartwissiana Stev. дуб Гартвиса, Hartwiss oak
Quercus iberica Stev. грузинский дуб, Georgian oak
Quercus ilex L. каменный дуб, holm oak
Quercus imeretina Stev. имеретинский дуб
Quercus longipes Stev. длинноножковый дуб
Quercus macranthera F. et M. крупнопыльниковый дуб, восточный дуб, Caucasian oak
Quercus mongolica Fisch. монгольский дуб, Mongolian oak
Quercus pedunculiflora C.Koch ножкоцветный дуб
Quercus petraea (Matt.) Liebl. сидячецветный дуб, скальный дуб, зимний дуб, durmast oak, sessile oak
Quercus phellos иволистный дуб, willow oak
Quercus pontica Koch понтиийский дуб, Pontine oak
Quercus pubescens Willd. пушистый дуб, downy oak
Quercus robur L. черешчатый дуб, стебельчатый дуб, летний дуб, pedunculate oak
Quercus rubra L. красный дуб, red oak
Quercus sessiliflora Salisb. сидячецветный дуб, скальный дуб, зимний дуб, durmast oak, sessile oak
Quercus suber L. пробковый дуб, cork oak

R

Rhamnus L. крушина, жестер, жостер, buckthorn
Rhamnus alaternus L. вечнозеленая крушина, evergreen buckthorn
Rhamnus cathartica L. слабительная крушина, common buckthorn
Rhamnus frangula L. ломкая крушина, alder buckthorn
Rhododendron L. рододендрон, rhododendron

Rhododendron camtschaticum Pall. камчатский рододендрон, Kamchatka rhododendron
Rhododendron caucasicum Pall. кавказский рододендрон, Caucasian rhododendron
Rhododendron dauricum L. даурский рододендрон, Dahurian rhododendron
Rhododendron luteum Sweet желтый рододендрон, yellow rhododendron
Rhododendron ponticum L. понтийский рододендрон, Pontine rhododendron
Rhus L. сумах, sumac
Rhus coriaria L. дубильный сумах, tanners' sumac
Rhus typhina L. уксусник, уксусное дерево, пушистый сумах, оленерогий сумах, staghorn sumac
Rhus verniciflua Stokes лаковое дерево, varnish tree
Ribes L. смородина, currant
Ribes alpinum L. альпийская смородина, alpine currant
Ribes aureum Pursh. золотиситая смородина, golden currant
Ribes dikuscha Fisch. смородина дикуша
Ribes grossularia L. крыжовник, gooseberry
Ribes nigrum L. черная смородина, blackcurrant
Ribes procumbens Pall. моховка
Ribes pubescens Hedl. пушистая смородина
Ribes rubrum L. красная смородина, redcurrant
Robinia pseudoacacia L. белая акация, лжеакация, black locust
Rosa L. роза, шиповник, rose
Rosa canina L. роза собачья, dog rose
Rubus arcticus L. поленика
Rubus caesius L. ожина, сизая ежевика, dewberry
Rubus chamaemorus L. морошка, cloudberry
Rubus idaeus L. малина, raspberry
Rubus nessensis W. Hall кумяника
Rubus saxatilis L. костяника
Ruscus aculeatus L. иглица, butcher's broom

S

Salix L. ива, willow
Salix acutifolia Willd. верболоз, краснотал, красная верба, остролистная ива, красная шелюга, willow
Salix alba L. серебристая ива, белая ива, ветла, белолоз, willow, white willow
Salix amygdalina L. миндальнолистная ива, almond willow
Salix arbuscula L. деревцовидная ива
Salix aurita L. ушастая ива
Salix babylonica L. плакучая ива, бабилонская ива, weeping willow
Salix caprea L. козья ива, ракита, бредина, goat willow, sallow
Salix caspica Pall. каспийская ива, Caspian willow
Salix cinerea L. серая ива, пепельная ива, grey willow
Salix daphnoides Vill. шелюга желтая, верба, willow
Salix fragilis L. ломкая ива, crack willow
Salix glauca L. сизая ива, blue willow
Salix hastata L. копьевидная ива, hastate willow
Salix herbacea L. травянистая ива, dwarf willow
Salix lanata L. мохнатая ива, woolly willow
Salix livida Whlb. синеватосерая ива
Salix myrsinites L. миртолистная ива
Salix myrtilloides L. черничная ива
Salix nigricans Sm. чернеющая ива
Salix pentandra L. пятитычинковая ива, чернолоз, чернотал, bay willow
Salix phylicifolia L. двуцветная ива
Salix polaris Wg. полярная ива, polar willow
Salix purpurea L. тал, талина, пурпурная ива, желтолозник, тальник, purple osier, purple willow
Salix pyrolifolia Ldb. грушанколистная ива

Salix reptans Rupr. ползучая ива, creeping willow
Salix reticulata L. сетчатая ива, reticulate willow
Salix rossica Nas. русская ива, Russian willow
Salix sibirica Pall. сибирская ива, Siberian willow
Salix tenuifolia Turcz. тонколистная ива
Salix triandra L. трехтычинковая ива, лоза, лозина, белолоз, белотал, willow
Salix viminalis L. верболоз,белотал, корзинная лоза, корзиночная ива, прутовидная ива, common osier
Salsola L. солянка, черкез
Salsola arbuscula Pall. баялыш
Salsola paletzkiana Litw. кара-черкез
Salsola rigida Pall. кейреук
Salsola subaphylla чогон
Sambucus L. бузина, elder
Sambucus ebulus L. зеленик, danewort
Sambucus nigra L. бузина черная, common elder
Sambucus racemosa L. красная бузина, red elder
Sequoia Endl. секвоя, sequoia
Sequoia gigantea Decne. мамонтовое дерево, гигантская секвоя, giant sequoia
Sorbus L. рябина, rowan, whitebeam
Sorbus aria Crantz круглолистная рябина, whitebeam
Sorbus aucuparia L. обыкновенная рябина, rowan, mountain ash
Sorbus domestica L. крупноплодная рябина, садовая рябина, true service tree
Sorbus torminalis Crantz рябина глоговина, глоговина, лечебная берека, wild service tree
Staphylea L. клекачка, bladdernut
Staphylea colchica Stev. колхидская клекачка, Colchis bladdernut
Staphylea pinnata L. перистая клекачка, pinnate bladdernut
Symphoricarpos rivularis Suks. снежеягодник, снежная ягода, snowberry
Syringa vulgaris L. сирень, lilac

T

Tamarix L. тамарикс, гребенщик, tamarisk
Tamarix elongata Ldb. вытянутый гребенщик
Tamarix gracilis Willd. изящный гребенщик, graceful tamarisk
Tamarix hispida Willd. жестковолистный гребенщик, опушенный гребенщик, Kashgar tamarisk
Tamarix laxa Willd. рыхлый гребенщик, limp tamarisk
Tamarix ramosissima Ldb. ветвистый гребенщик, branchy tamarisk
Tamarix tetrandra Pall. четырехтычинковый гребенщик, tetrandrous tamarisk
Taxodium distichum Rich. болотный кипарис, swamp cypress
Taxus L. тисс, yew
Taxus baccata L. ягодный тисс, европейский тисс, common yew
Taxus cuspidata Sieb. et Zucc. восточный тисс, Japanese yew
Tectona grandis L. тиковое дерево, тик, teak
Thuja L. туя, arbor-vitae
Thuja orientalis L. восточная туя, Oriental arbor-vitae
Tilia L. липа, linden, lime
Tilia amurensis Rupr. амурская липа, Amur lime
Tilia caucasica Rupr. кавказская липа, Caucasian lime
Tilia cordata Mill. мелколистная липа, small-leafed linden
Tilia euchlora K. Koch крымская липа, Crimean linden
Tilia mandshurica Rupr. маньчжурская липа, Manchurian linden
Tilia platyphyllos Scop. крупнолистная липа, large-leafed linden
Tilia rubra DC. красная липа, red lime
Tilia sibirica Bayer сибирская липа, Siberian linden
Tilia tomentosa Moench войлочная липа, серебристая липа, silver linden

U

Ulmus L. илмь, вяз, берест, elm

Ulmus carpinifolia Gled. берест, карагач, лиственный вяз, листоватый вяз, smooth elm

Ulmus densa Litv. густой вяз

Ulmus foliacea Gilib. берест, карагач, лиственный вяз, листоватый вяз, smooth elm

Ulmus glabra Huds. горный илмь, шершавый вяз, wych elm

Ulmus laevis Pall. гладкий вяз, European white elm

Ulmus parvifolia Jacq. мелколистный вяз, Chinese elm

Ulmus pinnato-ramosa Dieck. карагач, перистоветвистый вяз, Siberian elm

Ulmus pumila L. приземистый вяз, карагач, ильмовник, Siberian elm

Ulmus pumila L. var. ***arborea*** Litv. перистоветвистый вяз, мелколистный вяз, Siberian elm

Ulmus scabra Mill. горный илмь, шершавый вяз, wych elm

Ulmus suberosa Moench пробковый вяз

V

Vaccinium L. черника

Vaccinium arctostaphylos L. кавказская черника

Vaccinium myrtillus L. черника, bilberry, whortleberry

Vaccinium uliginosum L. голубика, bog whortleberry

Vaccinium vitis-idaea L. брусника, cowberry, red whortleberry

Viburnum lantana L. гордовина, wayfaring tree

Viburnum opulus L. калина, guelder rose

Z

Zelkova carpinifolia Dipp. граболистная дзельква, Caucasian elm

Zizyphus jujuba Mill. унаби, common jujube

Printed and bound by CPI Group (UK) Ltd, Croydon, CR0 4YY

16/09/2026

14963431-0002